Mouse Genetics and Transgenics

The Practical Approach Series

SERIES EDITOR

B. D. HAMES
Department of Biochemistry and Molecular Biology
University of Leeds, Leeds LS2 9JT, UK

See also the Practical Approach web site at **http://www.oup.co.uk/PAS**

★ indicates new and forthcoming titles

- Affinity Chromatography
- Affinity Separations
- Anaerobic Microbiology
- Animal Cell Culture (2nd edition)
- Animal Virus Pathogenesis
- Antibodies I and II
- Antibody Engineering
- Antisense Technology
- ★ Apoptosis
- Applied Microbial Physiology
- Basic Cell Culture
- Behavioural Neuroscience
- Bioenergetics
- Biological Data Analysis
- Biomechanics – Materials
- Biomechanics – Structures and Systems
- Biosensors
- ★ Caenorhabditis Elegans
- Carbohydrate Analysis (2nd edition)
- Cell-Cell Interactions
- The Cell Cycle
- Cell Growth and Apoptosis
- ★ Cell Growth, Differentiation and Senescence
- ★ Cell Separation
- Cellular Calcium
- Cellular Interactions in Development
- Cellular Neurobiology
- Chromatin
- ★ Chromosome Structural Analysis
- Clinical Immunology
- Complement
- ★ Crystallization of Nucleic Acids and Proteins (2nd edition)
- Cytokines (2nd edition)
- The Cytoskeleton
- Diagnostic Molecular Pathology I and II
- DNA and Protein Sequence Analysis
- DNA Cloning 1: Core Techniques (2nd edition)
- DNA Cloning 2: Expression Systems (2nd edition)

DNA Cloning 3: Complex Genomes (2nd edition)
DNA Cloning 4: Mammalian Systems (2nd edition)
★ DNA Microarrays
★ DNA Viruses
Drosophila (2nd edition)
Electron Microscopy in Biology
Electron Microscopy in Molecular Biology
Electrophysiology
Enzyme Assays
Epithelial Cell Culture
Essential Developmental Biology
Essential Molecular Biology I and II
★ Eukaryotic DNA Replication
Experimental Neuroanatomy
Extracellular Matrix
Flow Cytometry (2nd edition)
Free Radicals
Gas Chromatography
Gel Electrophoresis of Nucleic Acids (2nd edition)
★ Gel Electrophoresis of Proteins (3rd edition)
Gene Probes 1 and 2
★ Gene Targeting (2nd edition)
Gene Transcription
Genome Mapping
Glycobiology
Growth Factors and Receptors
Haemopoiesis
★ High Resolution Chromotography
Histocompatibility Testing
HIV Volumes 1 and 2
★ HPLC of Macromolecules (2nd edition)
Human Cytogenetics I and II (2nd edition)
Human Genetic Disease Analysis
★ Immobilized Biomolecules in Analysis
Immunochemistry 1
Immunochemistry 2
Immunocytochemistry
★ *In Situ* Hybridization (2nd edition)
Iodinated Density Gradient Media
Ion Channels
★ Light Microscopy (2nd edition)
Lipid Modification of Proteins
Lipoprotein Analysis
Liposomes
Mammalian Cell Biotechnology
Medical Parasitology
Medical Virology
MHC Volumes 1 and 2
★ Molecular Genetic Analysis of Populations (2nd edition)
Molecular Genetics of Yeast
Molecular Imaging in Neuroscience
Molecular Neurobiology
Molecular Plant Pathology I and II
Molecular Virology
Monitoring Neuronal Activity

★ Mouse Genetics and Transgenics
Mutagenicity Testing
Mutation Detection
Neural Cell Culture
Neural Transplantation
Neurochemistry (2nd edition)
Neuronal Cell Lines
NMR of Biological Macromolecules
Non-isotopic Methods in Molecular Biology
Nucleic Acid Hybridisation
★ Nuclear Receptors
Oligonucleotides and Analogues
Oligonucleotide Synthesis
PCR 1
PCR 2
★ PCR 3: PCR In Situ Hybridization
Peptide Antigens
Photosynthesis: Energy Transduction
Plant Cell Biology
Plant Cell Culture (2nd edition)
Plant Molecular Biology
Plasmids (2nd edition)
Platelets
Postimplantation Mammalian Embryos
★ Post-translational Processing
Preparative Centrifugation
Protein Blotting
★ Protein Expression
Protein Engineering
Protein Function (2nd edition)
Protein Phosphorylation (2nd edition)
Protein Purification Applications
Protein Purification Methods
Protein Sequencing
Protein Structure (2nd edition)
Protein Structure Prediction
Protein Targeting
Proteolytic Enzymes
Pulsed Field Gel Electrophoresis
RNA Processing I and II
RNA-Protein Interactions
Signalling by Inositides
★ Signal Transduction (2nd edition)
Subcellular Fractionation
Signal Transduction
★ Transcription Factors (2nd edition)
Tumour Immunobiology
★ Virus Culture

Mouse Genetics and Transgenics

A Practical Approach

Edited by

IAN J. JACKSON,
MRC Human Genetics Unit,
Edinburgh

and

CATHERINE M. ABBOTT,
Department of Medical Sciences,
University of Edinburgh

OXFORD
UNIVERSITY PRESS

Great Clarendon Street, Oxford OX2 6DP

Oxford University Press is a department of the University of Oxford
and furthers the University's aim of excellence in research, scholarship,
and education by publishing worldwide in

Oxford New York

Athens Auckland Bangkok Bogotá Buenos Aires Calcutta
Cape Town Chennai Dar es Salaam Delhi Florence Hong Kong Istanbul
Karachi Kuala Lumpur Madrid Melbourne Mexico City Mumbai
Nairobi Paris São Paulo Singapore Taipei Tokyo Toronto Warsaw

and associated companies in Berlin Ibadan

Oxford is a registered trade mark of Oxford University Press

Published in the United States
by Oxford University Press Inc., New York

A catalogue record for this book is available from the British Library

Library of Congress Cataloging in Publication Data
(Data available)

ISBN 0-19-963709-1 (Hbk)
0-19-963708-3 (Pbk)

Typeset by Footnote Graphics,
Warminster, Wilts
Printed in Great Britain by Information Press, Ltd,
Eynsham, Oxon.

Preface

There is always the temptation, when prefacing a book like this, to use it as a mark in time; to view the history of the field and to speculate where it will go. As the end of the 20th Century (another arbitrary mark in time!) approaches we can use this as further justification; in fact mouse genetics is roughly the same age as the century.

When William Bateson published his book on 'Mendel's Principles of Inheritance' in 1902 the subject did not yet have a name. Bateson himself invented the term 'Genetics' and when offered, in 1908, the Chair of Biology in Cambridge he took it on the understanding that it was to study this new field of Genetics. He and two others in the first decade of this century, Cuenot and Castle, all looked at coat colour genes, and are usually credited with the first studies of mouse genetics. Genetics requires variation that can be followed through generations, and the earliest examples of variants were mutant phenotypes. Among these were a number of coat colour mutations, published in 1913, the short-ear mutation published in 1921, rodless retina (1924) and the inner-ear mutation shaker-1 (1929). In the 1980s and 90s each of the genes in which these mutations occur have been cloned and sequenced as well as many other mutant loci. But the search for new phenotypes continues, and has been accelerated in recent decades by chemical mutagenesis (Chapter 8).

In 1915 Haldane showed that the inheritance of the chinchilla and pink-eyed-dilute coat colour mutations was not independent and thus made the first demonstration of linkage of mammalian autosomal genes, and identified what was known as linkage group I. Over the next 20 years only 9 more mutant loci were identified and placed onto a total of 5 linkage groups. By the end of the 1970s there were over 400 genes on the linkage map, and many of these were identified not by mutant phenotype but by variation in enzyme activity or protein isoforms. Once the technology developed to permit following the inheritance of silent, variant DNA markers, and once microsatellites were discovered to be an easily accessible source of such variation the way was open for large-scale efforts to generate assays for the many thousands of mapped loci now available. In a remarkably short time microsatellites have become the method of choice for mapping phenotypes (Chapter 4) and genomes (Chapter 5).

DNA markers have now been generated at such a high density that genetic mapping will not suffice to separate them; we have virtually saturated the linkage map. Many markers can now only be mapped relative to each other by physical means such as yeast artificial chromosome clone maps (Chapter 5), and as this type of mapping no longer requires inherited variation, any DNA marker can be mapped in this way. These physical maps are the logical progression from cytogenetic maps. Analysis of mouse chromosomes is relatively recent; they were only distinguished from each other in the early 1970s, and once each linkage group was assigned to a particular chromosome, then in situ hybridisation (initially using radioactive probes) could link the physical and genetic maps. The last decade has seen a revolution in molecular cytogenetics. The use of flourescent probes has greatly improved the sensitivity and resolution of in situ mapping. Chromosome-specific labelled paints have facilitated the identification of individual chromosomes, and can indicate the presence of translocations and rearrangements not otherwise detectable using the old techniques (Chapter 6).

Throughout the late 1970s and 1980s DNA cloning technology shifted the emphasis in genetics from phenotype-based to gene-based. Cloning began by isolating genes known to encode particular proteins, moved on to identifying homologues of genes from other species and recently has included the substrates for mass analysis of expressed sequences. One part of the characterisation of newly isolated genes is to determine their expression pattern. Finding where transcripts are localised not only on a broad organ basis but down to the single cell level by in situ hybridisation became possible in the 1980s, at first using radioactive probes (which is still the method of choice for certain stages) and later using non-radioactive probes and whole mount hybridisation (Chapter 3). Another key step in analysis of genes is assaying for function after insertion into mouse embryos by transgenesis. The technique of transgenesis changed little from its inception in the late 1970s until recent developments using large DNA fragments cloned in YACs and BACs have given more reliable expression of the transgene and have expanded the possibilities for the use of this technology to explore gene function (Chapter 9). The isolation of embryonic stem cells at the beginning of the 1980s, and the first gene targeting results a few years later were the first step towards a reverse genetics; where a mutant phenotype was produced after a gene had been isolated. ES cell technology has grown in sophistication and now there are numerous methods for making subtle and conditional mutations in any gene (Chapter 10).

The phenotypes that we observe in mice are not only genetic in origin, but are due to the interaction between genotype and the environment. Attempts to standardise the environment in which mice are kept have gone on for many decades, which have culminated in the various barrier and isolation systems in use today (Chapter 1). Such clean environments may ease the transfer of animals between labs, and the defined health status of the mice improves the comparison of results between experiments. Movement of clean animals is made easier by (and in fact may absolutely require) transfer of embryos or sperm; and transportation of these is best achieved after freezing. Mouse embryos were first successfully recovered after freezing in 1972 by Whittingham and by Wilmut. This technology has enabled archiving of novel mouse mutations in liquid nitrogen, which eases demand for space in mouse facilities, provides a safeguard against catastrophic loss of stocks and is a convenient way of transporting stocks. It is suprising that mouse sperm cryopreservation has only become viable in the last few years; but this may well become the technology of choice in the future (Chapter 2).

The accumulation of a huge amount of data, such as microsatellite mapping and physical clone mapping, necessitated a change in the way data is published. Access to these data has only been possible through the use of the Internet, and the World Wide Web in particular. The Internet has also influenced the way labs working in mouse genetics communicate. An informal publication, Mouse News Letter, had previously been used as a means of communicating unpublished, prepublished or other information, such as lists of resources. This means of communication became increasingly difficult as participation in the field grew, and expectations changed. However, much of the function of Mouse News Letter is now performed by large database repositories and by electronic mail groups (such as MGI-list) which disseminate knowledge in an informal way (Chapter 7).

So, where is mouse genetics going in the new century? Bateson in his Inaugural Lecutre at Cambridge on 'The Methods and Scope of Genetics' (1908) concluded that 'in the disentaglement of the properties and interaction of' [the genes] 'we must call to our aid physiological chemistry. . . . Such work can only be conducted by those who have the good fortune to be able to count upon continual help and advice from specialists in the various branches of zoology and physiology' These thoughts, from over 90 years ago, neatly sum up what has been called the post-genomic era; where attention will return again to phenotypic analysis. We are going to have quite soon the complete sequence of the human and mouse genomes, which is going to bring about radical changes in the way we do genetics. The massive task of assembling and annotating all of the sequence information will only be the first step. We will all have to become expert in handling the information, and in using the data from other species. Comparisons between mouse, human and other genome sequences will clearly facilitate the analysis of functional regions. It is highly likely that we will accumulate a bank of knock-outs corresponding to every mouse gene, probably by gene trapping, and there is no doubt that techniques for making conditional knock-outs will improve. From the phenotype driven side we can expect a large resource of more subtle, or at least unexpected, phenotypes arising from the various large-scale mutagenesis programmes which are already in progress. The genome sequence will greatly facilitate the identification of mutant genes corresponding to these mapped phenotypes. Functional analysis, in particular of human genes, will continue to be carried out using transgenesis with improved control over sites and level of expression. All these new mutants lines will be need to be housed and stored, whether as live animals, as frozen sperm or embryos or even as frozen ES cells. We can look forward to detailed gene expression databases containing a complete description of expression patterns at all stages of development. Some of this data will come from conventional in situ hybridisation, but will be stored and analysed in much more sophisticated ways as 3D reconstruction techniques provide standardised, computer-based atlases of embryonic development onto which expression can be mapped. Other expression data will come from high-throughout analysis of many (even all) genes using hybridisation of fluorescent probes to microarrayed cDNAs or oligonucleotides on glass slides.

Finally, we should say that this will not be the end of mouse genetics. The contribution of quantitative trait loci to phenotype is only just beginning, and can only be analysed by linkage analysis. Elucidating the genetic contribution to complex and common diseases of humans will be one of the major challenges of the new century, and the mouse model system will be key in finding these genes, and establishing their functional significance through transgenesis.

The mouse is the pre-eminent model for human genetics and we will see an increasing number of new workers coming to the field in search of answers to specific questions raised by human genetics. We anticipate that this book will continue to be a useful resource for current and future mouse geneticists for many years to come.

Edinburgh
August 1999

Ian J. Jackson
Cathy Abbott

Contents

List of Contributors xxi

Abbreviations xxiii

1. Mouse care and husbandry 1

Colin M. Hetherington, Brendan Doe and Donald Hay

1. Introduction 1

2. Housing 1
Types of facilities 1
Conventional accommodation 1
Barrier accommodation or specified pathogen free 1
Germ-free (axenic) and gnotobiotic accommodation 2
Maintenance of pathogen-free animals 2
Breeding performance and production 2
Interpretation of data 2
Collaborative experiments 2
Health screening 3
Quarantine 3
Flexible film isolators 4
Individually ventilated cage (IVC) racks 4
Fumigation 5

3. Husbandry 8
Caging 8
Bedding 9
Diet 9
Water 10
Environmental control 10
Environmental enhancement 10

4. Animal identification 11
Ear punch 11
Toe clip 11
Ear tags 11
Tail and paw tattooing 11
Electronic tags 14

5. Record keeping 16

6. Breeding systems 16
Timed mating 17
Pseudopregnancy and vasectomy 18
Superovulation 19

7. Rederivation 20
8. Euthanasia 24
9. Transportation 24
10. Legislation 24
Acknowledgements 25
References 25
Further reading 25

2. Cryopreservation and rederivation of embryos and gametes 27

Peter H. Glenister and William F. Rall

1. Introduction 27
2. Proliferation of mouse models 27
Long-term maintenance of mouse models 28
Cryopreservation and banking of mouse germplasm: resources for future research 28
3. Principles of cryopreservation 29
Controlled freezing 30
Vitrification 32
4. Skills required for embryo and gamete cryopreservation 33
5. Special equipment required for cryopreservation 34
6. Cryopreservation and recovery of mouse embryos 34
Source of embryos and preferred developmental stages 35
Freezing embryos using propylene glycol as cryoprotectant 36
Recovery of embryos and live mice from embryos frozen according to *Protocol 2* 39
Freezing embryos using glycerol as cryoprotectant 40
Vitrification of embryos in vitrification solution VS3a 42
Recovery of embryos and live mice from embryos cryopreserved according to *Protocols 4* and *5* 44
7. Cryopreservation and recovery of mouse gametes 46
Cryopreservation of mouse metaphase II oocytes 47
Recovery of oocytes frozen using *Protocol 7* 48
Cryopreservation of mouse sperm 49
In vitro fertilization of mouse oocytes 51
8. Hints and tips 53
Embryo cryopreservation 53
Sperm cryopreservation and *in vitro* fertilization 54

9. Long-term storage of embryos and gametes 55
Expected overall survival of cryopreserved embryos and gametes 55

10. Genetic resource banks 57

Acknowledgements 58

References 58

3. Spatial analysis of gene expression 61

Stefan G. Nonchev and Mark K. Maconochie

1. Introduction 61

2. Dissection of post-implantation embryos 62

3. Whole-mount non-radioactive *in situ* hybridization 63
Precautions for RNA work 63
Theory of *in situ* hybridization 64
Generation of riboprobe 65
Fixation and pre-treatment of embryos 67
Hybridization and removal of non-specific hybrids 68
Immunodetection of signals 69
Troubleshooting 71

4. Reporter transgenes 72
General considerations 72
Introduction and uses of reporter genes 72
Detection of β-galactosidase activity 73
Detection of alkaline phosphatase activity 74
Use of the green fluorescent protein (GFP) in transgenesis 75

5. Multiple and combined detection systems 77
Whole-mount staining of early (8.0–12.5 dpc) embryos 77
Multiple detection on cryostat and paraffin sections 78

6. Whole-mount skeletal analysis 83

Acknowledgements 85

References 85

4. Mapping phenotypic trait loci 87

Benjamin A. Taylor

1. Introduction 87

2. Rationale for mapping 87
Genetic definition of trait locus 87
Identification of candidate genes 87
Genetic manipulation 87

3. Background information for mapping phenotypic trait loci of the mouse 88
Definitions 88
Mouse genome 89
Mapping basics 89

4. A backcross mapping experiment: a model for simple mapping experiments 89
Identifying the phenotypic variant 89
Choosing a tester strain 90
Mating mutant and tester stock to produce F1 progeny 92
Producing backcross progeny 92
Classifying progeny for the target locus 93
Extracting a DNA sample 93
A linkage testing plan 93
Genotyping a subset of backcross progeny for selected markers 95
Recording data 96
Scanning data for linkage 96
Finding markers that flank the target 97
Finding closer markers 97
Analysing the complete data set to determine gene order and distance 97
Reporting linkage data 98
Size of cross 99
Markers for fine mapping 100

5. Incomplete penetrance, phenocopies and reduced viability 100

6. Mapping by intercross 101

7. QTL mapping, including the use of derivitive strains 102
General nature of QTL mapping 102
QTL mapping using strain crosses 103
Strain selection 103
Backcross versus intercross 103
Number of progeny 104
Marker spacing 104
Selective genotyping 105
Significance thresholds 106
Confidence intervals 107
Fine mapping QTLs 107
Selective phenotyping 107
Progeny testing 108
Congenic strain construction 108
Advanced intercross populations 108
Derivative inbred strains for QTL mapping 109
Recombinant inbred strains 109
Congenic strains 111
Recombinant congenic strains 111
Consomic strains 112

8. Markers 112
9. DNA pooling 114
Acknowledgements 118
References 118

5. Mapping genomes 121

Paul Denny and Stephen D. M. Brown

1. Introduction 121
2. Applications of genetic and physical mapping 122
3. Genetic mapping 124
Types of genetic marker 124
Genotyping using silver-staining of SSCP gels 124
Genotyping using fluorescently-labelled dCTP incorporation into PCR products from SSLP markers, for analysis on ABI sequencers 126
Linking different maps together 128
Generating new genetic markers in specific genomic regions 129
Criteria for making the decision to construct physical maps 131
4. Physical mapping 131
Introduction to physical mapping 131
Primary contig generation 132
Clone identification using hybridization assay 134
Chromosome walking/gap filling 136
Isolating insert end sequences 136
Validation of clone contigs 138
Acknowledgements 140
References 140

6. Mouse cytogenetics and FISH 143

1. Introduction 143

6A. Analysing mouse chromosomal rearrangements with G-banded chromosomes 144

Ellen C. Akeson and Muriel T. Davisson

1. Introduction 144
2. Metaphase chromosomes 144

3. Stimulation of peripheral mouse lymphyocytes by phytohaemagglutinin and lipopolysaccharide 148
Phytohaemagglutinin 148
Lipopolysaccharide 149

4. G-bands 149

5. Identifying and karyotyping mouse chromosomes 151

6. Recognizing aberrant chromosomes 152

Acknowledgement 153

References 153

6B. Fluorescent in situ hybridization (FISH) to mouse chromosomes 154

Margaret Fox and Sue Povey

1. Introduction 154

2. FISH and chromosome identification 154
Repeat DNA 154
Mouse chromosome paints 155
P1 probes 155

3. Obtaining mouse chromosomes 155

4. Probes and labelling 158
DNA 158
Labelling by nick translation 158
Chromosome paints and direct labelling 159

5. FISH 160
Probe preparation 160
Hybridization 162

6. Analysis and microscopy 167

References 169

7. Electronic tools for accessing the mouse genome 171

Janan T. Eppig

1. Introduction 171

2. Databases of genomic information for the mouse 171
The Mouse Genome Database (MGD) 171
The MRC Mammalian Genetics Unit 175

3. Specific data sets for genetic, radiation hybrid and physical mapping 175
Genetic data 175
The Whitehead Institute/MIT map 175
EUCIB 175
Jackson Laboratory backcross 176
Copeland-Jenkins interspecific backcross 176
M. Seldin interspecific backcross 176
Radiation hybrid data 176
The Jackson Laboratory mouse radiation hybrid database 177
The EBI radiation hybrid database 177
Physical mapping data 177
Genome-wide physical map 177
X-Chromosome physical map 177

4. Gene expression data 178
The Gene Expression Database (GXD) 178
The mouse 3D atlas 178

5. Databases of transgenics, knock-outs and other induced mutations 179
The transgenic/targeted mutation database (TBASE) 179
Mouse Knockout and Mutation Database (MKMD) 179
Induced mutant resource (IMR) 179

6. Animal resources lists 180
International mouse strain resource (IMSR) 180
The Jackson Laboratory: JAX mice IMR, MMR, DNA resource 180
European Mouse Mutant Archive (EMMA) 181

7. mgi-list, an electronic bulletin board for the mouse community 181

8. Summary 181

References 182

8. Mutagenesis of the mouse germline 185

Monica J. Justice

1. Introduction 185

2. Mouse mutagenesis 185
Spontaneous mutations 185
Systems for assessing mutation rate 186
Reporter genes 186
Specific locus test 187
Dominant phenotype assays 188
Other assays 188
Developing a mouse mutant resource 188
Determining functional complexity of genomic regions 189

Human disease models 189
Allelic series 190
Unravelling biochemical or developmental pathways 190

3. High-efficiency mutagenesis with *N*-ethyl-*N*-nitrosourea 191
Mode of action 191
ENU: chemical properties, stability and half-life 191
Types of DNA lesion caused by ENU 191
Effects on protein products 192
Induction of mutations in the mouse germline 192
Doses and treatment protocols 192
Genetic screens to isolate mutations 196
Dominant mutations 196
Single-locus screens for recessive mutations 197
Three-generation breeding scheme for recessive mutations 197
Two-generation breeding scheme using deletions 199
Modifiers and sensitized pathways 200

4. Practical considerations for ENU mutagenesis 201
Breeding considerations 201
Male rotations 201
Gamete sampling for spermatogonial stem-cell mutagenesis 203
Strain background effects 206
Inbred strains to use for mutagenesis 206
F1 hybrids 207
Other observations 208
Rate of recovery to fertility 208
Mutation rates for different loci 208
Cancer susceptibility and lifespan 208

5. Other mutagens 209
Radiation mutagenesis 209
X-rays: treatment and mutations recovered 209
Other types of radiation 210
Chlorambucil 210

6. Future prospects 211
DNA repair 211
Sequence-based screening for lesions 211
The future of mutagenesis 211

References 212

9. Generation of transgenic mice from plasmids, BACs and YACs 217

Annette Hammes and Andreas Schedl

1. Introduction 217
Principles and general considerations 217

Difficulties and limitations 218
Construct design 219
Perspectives 220
Factors influencing the efficiency of transgenesis 220
Choice of mouse strains 221

2. DNA isolation 221
Plasmid DNA 221
YAC DNA 223
BAC DNA 226
Testing the DNA concentration and quality 227

3. The microinjection set-up 228
Location and design of the injection table 229
Microscope 230
Micromanipulators 230
The holding pipette 231
Microinjection needles 231

4. The microinjection experiment 232
Superovulation and isolation of fertilized oocytes 232
Microinjections 234
Timing of injections 234
Injections 234
Oviduct transfer 236
Pseudopregnant females 236
Preparation of the transfer pipette 237
Oviduct transfer 238

5. Analysis of transgenic founders 240
Tail tipping and ear punching 243
Isolation of genomic DNA from tail biopsies 243
Southern blot analysis 245

References 245

10. Directed mutagenesis in embryonic stem cells 247

Antonius Plagge, Gavin Kelsey and Nicholas D. Allen

1. Introduction 247

2. Basic elements of construct design 249
Replacement versus insertion 249
Regions of homology 250
The mutation 251
Components of the targeting cassette 251
Enrichment for targeted events 252
Screening strategies 253
Alternative approaches for construct design 254

3. The use of site-specific recombinases in targeting 254
Excision of heterologous DNA from a targeted locus 255
Excision from ES-cell clones with recombinase expression plasmids 256
Excision in mice using transgenic mice expressing recombinase 257
Oocyte microinjection of recombinase 257
Use of recombinases to generate conditional somatic knock-outs 257
Spatially restricted knock-outs using tissue specific promoters to drive recombinase expression 258
Temporally regulated knock-outs using inducible recombinases 259
Knocking-in: recombinase mediated integration into chromosomally positioned target sites 262

4. Subtle mutations without site-specific recombinases 263
Double replacement 264
'Hit-and-run' 265

5. Chromosome engineering in ES cells 265
Deletions, inversions and duplications 265
Translocations 268

6. Generating, analysing and maintaining knock-out mice 268
ES-cell culture and pluripotency 268
Analysis of ES-cell potency in chimeras with tetraploid embryos 269
Equipment for injection of ES cells into embryos 269
Host embryos 271
Blastocyst injection 272
Morula injection 275
Morula aggregation 277
Transfer of embryos to the uterus 278
Genetic background 279
Maintenance of mutations as co-isogenic and congenic lines and mutant analysis in a hybrid background 280
Chimera analysis 281

Acknowledgements 282

References 282

List of suppliers 285

Index 290

Contributors

ELLEN C. AKESON
The Jackson Laboratory, 600 Main Street, Bar Harbor, ME 04609, USA.

NICHOLAS D. ALLEN
The Babraham Institute, Babraham Hall, Babraham, Cambridge CB2 4AT, UK.

STEPHEN D. M. BROWN
MRC Mouse Genome Centre, Harwell, Oxfordshire OX11 0RD, UK.

MURIEL T. DAVISSON
The Jackson Laboratory, 600 Main Street, Bar Harbor, ME 04609, USA.

PAUL DENNY
MRC Mouse Genome Centre, Harwell, Oxfordshire OX11 0RD, UK.

BRENDAN DOE
MRC Human Genetics Unit, Western General Hospital, Crewe Road, Edinburgh EH4 2XU, UK.

JANAN T. EPPIG
The Jackson Laboratory, 600 Main Street, Bar Harbor, ME 04609, USA.

MARGARET FOX
MRC Human Biochemical Genetics Unit, University College London, Wolfson House, 4 Stephenson Way, London NW1 2HE, UK.

PETER H. GLENISTER
MRC Mammalian Genetics Unit, Harwell, Oxfordshire OX11 0RD, UK.

ANNETTE HAMMES
Max-Delbrück-Centrum for Molecular Medicine, Robert-Rössle-Strasse 10, 13122 Berlin-Buch, Germany.

DONALD HAY
Biomedical Research Facility, Western General Hospital, Crewe Road, Edinburgh EH4 2XU, UK.

COLIN M. HETHERINGTON
University of Oxford, Biomedical Sciences, John Radcliffe Hospital, Oxford OX3 9DU, UK.

MONICA J. JUSTICE
Department of Molecular and Human Genetics, Baylor College of Medicine, One Baylor Plaza, Houston, TX 77096, USA.

GAVIN KELSEY
The Babraham Institute, Babraham Hall, Babraham, Cambridge CB2 4AT, UK.

MARK K. MACONOCHIE
MRC Mammalian Genetics Unit, Harwell, Oxfordshire OX11 0RD, UK.

STEFAN G. NONCHEV
Université Joseph Fourier, Institut Albert Bonniot, Domaine de la Merci, 38706 La Tronche Cedex, France.

ANTONIUS PLAGGE
The Babraham Institute, Babraham Hall, Babraham, Cambridge CB2 4AT, UK.

SUE POVEY
MRC Human Biochemical Genetics Unit, University College London, Wolfson House, 4 Stephenson Way, London NW1 2HE, UK.

WILLIAM F. RALL
Genetic Resource Section, Veterinary Resources Program, Office of Research Services, NIH, Building 14F, Room 101, Bethesda, MD 20892-550, USA.

ANDREAS SCHEDL
Max-Delbrück-Centrum for Molecular Medicine, Robert-Rössle-Strasse 10, 13122 Berlin-Buch, Germany.

BENJAMIN A. TAYLOR
The Jackson Laboratory, 600 Main Street, Bar Harbor, ME 04609, USA.

Abbreviations

AGT	O^6-alkylguanine–DNA alkyltransferase
AP	alkaline phosphatase
ASG	acetic-saline-Giemsa
BAC	bacterial artificial chromosome
BCIP	5-bromo-4-chloro-3-indolyl phosphate
BSA	bovine serum albumin
CAT	chloramphenicol acetyl transferase
CHAPS	3-[(3-cholamidopropyl)dimethylammonio]-1-propane-sulfonate
CI	confidence interval
CL	confidence limit
Con-A	concanavalin A
DAPI	4′,6-diamidino-2-phenylindole
DEPC	diethyl pyrocarbonate
DIG	digoxigenin
DLT	dominant-lethal test
DMSO	dimethylsulfoxide
dpc	days *post coitum*
DTT	dithiothreitol
ENU	*N*-ethyl-*N*-nitrosourea
ES cell	embryonic stem cell
EUCIB	The European Collaborative Interspecific Backcross
FCS	fetal calf serum
FISH	fluorescent *in situ* hybridization
FITC	fluorescein isothiocyanate
GFP	green fluorescent protein
GXD	gene expression database
hCG	human chorionic gonadotrophin
HTT	heritable translocation test
ICM	inner cell mass
ICSI	intracytoplasmic sperm injection
IVC	individually ventilated cage
IVF	*in vitro* fertilization
LET	linear energy transfer
LH	luteinizing hormone
LIF	leukaemia inhibitory factor
LINE	long interspersed sequence
LPS	lipopolysaccharide
MGD	mouse genome database

NBT	nitroblue tetrazolium
NP40	Nonidet P40
PAC	P1 artificial chromosome
PBS	phosphate-buffered saline
PAP	placental alkaline phosphatase
PFGE	pulsed field gel electrophoresis
PHA	phytohaemagglutinin
PLAP	placental alkaline phosphatase
PMSG	pregnant mare serum gonadotrophin
PVA	polyvinyl alcohol
QTL	quantitative trait locus
RC	recombinant congenic
RFLP	restriction fragment length polymorphism
RFLV	restriction fragment length variation
RH	radiation hybrid
RI	recombinant inbred
RT-PCR	reverse transcription–polymerase chain reaction
SINE	short interspersed sequence
SLT	specific locus test
SNP	single nucleotide polymorphism
SPF	specific pathogen free
SSLP	simple-sequence length polymorphism
STS	sequence-tagged site
TAIL-PCR	thermal asymmetric interlaced PCR
TESPA	3-aminopropyl triethoxysilane
X-Gal	5-bromo-4-chloro-3-indolyl b-D-galactopyranoside
UCL	upper confidence limit
YAC	yeast artificial chromosome

1

Mouse care and husbandry

COLIN M. HETHERINGTON, BRENDAN DOE and DONALD HAY

1. Introduction

This chapter illustrates methods for the care and husbandry of mice. It outlines various types of animal accommodation and methods of identification, record keeping, breeding systems, timed mating, superovulation and rederivation.

2. Housing

Mice may be housed under of a variety of conditions. The type of accommodation and husbandry will depend on the nature of the animal experimentation and financial considerations. The more sophisticated the methods used, the greater the cost of building, equipping and running the facility. A balance also must be struck between ease of access and working and the level of pathogen barrier used.

2.1 Types of facilities

2.1.1 Conventional accommodation

A basic facility with a minimum of procedures to prevent the introduction of pathogens is known as a 'conventional facility'. This type of facility has no physical barrier; animals housed in such units can be host to a wide range of harmful organisms, but do not necessarily show clinical symptoms of their exposure. Conventional units, however, can still maintain a relatively low level of pathogen contamination if the husbandry and management methods are under rigorous control.

2.1.2 Barrier accommodation or specified pathogen free

A barriered facility is one where precautions are taken to avoid the introduction of pathogens. In some facilities the aim is to keep animals free from a specific list of pathogens which may interfere with your research. These facilities are referred to as specific pathogen free (SPF). The list of pathogens might include endoparasites and ectoparasites, such as mites, worms, protozoa, certain bacteria, mycoplasma and viruses.

The barrier may take a number of forms. If designed correctly the whole building can be used as the barrier, or more local containment facilities can be used (see Section 2.3). A number of different procedures and routines can be adopted and can include the requirement for staff to shower on entering the animal accommodation and to wear special clothing, including hats, face masks and gloves. It is important that all materials required in a barriered unit are free of pathogens before they are passed into the unit. Barriered units may be equipped with double-ended autoclaves, fumigation chambers, cage washers and dunktanks to allow entry of materials.

2.1.3 Germ-free (axenic) and gnotobiotic accommodation

Animals that are free of all micro-organisms are known as germ free or axenic. If they have a specified flora they are known as gnotobiotic ('known flora'). Such animals must be maintained in positive pressure isolators (see Section 2.3.1) or individually ventilated cage (IVC) racks (see Section 2.3.2) and all materials used, including diet, bedding and water must be sterile.

2.2 Maintenance of pathogen-free animals

There may be several reasons for wanting to eliminate exposure of research animals to pathogens. A summary of possible effects is given below.

2.2.1 Breeding performance and production

In addition to increased mortality, infected or contaminated animals may suffer a reduction in the number of breeding pairs producing offspring, and a reduction in number of offspring born or weaned from each productive pair. This results in an increase in the number of animals that are required for the work, with a consequent increase in both the length of time required for the research and an increase in research costs.

2.2.2 Interpretation of data

Commonly, infectious diseases cause few, if any, obvious signs of clinical disease in mice but may have more insidious effects on the host immunological system which potentially invalidates research data. Furthermore, if mice do demonstrate clinical signs of disease, this may be misinterpreted by researchers as an expected phenotype, particularly with disease model or mutant animals. A disease-free mouse is altogether a better experimental model.

2.2.3 Collaborative experiments

With an increase in barrier facilities, there is reluctance to accept animals from other units unless these units can provide animals of the highest health quality. This has implications for collaborative experiments, leading to periods in quarantine for animals or rederivation of lines before they can be used, with time and financial consequences.

2.2.4 Health screening

Monitoring the health status of animals in a unit, whether it be full barrier or conventional status, is a necessity so that one can determine the microbiological profile of the residents. This is done by means of serology testing for viruses, as well as bacteriology, mycoplasmology, mycology and parasitology tests on the mice. It acts as a method of quality control, in assuring that given operational procedures are being maintained.

Screening can be carried out by the following methods:

(a) Proportional selection of animals from existing stock are sent for screening.
(b) Sentinel animals, which are animals of known microbiological status, are purchased and housed in strategic positions on racks in the rooms for a given period of time. They are then sent off for screening. This method is useful in that potentially valuable animals do not have to be killed.
(c) Serum samples from specific animals are sent for testing or tested using commercially available kits. Such tests provide information on viral contaminants only.

The companies who operate these services offer various packages, from complete overall screening to selective screening for specific organisms. Some are members of approved schemes such as the Laboratory Animals Breeders Association Accreditation Scheme (LABAAS) (Harlan UK) Such schemes give a recommended list of pathogens to be tested and the frequency with which this should be carried out.

2.3 Quarantine

The greatest risk to animal health is the introduction of animals of a lower health status. Many barrier units will only accept animals from established commercial breeders who have rigorously pathogen-tested colonies. Before animals are introduced into a facility it is essential that their health status is known in order to ensure that pathogens currently absent from the facility are not introduced inadvertently. A current detailed health report from the supplier should always be obtained but it should be remembered that a health report only provides details of the health status of the animals actually screened and thus reflects the health status of the colony only on the date they were screened. The proportion of animals in a colony infected with certain pathogens may be low and, thus, if the screening sample is small, such pathogens may not be detected. The best advice for maintenance of a high pathogen-free facility is never to accept animals of uncertain health status and to use quarantine and rederivation to introduce animals.

The ability to quarantine imported animals is essential. In addition, precautions must be taken to stop animals becoming exposed to pathogens during transit. This can be achieved by the use of the correct type of transport boxes. Animals should be quarantined on arrival and their health status determined by health screening (see Section 2.2.4). If separate quarantine rooms are not available then local containment facilities can be employed. Note that at the present time, mice imported into the UK from any source are subject to the Rabies (Importation of Dogs, Cats and Other Mammals) Order, 1974, amended 1994, and require an importation licence and rabies quarantine. The Home Office inspector for the animal facility should be consulted.

2.3.1 Flexible film isolators (supplier: Moredun Isolators, Harlan UK)

These consist of a frame surrounded by a sealed, transparent, flexible canopy. The isolators have their own ventilation system with HEPA filters on both inlet and exhaust. Isolators can be run under positive pressure to protect the animals within the isolator from contamination or negative pressure to protect the animals within your facility from those held in the isolators. Movement of materials is via an entry/exit port. This is a double-ended porthole which has openings at both entry and exit via a PVC cover held in place by a rubber band and steel clamp. When using the isolator under negative pressure for quarantine purposes, anything leaving the isolator must be thoroughly decontaminated by using an appropriate disinfectant such as Virkon (BDH), Alcide (Arrowmight Biosciences) or 1% peracetic acid (Solvay Interox). Conversely if using the isolator under positive pressure then items entering must be sterilized before passing into the internal work space. Access to the internal workspace is by gloved sleeves welded to panels of the canopy. Some models are available that have access to the internal workspace by a PVC half suit which has its own air supply (see *Figure 1*). Isolators make excellent local containment facilities and can reduce the running costs of a unit as they avoid the need for air handling units with High Efficiency Particulate Air (HEPA)-filtered air supply. Their limitations are that they are labour intensive and once you have introduced a batch of animals another experimental group cannot be introduced until the first batch has been health screened and removed from the isolator. Isolators will need to be fumigated between emptying and restocking. This can be performed successfully with 1% peracetic acid or Alcide.

2.3.2 Individually ventilated cage (IVC) racks (supplier: Techniplast UK, Charles River)

These racks consist of filter-covered cages which have HEPA-filtered air supplied to them individually from a fan unit (*Figure 2*). Extract air is also HEPA filtered. The racks can run under both positive and negative pressure depending on whether you are trying to protect existing stock or maintain animals in isolation. These racks have several advantages:

- different experimental groups can be introduced without risk of compromising existing animals in the rack

Figure 1. Half-suit isolator for local containment (photograph supplied by Moredun Isolators).

- they can be used as an alternative to room air handling units and therefore reduce the capital investment and running costs
- they provide the animals with a superior micro-environment by reducing levels of ammonia, carbon dioxide and relative humidity.
- as a result of the air flow through the cage, the frequency of changes of bedding and cages can be reduced.

A limitation is the need for a change over station which has a HEPA filtered air supply so that animals or existing stock are not compromised during change of bedding, diet etc.

2.4 Fumigation

Animal rooms and new buildings can be fumigated before being stocked or restocked with animals. In barrier facilities all supplies which cannot be autoclaved need to be fumigated in other ways before entry into the facility.

Figure 2. Individually ventilated cage rack for local containment (photograph supplied by Techniplast UK).

Protocol 1. Formaldehyde fumigation (for fumigation of stock rooms/buildings)

WARNING: formaldehyde fumigation is a hazardous procedure. It is imperative that one is not exposed to the fumes.

Equipment and reagents

- Formaldehyde solution 40% (w/v) or paraformaldehyde pills
- Urn with boil-dry cutout or thermal fumigator and fogger
- Full protective clothing including self-contained breathing apparatus
- Formaldemeter (PPM Ltd)
- Bacillus spore strips (H. W. Anderson)
- Disinfectant with good bacterial and viricidal properties, such as Virkon (BDH), Alcide (Arrowmight Biosciences) or 1% peracetic acid (Solvay Interox)
- Neutral detergent such as Teepol or Tego 2000 (Arrowmight Biosciences)

Method

1. Ensure that all surfaces are free from dirt, organic matter and grease. Formaldehyde has poor penetrating powers.
2. Remove any equipment that is sensitive to formaldehyde or high humidity or seal it in waterproof bags or containers after sterilizing by a suitable method.
3. Spray under the feet of any equipment with disinfectant.

4. If possible, increase the air temperature to 24°C and relative humidity to 70%. The efficiency of formaldehyde fumigation is increased by high temperature and humidity.
5. Disable the fire alarm system.
6. Place spore strips around the area.
7. Prepare the formaldehyde solution: add 100 ml of 40% (w/v) formaldehyde to 900 ml water for every 30 m^3 fumigation volume and place in the urn, or: place 3.2 g of paraformaldehyde pills per m^3 fumigation volume in the thermal fumigator. It must be arranged so that the urn or thermal fumigator can be switched on and off without entering the area to be fumigated and preferably sited in such a position as to be visible through a window.
8. Label all entry points with hazard warning signs and if possible lock the doors.
9. Switch off ventilation plant and label it clearly, indicating why it is switched off. The ventilation system in the plant room should be labelled with hazard warning signs.
10. If paraformaldehyde is to be used and the relative humidity is not 70%, fog the area with water to raise the humidity.
11. Seal the doors with adhesive tape.
12. Switch on the thermal fumigator or urn.
13. Leave for 12 h.
14. Switch on the ventilation plant extract.
15. Enable the fire alarm system.
16. Wait at least 24 h before entering; wear protective clothing and self-contained breathing apparatus.
17. Confirm the absence of formaldehyde with a formaldemeter . Levels below 2 p.p.m. (0.5 mg/m^3) are safe.
18. All surfaces should be washed down using a neutral detergent such as Teepol or Tego 2000. When doing this wear a full-face respirator and protective clothing.
19. Confirm effective fumigation by culturing spore strips (see supplier's instructions); the absence of growth from these strips indicates that fumigation was effective.

Small items not suitable for autoclaving can be passed into a barrier facility through a fumigation chamber and sterilized with ethylene oxide.

Protocol 2. Ethylene oxide fumigation (for fumigation of consummables, electrical items and paper)

WARNING: Ethylene oxide fumigation is a hazardous procedure. It is imperative that one is not exposed to the fumes.

Equipment and reagents

- Anprolene (ethylene oxide) vials (H. W. Anderson)
- Liner bags and seals (H. W. Anderson)
- Exposure indicators (H. W. Anderson)
- Spore strip with culture media (H. W. Anderson)
- Disinfectant with good bactericidal and viricidal properties such as Virkon (BDH), Alcide (Arrowmight Biosciences) or 1% peracetic acid (Solvay Interox)
- Fumigation chamber or sterilizer with air extract system.

Method

1. Ensure that the items to be sterilized are free of dirt, organic matter and grease.
2. Place them in a liner bag with an Anprolene vial, exposure indicator and spore strip. The exposure indicator and spore strip should be placed in the most inaccessible part of the load to confirm penetration of the gas at an adequate concentration.
3. Seal the bag with twist seal or heat sealer.
4. While holding the Anprolene vial through the bag, break the pre-scored neck of the vial.
5. Place the bag in a fumigation chamber or purpose-built sterilizer. If the item being sterilized is to be passed through a double-ended chamber or sterilizer to a barrier unit the outside of the bag and the chamber should be sprayed with a suitable disinfectant with good viricidal properties.
6. Activate the sterilization cycle.
7. Sterilize for 12 h.
8. Purge the chamber for at least 2 h to remove gas.
9. Confirm effective gas penetration by colour change of the exposure indicator and by lack of growth from a cultured spore strip.
10. Remove the items from bag and air any absorbent item for 24 h in a well ventilated room.

3. Husbandry

3.1 Caging

Mouse cages are available in various shapes, sizes and designs. The two most useful sizes are those with a floor area of 200 cm^2 or 1200 cm^2 and 12 cm high.

The former provides accommodation for breeding pairs and trios or for up to five adult mice, and the latter may be used for up to 20 adult mice, depending on their size. Boxes are available made of clear polycarbonate or opaque polypropylene. They may have solid or stainless steel grid bases and stainless steel lids. There is little to suggest which the mice prefer and cages are chosen to suit the management system of the facility, although cages with grid bases are not recommended as breeding cages. Cage lids may be fitted with filter tops to reduce the likelihood of the spread of pathogens. Cages may be held on wall-mounted racks, allowing for easy floor cleaning and the recapture of escaped animals, or on mobile racks.

3.2 Bedding

Bedding is usually woodchips or shavings. The wood used must be free of insecticides and herbicides. Allergies to wood dust can develop in staff and only dust-free bedding should be used. In addition consideration should be given to the need for staff to wear masks when handling dry bedding. Bedding may be sterilized by autoclaving or irradiation.

Shredded paper, tissues or purpose made material (e.g. from International Market Supplies) can be provided for nesting. This is also easily sterilized by autoclaving.

3.3 Diet

Nutrition is an important factor in the breeding and maintenance of mice in good health and fertility. Various commercial diets are available and formulated as breeding or maintenance diets (Harlan Teklad, Special Diet Services). Diets may be supplied as pellets of various shapes and sizes or powders. Expanded pelleted diets result in less wastage by the animals and these diets can also be classed as pasteurized. If a higher pathogen-free standard is required, diets designed to be autoclaved can be used, which are supplemented with nutrients to compensate for the inevitable degradation that occurs on autoclaving, and treated to prevent the pellets sticking together (Harlan Teklad). The physical changes to diet pellets brought about by autoclaving depends on the constitution of the diet and the sterilizing and drying cycle of the autoclave. A diet should always be tested in the autoclave that will normally be used to ensure that it is acceptable after sterilization. It may be necessary to spread the diet on trays in the autoclave to prevent pellets sticking together. Colour-change indicators are available for use in autoclaves which should be placed in the core of the load to determine that the diet or bedding has been exposed to sufficient heat for a sufficient length of time for sterilization to have taken place.

Diets sterilized by irradiation are also available. These diets are expensive and are generally only used for isolators or IVC racks. When using irradiated diets it important to ensure that the packaging is undamaged, which is normally done by confirming that the vacuum drawn when the diet was packed is maintained.

3.4 Water

Water can be provided by water bottles or by automatic watering systems. With either system it is vital that leaking water is avoided. There is a real risk of mice, especially young animals, drowning or dying from the effects of a soaking if a bottle or nozzle of an automatic system leaks. In most circumstances mains water is of adequate microbiological quality but in barriered units or where immunodeficient animals are maintained it may be necessary to use water that has been filtered, or autoclaved, or acidified by the addition of chlorine or 1% hydrochloric acid. Where microbiological quality of water is important then a complete change over of water bottles twice weekly is advised as water bottles quickly become contaminated from saliva, faeces and bedding.

3.5 Environmental control

Forced air ventilation is required to remove odours and allergens. The room temperature should be maintained at 21 ± 2°C and at a relative humidity of 55 ± 10%. Full air conditioning is recommended. It is important that air is distributed evenly throughout the room and that animals are not exposed to draughts. The number of air changes per hour required depends on the stocking density and the effectiveness of the air distribution within the room. Sixteen air changes per hour in a room containing 150–175 mice/m^2 of floor space are adequate provided the cages are cleaned regularly. The use of pre- and HEPA filters on the incoming air supply can be used to stop pathogen entry. In addition it is possible to balance the air supply so that animal rooms are positive in pressure to adjacent corridors. It should be noted that the microclimate in the cage may differ significantly from the room with significantly higher temperature, humidity and ammonia concentration in the cage.

Animal rooms should have no windows and a controlled diurnal cycle. The most commonly used lighting regimen is 12 h light/12 h dark (see also Section 6.1). Drains should be avoided in barriered animal rooms as these offer a source of pathogens via wild rodents. In barrier situations the plant room should be accessible for maintenance without entering animal accommodation and physically sealed to rooms below.

3.6 Environmental enhancement

It is now an accepted fact that animals living in confined areas require some form of stimulus to prevent boredom which can cause stress; mice are no exception. Various methods have been recommended to help reduce this phenomenon, they include the following:

- Try to avoid housing mice singly.
- Add materials that give the mice some diversion, such as paper, wool or

shavings, which encourage the animals to build nests. An improvement on shavings are 'nestlets' (International Market Supplies); these are compressed cotton fibre squares which are sterilized during manufacture. When these are placed in the cage, the mice will shred them and build nests. Cardboard tubes which encourage gnawing are also clean, safe and disposable (Lillico).

- Putting some diet cubes directly into the cage when adding clean bedding can be an enhancement, especially with newly weaned mice.
- Devices such as activity wheels and other cage furniture are available, but are not necessarily more effective than the nesting materials.

4. Animal identification

4.1 Ear punch

Mice may be identified by marking the ears with an ear punch. Ear punches that produce holes of various sizes are available commercially (International Market Supplies). The standard marking system is shown in *Figure 3*. It is often sufficient to simply mark one or both ears to enable mice within a cage to be identified individually. Ear marking cannot be used on mice younger than 10 days as their ears are too small. There is always the risk of an ear being torn or chewed, resulting in the loss of an animal's identification marks.

4.2 Toe clip

Mice may be identified by removing the terminal joint of one or more toes. This should only be done under local or general anaesthetic. The standard numbering system is shown in *Figure 3*. When combined with ear marking, numbers up to 9999 can be achieved. This system is used only rarely as more humane systems are available and in the UK the local Home Office inspector should be asked for comments before proceeding with this method.

4.3 Ear tags

Ear tags are available commercially (International Market Supplies). Animals less than 3 weeks old cannot be identified with tags as their ears are too small.

4.4 Tail and paw tattooing

Mice may be identified by tattooing the tail and/or the paws. Animals can be tattooed from birth; from days 1–7 tattooing can be done by a single operator but for older animals it is easier with one person holding the scruff of the animal whilst the other tattoos. Tattooing equipment for marking mice is available commercially (Animal Identification and Marking Systems (AIMS) or International Market Supplies).

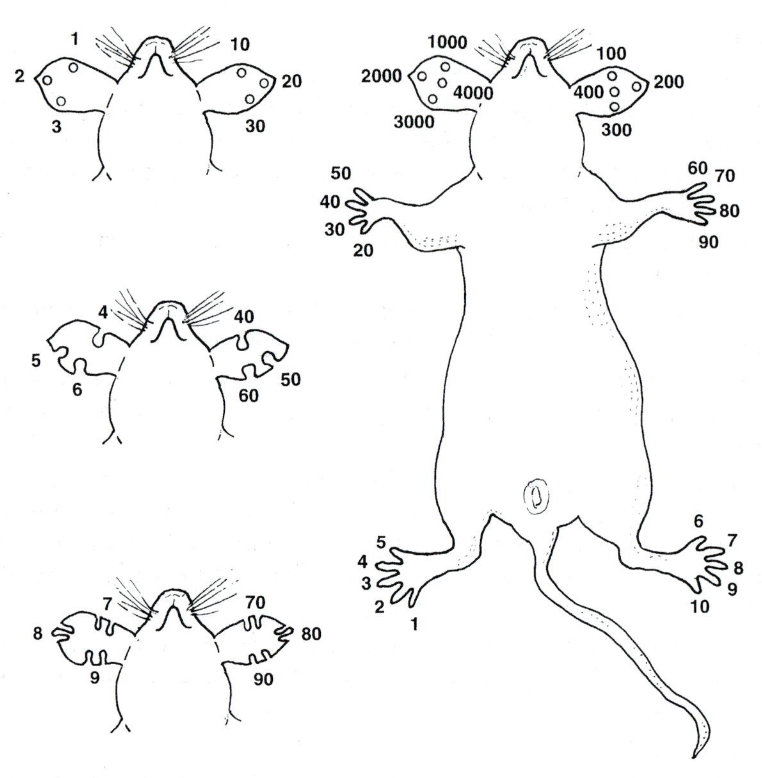

Figure 3. Left: standard ear-punch system for marking animals (ventral view). Animals can be marked from 1 to 99 with this system. Right: ventral view of a system using a combination of toe clips and ear punch to mark animals from 1 to 9999. Ear punches are combined to give the full range of hundreds and thousands.

A numbering system for paw tattoos is shown in *Figure 4*, and a mouse with a tail tattoo shown in *Figure 5*. Once applied, tattoos last indefinitely.

Protocol 3. Paw and tail tattooing

This protocol gives a general outline of the tattooing method. It is highly recommended that full training is undertaken to ensure that the animals are tattooed with minimal stress and that the tattoos are of a high quality.

Equipment and reagents

- Tattooing equipment (Animal Identification and Marking Systems or International Market Supplies)
- Permanent ink (Animal Identification and Marking Systems or International Market Supplies)
- Mouse restrainer (Animal Identification and Marking Systems)

Method

1. Check the needle with a magnifier or microscope to confirm that it is not damaged. A single-point needle should be used for paws and a three-point needle for tails. Damaged needles can cause bleeding and inflammation with increased phagocyte activity at the site of the tattoo leading to fading.
2. Load the needle so that the point protrudes about 1 mm from the tattoo gun.
3. Connect the gun to the power supply. The depth to which the needle penetrates the skin is adjusted by varying the power.
4. Shake the ink to ensure an even distribution of ink particles and therefore a dark permanent tattoo. Dispense the ink into a suitable receptacle.
5. When tattooing paws, pick up the mouse by the scruff if it is older than 7 days (*Figure 6*). If applying a tail tattoo, put the animal into a mouse restrainer which allows the tail to protrude for tattooing. The restrainer fits onto a tattooing platform which helps keep the restrainer in place during the procedure.
6. Apply tissue oil to the area to be tattooed with a cotton bud. The oil aids tattooing by lubricating the needle, so preventing staining to the surrounding area of the tattoo by excess ink .
7. Dip the tattoo needle in the ink.
8. Holding the gun perpendicular to the skin operate the foot switch and apply the needle to the area to be tattooed. If tattooing the tail, lay a specially designed tool over the area to be tattooed, and hold this down with one hand; a gap in the tool leaves an exposed area of the tail for tattooing.
9. When applying a tattoo to the tail, always tattoo from the base of the tail to the tip.
10. Wipe the tattooed area with a tissue to ensure that the ink has penetrated the skin. If not repeat steps 6–9. There should be no blood at the tattoo site. If bleeding occurs, reduce the power level and check that the needle is not damaged.
11. When tattooing mice of <15 g, do not use a tattoo that encloses a complete area, e.g. a '0', as this may cause the enclosed area of skin to die and the tattoo not to be permanent. Use '( )' rather than '0'. This is not a problem in older mice.
12. Clean the needle thoroughly by washing sequentially with skin cleanser, water and 95% ethanol before carefully wiping the tip. The needle should be inspected for damage and replaced if the needle tip is hooked or blunt.

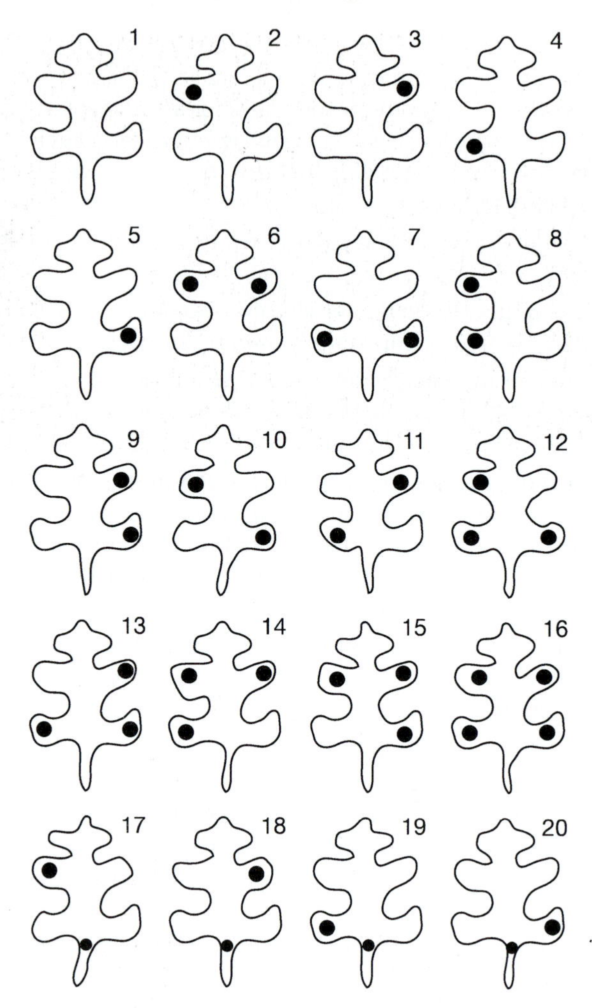

Figure 4. Schematic view of a marking system for paw tattoo marking. The cartoons represent the ventral view of a mouse indicating the locations of single ink dots on each paw or on the base of a tail which can be combined to give a number from 1 to 20.

4.5 Electronic tags

In these systems (such as the Trovan System from R. S. Biotech) an encapsulated transponder is injected into the animal subcutaneously via a needle inserted from an implanter. The transponser is about the size of a grain of rice. This requires the use of a 12G needle and is thus not suitable for young animals. Each transponder has a unique code which can be read at any time by a passing a scanner over the mouse; the unique number is shown on an LCD display on the reader. It is worth noting that electronic tags are relatively expensive, particularly if tagging all animals in a large facility.

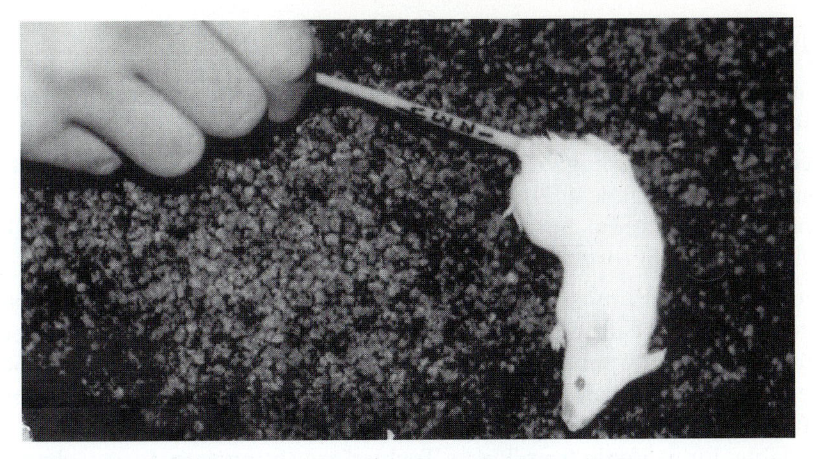

Figure 5. Mouse tail after tattooing (photograph supplied by AIMS).

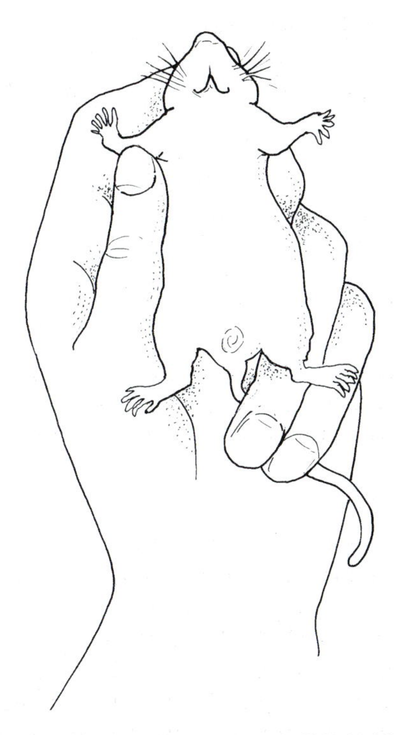

Figure 6. Mouse under restraint by 'scruffing'. The scruff is held between the index and middle fingers while the third and fourth fingers hold the tail.

5. Record keeping

Accurate record keeping is an important function in all modern animal facilities. It can broken down into three main areas:

- **General housekeeping:** recording information for maintenance charges, consumables, technical assistance, rental and use of unit equipment. This type of information can be kept accurately using any PC spreadsheet software, e.g. Microsoft Excel.
- **Breeding records:** most units have their own method of keeping breeding records. Each local unit may have different requirements for the type of information they want to record, but manual record keeping can be done by use of breeding record cards and pedigree charts. Specialized computer software specifically designed for keeping records of mouse colonies is also available, for example from Locus Technology (http://www.locustechnology.com). Software for keeping animal breeding records has recently been developed at the MRC Human Genetics Unit and is available for downloading over the Internet (http://www.hgu.mrc.ac.uk/Softdata/Lams/). However, at the moment there is no universally accepted standard. A standardized system would have the advantage that accurate information could be transferred between units involved in collaborative work associated with breeding systems. The use of correct nomenclature for genes, mutations, transgenes and strains does aid in communication between units. The Mouse Genome Database (http://www.informatics.jax.org/) maintains lists of genes and strains and has guidelines for nomenclature (see Chapter 7).
- **Experimental returns:** in many countries experimental work is covered by government legislation and it is important that accurate records are kept of animals used in experimental procedures. This information can be recorded by manual means, or on computers.

6. Breeding systems

Mice will normally begin to breed from 6 to 8 weeks of age and continue for about 200 days. There are, however, significant strain variations (see ref. 1). The gestation period of the mouse is 19–20 days and parturition is often associated with postpartum oestrus. Thus if a male is housed with a pregnant female, they will mate immediately after the birth of the litter and the female will become pregnant while feeding the newly born litter. Concurrent lactation and pregnancy can result in delayed implantation (i.e. embryos remain viable but blocked at the blastocyst stage for a number of days), resulting in a prolonged gestation.

Mice are commonly bred in pairs, trios (one male with two females) or harems (one male with more than two females). Weekly rotation of a male mouse through cages containing pairs of females may be used to optimize the yield of offspring from a single male. It is not advisable to put more than one male in a breeding cage as the males will fight. Pair mating is optimal for the maximum production of offspring per female; other mating systems result in a higher number of animals weaned per cage but preweaning mortality is higher.

Mice are weaned at approximately 3 weeks of age and should be removed from the breeding cage before the birth of a subsequent litter. Male mice can be caged together from weaning, but if adult males from different cages are mixed they will often fight.

Breeding records should be kept to identify the animals to be used as replacement breeders, to maintain pedigree details and to ensure that unproductive animals can be identified and culled. The breeding system adopted will depend on the number of animals required and whether they are of an inbred or outbred strain. The breeding nucleus of an inbred strain must be maintained by full-sib mating (brother × sister) while an outbred strain must be maintained in such a way as to minimize inbreeding (2).

6.1 Timed mating

The oestrous cycle of the mouse is normally 4 days and females are receptive to mating only at oestrus.

If a regular supply of mated animals is required it is sufficient to set up a series of cages with one male and three females in each. When determining how many animals to mate up, one should bear in mind the fact that one in four females should be in oestrus on any given day. If a group of mated animals is required for a particular day it is possible to increase the probability of mating using the Whitten effect (3). When a group of female mice is placed with a male their oestrous cycles become synchronized and the majority of animals will mate on the third night. A male is caged on one side of a barrier, or in a small cage within a cage, separating him from the females for 2 days, and the animals are allowed to mate on the third day. Alternatively it is possible to detect females in oestrus by observing changes in the appearance of the vagina and by mating only those animals that are in oestrus. At oestrus the vagina is gaping and the tissue is pink and slightly moist with pronounced striations.

When mice mate, the ejaculate of the male coagulates in the female's vagina, producing a solid copulation plug. By checking for the presence of a plug it is possible to determine whether the animal has mated. Sometimes the plug is easily visible but normally a blunt metal probe is used to detect its presence in the vagina. The plug remains in the vagina for varying lengths of time, up to 18 h. The sooner the animals are checked after mating the less the likelihood of the plug falling out prior to checking.

Mice normally mate during the middle of the dark period of the diurnal cycle. Various systems of defining the time of mating and the age of the embryos are in use. The day of finding the vaginal plug may be designated day 0, day 0.5, day 1 or the first day of pregnancy.

The time of mating can be changed relative to the working day by changing the diurnal light/dark cycle. By completely reversing the cycle mating will occur around midday. Changing the diurnal cycle allows different developmental stages of embryos to be available during normal working hours. If animals are brought in from a place with a different light/dark cycle a few days should be allowed for the mice to acclimatize.

6.2 Pseudopregnancy and vasectomy

Pseudopregnant animals are produced by mating animals as described in Section 6.1 but the mating is with vasectomized males or with genetically sterile animals, for example males carrying a chromosomal translocation such as *T(7;19)145H* ('T145'). It is easiest to vasectomize mice when they are about 6 weeks of age before they become fat. The males to be vasectomized should be of a strain or genotype with a good breeding performance, such as CBAB6F1 (offspring of CBA × C57BL6) or an outbred strain. Vasectomized mice can also be purchased from commercial breeders.

Protocol 4. Vasectomy

Equipment and reagents

- Male mice 4–6 weeks old (it is easier to vasectomize young mice as there is less fat)
- Surgical instruments: scissors (12 cm) and two pairs of toothed forceps (10 cm)
- Michel suture clips (≤12 mm) and applicator (Interfocus Ltd.), or suture thread and needle
- Suture clip remover (I.M.S.)
- General anaesthetic, e.g. Hypnorm and Hypnovel (Vetdrug). Make up by adding 2 vols of water for injection to 1 vol. of Hypnorm, then add 1 vol. of Hypnovel (10 mg/5 ml). Under no circumstances add the undiluted drugs together, as this causes crystallization. Store indefinitely at room temperature.

Method

1. Sterilize surgical instruments by autoclaving or use a glass bead sterilizer such as Sterilquartz (International Market Supplies).
2. Anaesthetize the mouse by injection of Hypnorm/Hypnovel intraperitoneally, 10 ml/kg (0.2 ml/20 g mouse) with a 25G needle. The anaesthesia lasts 20–40 min.
3. Make a single transverse incision in the skin on the ventral surface anterior to the coagulating glands.
4. Make an incision through the body wall at one end of the skin incision.
5. Locate the vas deferens by manipulating the fat pad, preferably without exteriorizing the testis.
6. Remove a short length (2–3 mm) of the vas deferens.
7. Repeat for the other side.

8. Close the skin incision with Michel clips (≤12 mm) or suture thread. There is no need to suture the body wall.
9. Remove suture clips after 12 days.
10. Check that the mouse mates but is sterile by mating with females, confirming that mating has occurred by the presence of vaginal plugs and checking for the absence of embryonic implantation sites in the uterus of the females on day 8 or later.

6.3 Superovulation

Large numbers of pre-implantation embryos may be obtained by injecting gonadotrophins prior to mating. The successful induction of superovulation depends on several variables including age, weight and strain of mouse and the times of injection. Some strains or genotypes of mice superovulate well while others do not. The best age for superovulation is usually in the range of 3–5 weeks. The optimum age varies with genotype and the optimum period lasts between 4 and 6 days. Body weight is often a better indication of development in certain strains and, if commercially bred animals are used, may be the only measure of developmental age. Trial and error may be needed to optimize superovulation of a particular strain.

The time of administration of the gonadotrophins relative to the light:dark cycle of the mouse room affects both the number of eggs ovulated and their developmental uniformity. A 46–48 h interval between the pregnant mare serum gonadotrophin (PMSG) and human chorionic gonadotrophin (hCG) injections will usually be found optimal. Ovulation normally occurs 10–13 h after injection of hCG. To obtain the best synchronization and yield it is important that the hCG is administered before the release of endogenous luteinizing hormone (LH) which is regulated by the diurnal cycle. It is generally assumed that endogenous LH release occurs 15–20 h after the midpoint of the second dark period following PMSG administration.

Protocol 5. Superovulation

Equipment and reagents

- 1 ml syringe with 25G needle
- Pregnant mare serum gonadotrophin (PMSG), 5–10 IU/ml (Intervet)
- Stud males of proven fertility[a]
- Human chorionic gonadotrophin (hCG), 5–10 IU/ml (Intervet)
- Female mice to superovulate

Method

1. Ensure that the diurnal cycle for the animal room is set to 12 h dark: 12 h light. In this example the light period is from 08:00h to 20:00h.
2. Inject female mice of the correct age for the genotype with 5–10 IU PMSG intraperitoneally at about 14:00h.

Protocol 5. *Continued*

3. Around 46–48 h later, i.e. at about 12:00–14:00h, inject mice with 5–10 IU hCG intraperitoneally.
4. Cage each female singly with a stud male overnight. Males normally will mate with only one female so they should be set up as pairs.
5. The following morning check for the presence of a vaginal plug.

[a] Note that the mating performance of males varies with genotype. They should not be used on consecutive days and it is important to keep a record of their mating success. Those that do not mate on a regular basis should be replaced.

7. Rederivation

Embryo transfer and Caesarean section are methods of rederiving mouse stocks when it is necessary to improve a colony's health status or to bring animals into a barrier facility. Embryo transfer is the preferred method for 'cleaning up' experimental animals that have pathogens. It has the advantage over Caesarean rederivations of eliminating organisms that may infect the fetus by crossing the placenta. A method for embryo transfer from a conventional to a barrier unit is described below. If animals are to be imported into a barrier unit from a distant site, it may be preferable to import them as frozen embryos or as frozen sperm, depending on circumstances (see Chapter 2). The ease of rederiving strains depends to some extent on the success of superovulating the females. If strains of defined genotype are to be rederived, then it will be important to use females of that strain. On the other hand if the genotype is less important (as may be the case with transgenic mice, for example) then females of a strain known to superovulate well can be mated with males carrying the desired gene or transgene.

A protocol for Caesarean rederivation is also described for use when the expertise and equipment is not available for embryo transfer. It is essential to keep the rederived litters in isolation (see Section 2.3) to ensure that their health status is satisfactory before introducing them to an area housing mice of higher health status. The foster mothers can be used as sentinels in these cases.

Protocol 6. Pre-implantation embryo rederivation

Equipment and reagents

- Female mice made pseudopregnant by mating with a vasectomized male (*Protocol 4*) and mating confirmed by presence of a vaginal plug (see also Chapter 9, Section 4.3.1).
- Handling and culture media: M2 + 4 mg/ml bovine serum albumin (BSA) and MI6 + 4 mg/ml BSA. Embryo-tested M2 and M16 media and embryo-tested BSA can be obtained from Sigma. See also ref. 4.

- Pre-implantation embryos from superovulated female mice (see *Protocol 5*, and Chapter 9, *Protocol 5*)
- Embryo-tested mineral oil (Sigma)
- Thermos flask
- 1% Virkon disinfectant (BDH)

A. *Preparation in the conventional unit*

A pool of pre-implantation embryos for transfer to the barrier unit must be obtained. This is achieved by superovulating wild-type or genetically typed females (*Protocol 5*) and mating them to males of the line or strain to be transferred. This is timed so that on the day of passing embryos to the barrier unit the embryos are at either the one-cell or morula stage (either 0.5 or 2.5 days *post coitum* (dpc)). This means that after overnight culture in the barrier unit the embryos will be ready for transfer at the two-cell or blastocyst stage. At least 2 h before it will be required, place in a 37 °C incubator: a flask of 1% Virkon, a thermos flask and M2 medium + BSA

B. *Preparation in the barrier unit*

Pseudopregnant females should be available in order to receive two-cell or blastocyst-stage embryos

1. Place drops of M16 + BSA under oil (see Chapter 9, *Protocol 5*, Step 11) and equilibrate in a 37 °C CO_2 incubator at least 3 h prior to arrival of embryos. Place a second dish of drops of M2 + BSA under oil in the incubator.
2. When all the solutions have equilibrated to 37 °C, kill the females and dissect out their oviducts (see Chapter 9, *Protocol 5*, Steps 1–8). If dealing with one-cell embryos (0.5 dpc), treat them with hyaluronidase to remove the cumulus cells (see Chapter 9, *Protocol 5*, Step 9). If dealing with morula-stage embryos (2.5 dpc), flush the infundibulum of the oviduct through with 0.1–0.2 ml of M2 + BSA per oviduct into a watch glass or Petri dish. Do this by putting a 30–34 gauge Luer-mount needle onto a 1ml plastic syringe filled with M2 + BSA at 37 °C and inserting the needle into the infundibulum, holding it in place with a pair of forceps, and flushing the medium through, expelling the embryos at the cut end of the uterine horn.
3. After removing all the embryos from the oviduct, wash them through a couple of drops of M2 + BSA medium to remove any unwanted uterine material and put them into approximately 1 ml of 37 °C M2 + BSA in a Sterilin tube or similar. Ensure that the top is firmly sealed so that no disinfectant can enter the tube.
4. Put the 1% Virkon into the thermos flask and then place the tube of embryos into the flask. This sterilizes the outside of the tube at the same time as keeping the embryos at 37 °C. Then put the flask through the appropriate cycle in the sterilizing chamber (do not autoclave!) after spraying it with disinfectant.

Protocol 6. *Continued*

5. On arrival in the barrier, take the embryos through three to five washes of M2 + BSA medium and then 3–5 times in the pre-equilibrated M16+BSA, using a fresh pipette each time (a wash consists of moving the embryos from one drop of medium into the top of a fresh drop of medium and letting them fall to the bottom of the drop; a portion of the medium surrounding the embryos is then sucked up and discarded).
6. After the washing process is complete, culture the embryos overnight to develop to the two-cell stage or blastocysts. Embryos can then be replaced into an appropriately aged pseudopregnant female by oviduct transfer for two-cell embryos (see Chapter 9, *Protocol 7*), or uterine transfer for blastocysts (Chapter 10, *Protocol 7*).

It is worth noting that one-cell embryos can be sensitive to temperature fluctuations (5) and it may be best to adopt the morula/blastocyst regime if embryos have to remain in a fumigation chamber for a length of time when passing across a barrier.

Protocol 7. Rederivation by Caesarian section

A breeding colony of the required or higher health status is required to provide foster mothers. If a successful operation is to be carried out it is essential that mice with newborn litters are available to act as foster mothers.

Equipment and reagents

For operation on donor:

- Sterile scissors (12 cm) and toothed forceps (10 cm)
- 500 ml beaker containing disinfectant solution, for example, 1% Vircon (BDH)
- Container with watertight lid containing disinfectant solution, e.g. 1% Vircon, at 37 °C
- Hair clippers

In area with foster mother:

- Sterile scissors (12 cm) and toothed forceps (10 cm)
- Sterile tissues
- Dish to receive uterus

Method

Note: the complete procedure from cervical dislocation of the mother to delivery of the last pup should take less than 4–5 minutes.

1. By palpation, select the animal that is to undergo the Caesarian.[a]
2. Ensure that a suitable foster mother is available. Some strains of mice make better foster mothers than others. BALB/c mice make good mothers. It is also an advantage to chose a foster mother whose natural litter is different in colour from the litter to be fostered. The

foster mother's litter should be newborn—it is possible to use a mother with a 2–3 day old litter but the success rate will be lower.

3. Prepare a 50 ml container with a watertight lid containing 1% Vircon, at 37 °C.

4. Shave the abdomen of the pregnant mouse.

5. Kill the mouse by cervical dislocation and immediately dip the body in the disinfectant solution in the beaker. This will sterilize the outer surface of the animal and prevent hair blowing about. Blot the body to remove excess moisture and transfer it to a sterile operating surface.

6. Expose the body wall by grasping the skin of the abdomen immediately above and below the midline and pulling simultaneously towards the head and feet. This procedure can be made easier if a small midline incision is first made in the skin with a pair of sterile scissors. By peeling the skin back in this fashion the exposed body wall remains sterile and drapes are unnecessary.

7. With sterile scissors make a longitudinal incision in the body wall.

8. Hold the vagina with sterile forceps and cut caudally. Raise the uterus while continuing to hold the vagina and trim the mesentery, taking care not to puncture the uterine wall. In this way the uterus can be removed without it touching any surfaces. Provided the skin is peeled well away any areas which the uterus might touch should, in any case, be sterile.

9. Transfer the uterus to the disinfectant in the container and take it to the vicinity of the foster mother. This might be via a dunk tank or entry port to an isolator or simply to a clean isolation area.

10. Tip the uterus into a suitable dish, remove it from the disinfectant and blot lightly.

11. Place the uterus on a sterile absorbent surface and slit longitudinally along the anti-mesometrial surface. Remove the fetuses from their membranes. The umbilical cord should be crushed or torn rather than cut to reduce bleeding. Cauterization is unnecessary. If the fetuses are sufficiently mature, bleeding should be slight.

12. Dry the pups with tissue and stimulate them gently until they are pink and breathing regularly. If the pups are immature their skin will be sticky. Mature pups should squeak if their tails are lightly squeezed with a pair of forceps. Pups that do not squeak are unlikely to survive.

13. Transfer the pups to the cage of the foster mother. If the foster mother's own litter has a different eye or coat colour a few of the natural litter may be left with the fostered litter. The foster mother with the rederived litter should be held in isolation until their health status has been shown to be satisfactory. If some of the mother's

Protocol 7. *Continued*

natural litter have been left these can be used for health screening in addition to the foster mother.

[a] Timed matings can be used to determine the likely time of birth but palpation is always advisable owing to the normal spread in the time of parturition. A full-term pregnant mouse may be palpated using the thumb and forefinger. At around the 18th and 19th days of pregnancy the individual conceptuses can be felt as small spheres. Immediately before parturition the fetuses lie in an elongated position and can be moved slightly within the uterine lumen. This is when the Caesarean operation should be performed. An animal that has already started to give birth should not be used.

8. Euthanasia

Mice may be easily killed by cervical dislocation. To do this, pick up the animal to be killed by the base of the tail. Allow the animal to grip the bars of a cage top and, while pulling gently backwards by the tail, firmly pinch the neck; alternatively, hold a rod or ruler across the neck and push down while pulling the tail. We strongly recommend learning this procedure under supervision to ensure humane killing.

If large numbers of animals are to be killed, a euthanasia chamber can be used, consisting of a container linked to a cylinder of carbon dioxide. The animals should be exposed to a steadily rising concentration of carbon dioxide and not plunged into a high concentration. Avoid overcrowding in a lethal chamber and remove all carbon dioxide before immediate re-use of the chamber.

9. Transportation

Mice can be easily transported by road, rail and air. It is vital to ensure that before transferring, importing or exporting mice that you conform with local, national and international regulations. The boxes should be clearly labelled with the name and address of the sender and consignee, and a description of the contents (species, strain, age, sex and number). As a minimum requirement animals in transit should have comfortable secure containers, not be subjected to overcrowding, and have an adequate supply of food and water to sustain them for the period of their journey. When shipping animals by air, International Air Transport Association (IATA) Live Animal Regulations control the size of containers, and the type of materials that can be used for their construction.

10. Legislation

Experimental procedures on living animals in many countries are subject to national or local control. You must ensure that you adhere to any relevant legislation before embarking on animal experimentation.

Acknowledgements

We would like to thank Moredun Isolators for *Figure 1* and for use of their fumigation protocol. We would also like to thank Techniplast UK and Animal Identification and Marking Systems for photographs (*Figures 2* and *5* respectively), and the Photography Department of MRC Human Genetics Unit.

References

1. Silver, L. M. (1995) *Mouse Genetics*. Oxford University Press, Oxford.
2. Festing, M. F. W. (1987) In *The UFAW Handbook on the Care and Management of Laboratory Animals*, 6th edn (ed. T. Poole), p. 58. Longman, Harlow.
3. Whitten, W. K. (1956) *Journal of Endocrinology*, **14**, 160.
4. Hogan, B., Beddington, R., Costantini, F. and Lacy, E. (1994) *Manipulating the Mouse Embryo*, 2nd edn. Cold Spring Harbor Laboratory Press, Cold Spring Harbor, NY.
5. Pickering, S. J. and Johnson, M. H. (1987) *Human Reproduction*, **2**, 207.

Further reading

Poole, T. (ed.) (1987) *The UFAW Handbook on the Care and Management of Laboratory Animals*, 6th edn. Longman, Harlow.

Monk, M. (ed.) (1987) *Mammalian Development: a Practical Approach*. IRL Press, Oxford.

Coates, M. E. and Gustafsson, B. E. (eds) (1984) *The Germ Free Animal in Biomedical Research*. London Laboratory Animals Ltd.

Wolfensohn, S. and Lloyd, M. (1994) *Handbook of Laboratory Animal Management and Welfare*. Oxford University Press.

Flecknell, P. A. (1987) *Laboratory Animal Anaesthesia, an Introduction for Research Workers and Technicians*. Academic Press.

Code of Practice for the Housing and Care of Animals Used in Scientific Procedures (1989). HC107. HMSO, London.

Guidance on the Operation of the Animals (Scientific Procedures) Act 1986 (1990) HC182. HMSO, London.

The Humane Killing of Animals under Schedule 1 to the Animals (Scientific Procedures) Act 1986 (1997) HC193. HMSO, London.

2

Cryopreservation and rederivation of embryos and gametes

PETER H. GLENISTER and WILLIAM F. RALL

1. Introduction

This chapter describes optimized cryopreservation protocols for the long-term storage and maintenance of pre-implantation embryos and gametes from mouse models of human genetic disease, and of normal biological processes. It outlines the theory and practical application of cryopreservation procedures and discusses the quality control requirements to ensure the availability of mouse models for future study. Comprehensive reviews of the history, theory and application of germplasm cryopreservation are available elsewhere (1,2).

2. Proliferation of mouse models

Several scientific trends have increased the number of mouse models and placed an intolerable load on the financial and infrastructural resources needed to maintain them as conventional breeding colonies. First, during the past decade, molecular approaches have permitted the mouse genome to be manipulated in increasingly sophisticated ways. Exogenous DNA and genes can be inserted either randomly (transgenic) or in specific locations of the mouse genome (knock-out, knock-in). These procedures produce unique mice for research, and permit experiments to be repeatable and often very precise, two of the most important criteria for experimental design. As more and more genes are sequenced and cloned, the number of new knock-out, knock-in and transgenic mice will undoubtedly increase.

Second, several institutions are planning large mutagenesis programmes to reduce the 'phenotype gap' in the mouse by producing large numbers of mutants using chemical mutagens such as *N*-ethyl-*N*-nitrosourea (ENU) as described by Brown and Peters (3) and this volume, Chapter 8. An enormous number of new mutations will be generated and require long-term maintenance for future study. Finally, laboratories throughout the world are now preparing for the post-genome sequencing era when the emphasis will be on studying the role of gene structure and function in the context of the whole organism.

2.1 Long-term maintenance of mouse models

Although the creation of new models provides immediate opportunities for scientific study, the models may also prove useful in unexpected ways in the future. History indicates that many of the spontaneous and induced mutations discovered and characterized three decades ago are now in high demand, primarily due to an increased understanding of disease and normal biological processes. There is every reason to believe that many of today's mutations and genetically engineered strains will be similarly useful in the future.

Unfortunately, most institutions do not have the resources to maintain more than a small fraction of new mouse models as living colonies. Therefore, alternative strategies are needed to reduce the costs of maintaining these genotypes and ensure that the most promising animal models will be available for future study. In some cases, a bank of tissue, cell lines or DNA may be sufficient. However, maximum flexibility requires the ability to re-establish living animals.

Apart from space and economic considerations, other pitfalls can befall mouse colonies during long periods of conventional breeding. These include physical disasters such as fire, disease, breeding failure, accidental genetic contamination and inevitable degeneration—mutation and genetic drift. The protection of unique mouse models from these misadventures is an essential requirement for long-term maintenance of genetic resources.

2.2 Cryopreservation and banking of mouse germplasm: resources for future research

Until the mid-1970s, the only method to ensure the continued availability of mouse models was continuous breeding. The first reports of successful freezing of mouse embryos in 1972 by Whittingham *et al.* (4) and Wilmut (5) were immediately recognized as enabling technology for the economical long-term banking of infrequently used mouse stocks (6). This was followed in 1977 by the successful freezing of unfertilized mouse oocytes also by David Whittingham (7). In the intervening years, a large number of protocols for embryo and oocyte cryopreservation have been described using many cryoprotective agents and diverse cooling and warming rates. It is beyond the scope of this chapter to present or compare the relative merits of these protocols. Here, we present those protocols we have optimized and applied for many years in our respective laboratories.

Since mouse embryos and oocytes were successfully cryopreserved, it was natural for many laboratories to attempt to preserve mouse sperm. After all, spermatozoa of many species have been successfully cryopreserved and used for routine artificial insemination of farm animals and humans since the 1950s. For reasons that are still not entirely clear, mouse sperm remains difficult to cryopreserve. In the past 5 years, some progress has been made towards the goal of mouse sperm banking. At Harwell, we have focused our efforts on the protocol of Okuyama *et al.* (8), as improved and refined by Nakagata and Takeshima (9) and Sztein *et al.* (10). Although mouse sperm cryopreservation is still under development, we will present our current procedure as excellent results can be obtained from some hybrid and outbred males. Recent reports show that live mice can also be obtained by intracytoplasmic sperm injection (ICSI) of spermatozoa preserved without cryoprotectant (11) and also with freeze-dried sperm (12). This work is extremely interesting as the spermatozoa injected are technically dead. The ICSI procedure in mice is technically demanding and requires the use of expensive piezoelectric pipette injection systems, so whether these procedures become commonplace remains to be seen.

The ability to cryopreserve both embryos and gametes offers opportunities for innovative research. For example, novel combinations of banked gametes can be used to create embryos that may be difficult to obtain using conventional breeding. Also, if large numbers of an important transgenic or mutant stock are required from banked spermatozoa, many offspring can be produced in a matter of weeks using frozen sperm from a single male. At Harwell, recent results indicate that approximately 1000 pups can be produced from spermatozoa frozen from a single (C3H/HeH × BALB/c) F1 male following *in vitro* fertilization (IVF) using fresh (C3H/HeH × 101/H) F1 eggs (13). Unfortunately, similar success is not yet possible using gametes from most inbred strains (14).

Embryo banking will continue to be the most appropriate method for animal models that require strict pedigree breeding and when homozygosity or heterozygosity must be maintained (e.g. inbred strains and outbred stocks). 'Haploid' banking (oocytes, spermatozoa) may be appropriate when the preservation of only one or a few genes is needed. Specific examples include congenic and co-isogenic strains carrying spontaneous or induced mutations. Also, the archiving of sperm and DNA from the many thousands of male offspring produced in the large ENU mutagenesis programmes will also prove essential for future genotype evaluation screens (13).

Embryo cryopreservation, being relatively simple and well proven, will probably remain the method of choice in most laboratories. Frozen embryos can be distributed to other laboratories and successfully re-established by standard embryo transfer techniques. The IVF or artificial insemination procedures required to re-establish breeding colonies from cryopreserved sperm or oocytes are not yet as readily available. Procedures for the cryopreservation of mouse spermatozoa and subsequent IVF will be presented in this chapter, but not artificial insemination, as this technique is not routinely used by the authors.

3. Principles of cryopreservation

Studies of the mechanisms of cryopreservation and cryoinjury indicate that the properties of cells, such as size, shape and permeability to water and cryoprotectants, play a major role in determining the appropriate conditions for successful cryopreservation. These have been reviewed by Mazur (15) and Rall (16). Considerations that are unrelated to the osmometric behaviour of cells, such as the presence of special structural or functional features and susceptibility to toxicity and cold, may play equally important roles. Cellular properties vary widely and usually depend on the type of cell (embryo, spermatozoa, oocyte), stage of development, and species. Two cryopreservation approaches have developed over the past 20 years. The first, controlled freezing, evolved from studies on the osmotic behaviour of cells during the freezing process and the permeation and efflux of cryoprotectants. This knowledge was applied to the cryopreservation of embryos by Whittingham *et al.* in 1972 (4).

The second approach, vitrification, was proposed in the late 1930s from studies on the physics of freezing and solidification and was recently successfully applied to embryos by Rall and Fahy (17).

3.1 Controlled freezing

Successful controlled-rate freezing procedures are characterized by three features.

- Concentrations of 1–2 M of a permeating cryoprotectant (usually propylene glycol, glycerol, dimethylsulphoxide (DMSO), methanol or ethylene glycol) are added to the cell suspension prior to cooling.
- The cell suspension is cooled and frozen using controlled conditions, usually at a constant rate of cooling between –5 °C and at least –30 °C.
- The cells undergo a characteristic sequence of changes in their osmotic volume during the cryopreservation and recovery processes.

If the permeability properties of the cells are known, mathematical models are available to calculate the optimum conditions for cryopreservation and recovery of living cells (18,19). However, a knowledge of the osmotic properties of the cells does not always permit the immediate development of a simple cryopreservation protocol. Such models do not take into account all mechanisms of cryoinjury. Special features of some cells, such as high susceptibilities to toxicity and cold or extremes of cryoprotectant and water permeability, may complicate and, in rare instances, prevent successful cryopreservation. Fortunately, all of the pre-implantation stages of the mouse, including the mature oocyte, can be cryopreserved using one or more cryoprotectant, cooling and warming conditions (0.5 to >2000 °C/min), and a variety of containers (straw, ampoule, cryotube). Variability in the success of individual attempts of mouse embryo or oocyte cryopreservation may indicate problems with their starting quality. Poor-quality cells usually do not survive cryopreservation. Mouse spermatozoa have proven to be more challenging to cryopreserve, presumably due to their special features.

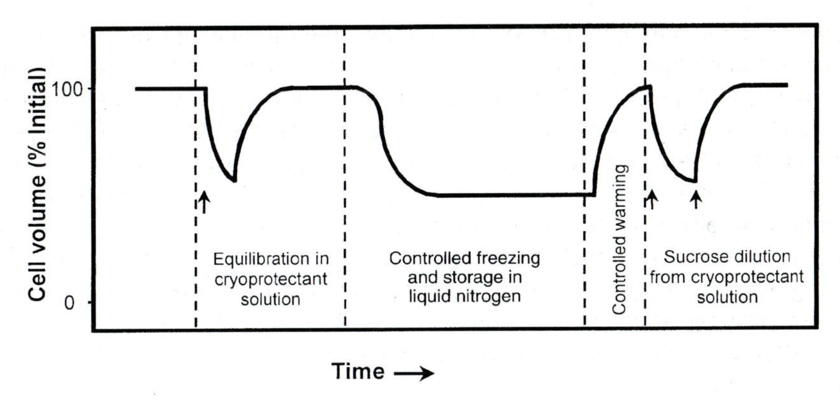

Figure 1. Diagrammatic representation of cell volume changes during each step of successful controlled freezing. These characteristic changes provide a useful guide for optimizing the steps of freezing for any embryo, gamete or cell. Figure reprinted from Rall (ref. 41) with modifications.

Regardless of the type of biological cells, successful cryopreservation by controlled freezing leads to a series of changes in the osmotic volume of the cells. The freezing (*Protocols 2*, *4* and *8*) and warming (*Protocols 3* and *6*) steps result in the series of changes shown in *Figure 1*.

Embryos remain at their normal isotonic volumes from the time of collection to transfer into the cryoprotectant solution. This produces a transient shrink–swell change in the volume of the cells. In the case of eight-cell mouse embryos transferred into propylene glycol or glycerol at room temperature, embryos will return to their isotonic volume within approximately 10–15 min. Seeding and controlled slow freezing results in the growth of ice crystals in the surrounding solution. Embryos respond to the changes in concentration of the suspending solution by a progressive shrinkage due to the exosmosis of water from their cytoplasm. Embryos gradually shrink to approximately 60% of their initial volume after slow freezing to –30°C. No change occurs in the volume of embryos during rapid cooling in liquid nitrogen or storage. Embryos swell during warming and thawing due to the influx of water into the cytoplasm as the ice melts and dilutes the extracellular solution. The embryos then shrink when mixed with a sucrose diluent and the intracellular cryoprotectant leaves the cytoplasm. Finally, embryos swell back to their initial volume when transferred into isotonic medium.

Mouse spermatozoa are a good example of the limitations of applying mathematical models to predict the appropriate cryopreservation conditions. The difficulty presumably results from the highly specialized nature of spermatozoa and unique features of their structure and physiology. For example, mouse spermatozoa are characterized by extreme sensitivity to osmotic damage (20) and mechanical stresses such as centrifugation (21). Despite these features, some success has been achieved using empirical approaches with novel cryoprotectants, including complex macromolecules such as bovine serum albumin (22), egg yolk (23) and skimmed milk (8,9), and high concentrations of non-permeating sugars, especially raffinose. The mechanisms by which these polymers and sugars protect spermatozoa during cryopreservation are poorly understood.

3.2 Vitrification

Vitrification refers to the physiochemical process by which liquids and solutions solidify into a glass during cooling without crystallization of the solvent or solute. Successful cryopreservation by vitrification is characterized by three features (2):

- no ice forms in the cell suspension during cooling, storage or warming;
- cells are osmotically dehydrated prior to cooling by controlled equilibration in a highly concentrated solution of cryoprotectants (usually >6 M);
- cells undergo a characteristic sequence of changes in their osmotic volume during the cryopreservation and recovery processes.

The volumetric behaviour of cells during the course of successful vitrification is very similar to that of controlled freezing procedures. This is a consequence of the fact that many of the osmotic factors affecting the success of controlled freezing also apply to vitrification. Although most vitrification protocols use rapid cooling (>1000 °C/min) by direct transfer into either liquid nitrogen or liquid nitrogen vapour, successful cryopreservation can also be achieved using relatively slow cooling and warming (5–100 °C/min).

Vitrification offers advantages over controlled freezing because a controlled-rate freezer is not required and cells are not subject to injury associated with chilling injury or with ice formation in the cytoplasm or solution surrounding the cells. Nearly all of the disadvantages of vitrification result from the need to dehydrate cells in a concentrated solution of cryoprotectants. Exposure to this so-called 'vitrification solution' must be carefully controlled to prevent toxicity and the extent of permeation of the cryoprotectant. This requirement for careful control of equilibration in cryoprotectant solutions (concentration, time and temperature) reduces opportunities for the batch cryopreservation of large numbers of embryos.

Regardless of the type of biological cells, successful cryopreservation by vitrification leads to a series of changes in the osmotic volume of the cells. The vitrification (*Protocol 5*) and warming (*Protocol 6*) steps result in the following changes. Embryos remain at their normal isotonic volumes from the time of collection to transfer into the cryoprotectant solution (*Figure 2*).

The three-step equilibration used for eight-cell embryos produces a series of changes in the volume of the cells. First, equilibration in 1.6 M glycerol results in the same transient shrink–swell change in the volume of the cells as seen in *Figure 1*. As before, embryos transferred into glycerol at room

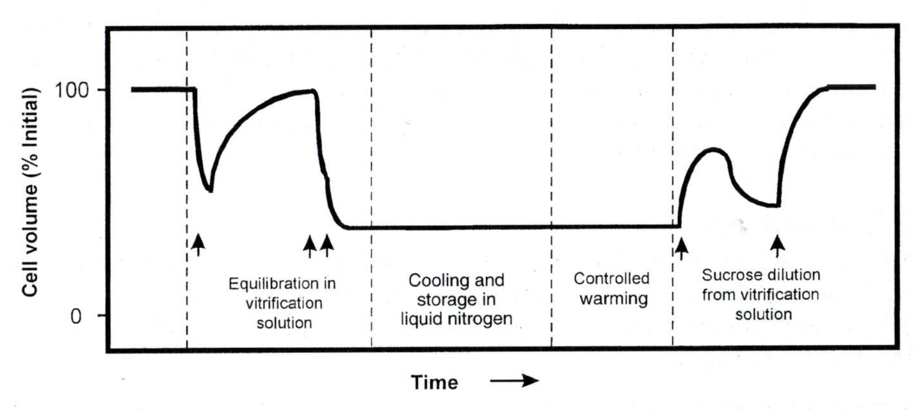

Figure 2. Diagrammatic representation of cell volume changes during each step of successful vitrification. These characteristic changes provide a useful guide for optimizing the steps of vitrification for any embryo, gamete, or cell. Figure reprinted from Rall (ref. 41) with modifications.

temperature return to their isotonic volume within approximately 15 min. Then, embryos shrink when transferred into each of the increasingly concentrated glycerol solutions. After 1 min exposure to the vitrification solution (VS3a; see *Protocol 5*), the shrunken embryo suspension is cooled rapidly to below –150 °C and then stored in liquid nitrogen.

There are no changes in volume of the cells during cooling, storage or warming. Embryos exhibit a complex series of changes in volume when glycerol is diluted using a sucrose solution. Initially, cells swell when placed in the sucrose diluent to restore osmotic equilibrium with less concentrated sucrose solution. Then, cells shrink as intracellular cryoprotectant leaves the cytoplasm. Finally, embryos swell back to their initial volume when transferred into isotonic medium. Note that in many respects, the osmometric behaviour of embryos cryopreserved by vitrification is similar to that during cryopreservation by controlled freezing.

4. Skills required for embryo and gamete cryopreservation

It is beyond the scope of this chapter to detail all the embryological skills needed for competent handling of embryos, gametes and the eventual transfer of embryos to pseudopregnant foster mothers. The same technical skills and equipment required to manipulate embryos in a mouse transgenic or IVF laboratory are needed for cryopreservation. All media and cryoprotectant solutions are filter-sterilized immediately after preparation and sterile technique is used for handling solutions and embryo suspensions. These basic techniques, as well as most of the media required, are described elsewhere in this volume and in ref. 24. A previous publication in the 'Practical Approach' series (25) also has excellent chapters on preparation of media, cryopreservation and IVF.

5. Special equipment required for cryopreservation

In general, the standard equipment required for mouse embryology, IVF and embryo transfer is supplemented with cryobiology equipment. All cryopreservation methods require vacuum flasks designed to hold liquid nitrogen, a thermocouple device for accurate temperature measurements, plastic insemination straws or cryotubes as the freezing container, and liquid nitrogen storage refrigerators.

If a slow cooling cryopreservation method for embryos or oocytes is chosen, the purchase of a controlled-rate freezing machine is recommended. Although some are expensive, a freezer is a worthwhile investment if routine embryo banking is contemplated (one or more freezing sessions per week). They yield reproducible cooling conditions and simplify and standardize the cryopreservation process. At Harwell and the National Institutes of Health (NIH), we have used the Planer Kryo 10 controlled-rate freezer with very satisfactory results. Planer now manufacture a more basic (and less expensive unit), the Kryosave Compact, which is suitable for cryopreservation of mouse embryos and oocytes.

The NIH embryo cryopreservation laboratory converted to BioCool, alcohol bath, controlled-rate freezers and plastic insemination straws 10 years ago. Alcohol baths are less expensive to purchase and operate (they do not consume liquid nitrogen), but are not appropriate for cooling rates of >3°C/min. Equivalent success can be achieved using other devices to control the rate of cooling, including home-made apparatus described by Wood *et al.* (23) in a previous volume in the 'Practical Approach' series. In some cases, minor adjustment in the procedures described below may be necessary for some types of freezers. Please consult us directly for advice.

Sperm freezing and embryo vitrification procedures usually do not require a controlled-rate freezer or special equipment other than a chamber in which a temperature gradient of liquid nitrogen vapour can be generated. A deep polystyrene box usually suffices.

6. Cryopreservation and recovery of mouse embryos

Here, we present three well-established protocols that have been used for many years with consistent and high yields of viable embryos from a variety of different mouse models. These include mutant, transgenic, knock-out, knock-in, inbred, congenic, co-isogenic, recombinant inbred and outbred strains as well as strains carrying complex chromosome anomalies. The first protocol, based on the method of Renard and Babinet (26), has been in routine use at the MRC Mammalian Genetics Unit (MGU), Harwell since 1989 (*Protocols 2* and *3*). Briefly, the embryos are frozen in plastic insemination straws, using 1.5 M propylene glycol and a slow cooling to –30 °C.

The second method employs a similar slow cooling regime but uses 1.5 M glycerol and slow cooling to –40 °C (27), and has been in routine use at the NIH since 1989 (*Protocols 4* and *6*). The third method is a rapid vitrification method using a concentrated mixture of glycerol and bovine serum albumin (*Protocols 5* and *6*).

6.1 Source of embryos and preferred developmental stages

All pre-implantation stages of the mouse embryo can be successfully frozen. Most laboratories prefer to freeze at about the eight-cell stage (e.g. day 2.5 *post coitum* (see ref. 24), sometimes referred to as 'day 3', when day 1 is defined as the day of finding a copulation plug) for the following reasons:

(a) These embryos are less sensitive to *in vitro* handling than one- or two-cell embryos.
(b) If one or two blastomeres are damaged during freezing and thawing, the surviving cells still have the potential to develop into a normal pup.
(c) Overall survival rates after cryopreservation are often higher than those obtained with earlier or later developmental stages.
(d) It is easy to score normal morphological appearance before freezing and after thawing. Abnormal or severely damaged embryos can be discarded.
(e) If desired, a period of *in vitro* culture can be employed to act as a preliminary viability assay before transfer of the post-culture blastocysts to the oviducts or uterine horns of pseudopregnant foster mothers.

Donor females are usually treated with superovulatory doses of gonadotrophic hormones to maximize the yield of embryos per female (Chapter 1, *Protocol 5*). There is no evidence that hormone treatment of females yields embryos that are less viable than those produced from natural ovulation. Care must be taken when scoring the embryos, as sometimes superovulation results in a proportion of eggs that do not fertilize. These unfertilized eggs often fragment and may resemble normal embryos. Fortunately, fragmented embryos are easily identified when placed into cryoprotectant solutions (they remain shrunken). When a mouse strain does not respond consistently to superovulation, embryos can be obtained from females during natural cycles or following post-partum mating.

Protocol 1. Collection of embryos for freezing

Equipment and reagents

- Pregnant female mice at day 2.5 *post coitum* (see Section 6.1).
- Medium M2 containing 0.4% (w/v) BSA (see Chapter 1, *Protocol 6*)[a]

Protocol 1. *Continued*

- Sterile instruments for dissection: scissors, #5 watchmakers forceps
- Sterile mouth pipettes drawn from Pasteur pipettes or other suitable glass capillaries (the internal diameter of the pipettes should be ~100 μm)
- 35 mm Petri dishes (tissue culture quality)
- 60 mm Petri dishes (tissue culture quality)
- 1 ml syringe
- Mouthpiece and aspirator tube assembly (e.g. Sigma)

Method

1. Dissect the oviducts from the donor females and place into ~20 μl drops of M2 or PB1 medium contained in a 35–100 mm Petri dish. With experience, four to six pairs of oviducts may be processed in an appropriate number of drops in one dish. Leave a few millimetres of uterus attached to the oviducts. For details of oviduct dissection, see ref. 24.
2. Flush the contents from the oviducts using a blunt tipped Hamilton needle attached to a 1 ml syringe containing M2 or PB1 medium. See refs 24 and 25.
3. When all the oviducts are flushed, collect the embryos using a mouth pipette and transfer them to a 35 mm Petri dish containing ~3 ml M2 or PB1 medium.
4. Evaluate the embryos carefully under the microscope (100×) and discard any that are abnormal, fragmented, unfertilized or lysed.
5. Transfer morphologically normal embryos to a fresh dish of M2 or PB1 medium using a new sterile pipette.
6. Wash the embryos by pipetting them up and down several times in the medium and wash them again through two more dishes of fresh medium using new sterile pipettes and sterile technique. A multi-well dish may be useful for the washing steps. Serial washing is important when embryos are recovered from animals infected with microbial pathogens or housed under 'conventional' animal holding conditions. The washing is thought to remove most pathogens with the exception of mycoplasma and viruses that integrate into the genome.
7. When the embryos have been scored and collected, they can be kept on the bench at room temperature in M2 or at 4°C in PB1 or M2 while preparations are made for freezing.

[a] We use BSA from ICN Flow (albumin, bovine crystallized). PB1 medium containing 0.4% (w/v) BSA (6) can be substituted for M2 medium. PB1 medium (ref. 6) is 136.87 mM NaCl, 2.68 mM KCl, 0.90 mM $CaCl_2$, 1.47 mM KH_2PO_4, 0.49 mM $MgCl_2{\cdot}6H_2O$, 8.09 mM Na_2HPO_4, 0.33 mM sodium pyruvate, 5.56 mM glucose.

6.2 Freezing embryos using propylene glycol as cryoprotectant

This is the MGU Harwell method based on the method of Renard and Babinet (26).

Protocol 2. Freezing eight-cell mouse embryos using propylene glycol as cryoprotectant

Equipment and reagents

- Planer programmable freezing machine or equivalent controlled-rate freezing apparatus (see Section 5)
- Medium M2 (Sigma) containing 0.4% (w/v) BSA (see Chapter 1, *Protocol 6*); we use BSA from ICN Flow (albumin, bovine crystallized)
- 1.5 M propylene glycol (Sigma) in M2 (0.6 ml propylene glycol + 4.4 ml M2), filter sterilized (using a 0.22 μm filter).
- 1.0 M sucrose in M2 (dissolve 1.71 g Analar grade sucrose in 5 ml M2 by gentle inversion), filter sterilized (0.22 μm filter).
- 1 ml plastic syringe
- Fine watchmaker's forceps
- Plastic insemination straws, 0.25 ml, 133 mm long (Planer or IMV).
- Nylon (or steel) cannula rod (e.g. from IMV or Planer)
- Plastic goblets (IMV or Planer) and aluminium canes (IMV or Planer) for storage in liquid nitrogen
- Electricians' wire labels for identifying the straws (W. H. Brady, Inc.)
- Equipment for sealing the straws, such as a heat sealer or 'Cristaseal' putty as used in haematology laboratories (Hawksley, UK)
- Small Dewar flask containing liquid nitrogen
- Eight-cell embryos held in M2

Method

1. Prepare the freezing apparatus. Cool and maintain at 0 °C.
2. Prepare the straws. Please refer to *Figure 3* for final appearance of straw.
   (a) Using a metal rod, push the cotton plug down the straw until the end of the plug is 75 mm from the open end of the straw.
   (b) Use a fine marker pen to make three guide marks on the straw to aid loading with solutions.
      (i) Mark 1: 20 mm from the end of the plug.
      (ii) Mark 2: 7 mm from mark 1.
      (iii) Mark 3: 5 mm from mark 2.
   (c) Label the straws with an appropriate code for the stock being frozen. Use wire labels or write on the straws using a very fine marker pen.
3. Fill the straws as follows using a 1 ml syringe attached to the straw:
   (a) Aspirate a column of 1 M sucrose to mark 3.
   (b) Aspirate a column of air so that the sucrose meniscus reaches mark 2.
   (c) Aspirate a column of 1.5 M propylene glycol so that the sucrose meniscus reaches mark 1.
   (d) Aspirate air until the column of sucrose reaches half way up the plug and forms a seal with the polyvinyl alcohol (PVA) powder in the straw. The straws are now ready for loading with embryos.

Protocol 2. *Continued*

4. Pipette the embryos into a dish of 1.5 M propylene glycol and equilibrate for 15 min at room temperature (20–25 °C). Examine the embryos after 5–10 min of exposure to the propylene glycol solution and discard those remaining shrunken or with abnormal morphology.
5. Pipette the embryos into the 1.5 M propylene glycol column of each straw. Approximately 20 embryos can be frozen in each straw and the loading can be visualized under the microscope if necessary. With experience, it is possible to pipette the embryos into the straw by eye, using a small air bubble behind the embryos in the pipette to act as a visible marker. When the air bubble enters the propylene glycol column, the operator knows that the embryos will have entered the straw.
6. The straws must now be sealed. This can be achieved by careful use of a heat sealer, or the end of the straw can be plugged with a commercially available putty such as 'Cristaseal' (see Section 8.1).
7. Place the straws into the chamber of the controlled-rate freezer and cool to –7 °C (the cooling rate is not critical). Wait for 5 min for the samples to equilibrate at –7 °C. The total time of exposure of embryos to propylene glycol at room temperature should be approximately 30 min and no more than 45 min.
8. Ice formation must now be induced in the medium surrounding the embryos to prevent supercooling during the slow cooling process that follows. This process is called 'seeding'. Cool the tips of a pair of watchmaker's forceps by dipping in liquid nitrogen. Touch the straws in the middle of the sucrose column with the cold forceps until ice is seen to form (see Section 8.1).
9. Quickly return the straws into the cooling chamber. Examine each straw 5 min later to confirm that ice has formed in the entire sucrose column and grown across the air bubble into the propylene glycol column. If any straws are not completely seeded, repeat step 8.
10. Cool at 0.3 °C/min to –30 °C, then quickly remove the straws from the cooling chamber and plunge them directly into a small Dewar flask of liquid nitrogen. Alternatively, the freezing machine can be programmed to cool at 10 °C/min to below –150 °C after –30 °C is reached. The Planer Kryo 10 can cool and hold samples at –160 °C for several hours if required, but this is unnecessary and consumes large quantities of liquid nitrogen.
11. Store the straws in plastic goblets on aluminium canes in suitable liquid nitrogen storage refrigerators. Many sizes of vessel and storage systems are available. We use 'cane' storage refrigerators designed for cattle semen. Consult your supplier.

6.3 Recovery of embryos and live mice from embryos frozen according to *Protocol 2*

Again, this is the MGU Harwell protocol, based on the method of Renard and Babinet (26).

Protocol 3. Recovery of embryos by rapid warming

Equipment and reagents

- Frozen straw(s) of embryos
- Water bath at 25°C
- Small Dewar flask of liquid nitrogen
- Nylon (or steel) cannula rod (IMV or Planer)
- Anaesthetic and instruments for embryo transfer (Chapter 9, *Protocol 7*)
- 0.5 day pseudopregnant female mice to receive the embryos (Chapter 1, Section 6.2)
- Medium M2 containing 0.4% (w/v) BSA (see *Protocol 1*)

Method

1. Transfer the appropriate straws from the storage refrigerator to a small Dewar flask of liquid nitrogen.
2. Using forceps, hold the straw near the wire markers for 40 sec in air and then in water at room temperature until the ice disappears. Handle the straw gently and do not agitate it in the water bath.
3. Wipe the straw with a tissue, then cut off the seal and also cut through the PVA plug leaving about half the cotton plug in place to act as a plunger.
4. Using a metal rod, expel the entire liquid contents of the straw into a 35 mm Petri dish so that the sucrose and propylene glycol columns mix together. Alternatively, holding the straw firmly and horizontally, cut off both seals entirely then allow the contents of the straw to run out into the dish. It may help to have a 100 μl drop of 1 M sucrose already in the dish to mix with the contents of the straw.
5. Wait for 5 min. The embryos will shrink considerably due to the presence of sucrose, which is a non-permeating solute. The sucrose assists dilution of the propylene glycol in two ways. Firstly, upon rapid warming, the embryos swell quickly due to a rapid intake of water. The osmotic action of the sucrose counteracts this swelling and reduces osmotic stress. Secondly, thawed blastomeres still contain propylene glycol. Again, the osmotic effect of the sucrose accelerates the efflux of propylene glycol from the cells and reduces osmotic swelling.
6. Transfer the embryos to a 200 μl drop of M2. They will rapidly take up water and assume normal appearance. Wait for 5 min, then examine the embryos closely under the microscope. Discard any that appear grossly abnormal or with more than two lysed blastomeres.

Protocol 3. *Continued*

7. Transfer the surviving embryos to fresh dish of M2 for a further 5 min.
8. Wash the embryos through four or more dishes of M2 using new sterile pipettes for each wash to ensure dilution of any pathogens.
9. Transfer normal or slightly damaged embryos to the oviducts of 0.5 day pseudopregnant recipients. We normally bilaterally transfer six to eight embryos into each oviduct. Alternatively, culture the embryos to the blastocyst stage in M16 (see Chapter 10, *Protocol 7*), and transfer to the uterine horns of 2.5 day pseudopregnant recipients. Always split the surviving embryos of a straw between two or more recipient females. If the embryos are very important, or there are few survivors, transfer fresh embryos (with a different coat colour) along with the thawed embryos, or transfer into a pregnant recipient.

6.4 Freezing embryos using glycerol as cryoprotectant

This is the slow freezing method routinely used for mouse and rat embryo banking at the Genetic Resource Section on the NIH campus in Bethesda, Maryland. It is based on the sucrose dilution procedure of Leibo (28), as modified by Rall (16).

Protocol 4. Freezing eight-cell mouse embryos using glycerol as cryoprotectant

Equipment and reagents

- Alcohol bath freezing machine (e.g. model BC-IV-40, FTS Systems)
- Small Dewar flask containing liquid nitrogen
- Straw rack (IMV)
- Plastic insemination straws, 0.25 ml, 133 mm long (Planer or IMV)
- Plastic insemination straws, 0.5 ml, 133 mm long (Planer or IMV), with plugs removed
- 1 ml syringe
- 35 mm Petri dishes
- Nylon (or steel) cannula rod (IMV or Planer)
- Fine tipped alcohol resistant pen (Marker II Superfrost, Precision Dynamics Corp.)
- Impulse heat sealer for plastic bags and sterilization pouches
- Ruler
- Adhesive labels (B-500 vinyl cloth wire label or Datab markers,W. H. Brady Co.)
- Goblets (IMV or Planer) and aluminium canes (IMV or Planer) for storage in liquid nitrogen
- PB1 medium containing 0.4% (w/v) BSA (6) (M2 medium containing 0.4% (w/v) BSA may be substituted); we use Bayer Pentex crystallized BSA
- Glycerol solution: 1.5 M glycerol in PB1 medium (or in M2 medium)
- Sucrose solution: 1.0 M sucrose in PB1 medium (or in M2 medium)
- Eight-cell embryos held in PB1 at room temperature or 4°C (or in M2 medium)

Method

1. Prepare straws:
   (a) Push the plug into the straw with a metal rod until the length from the open end to the plug equals 11 cm (note the length of plug may vary from batch to batch).

(b) Label the straws with an appropriate code for the embryo being frozen. Use wire labels or write on the straw with a fine tipped, alcohol-resistant marker pen.

(c) Mark each straw with guides for loading cryoprotectant solutions using a marker pen. Mark each straw at 0.8, 1 and 6.8 cm from open end of straw.

(d) Attach the straw to a 1 ml syringe and aspirate sucrose solution until the column reaches the third mark. Then aspirate air until the first mark is reached and wipe excess sucrose from outside of straw. Aspirate glycerol solution until the column reaches the second mark, air to the first mark, cryoprotectant solution to the second mark, and finally air until the powder in the plug is reached. Please refer to *Figure 3* for final appearance of straw.

(e) Heat seal the plugged end of the straws.

(f) List strain, date and other information on Datab label. Wrap the label around one end of a 0.5 ml plastic insemination straw and slip the other end over the heat seal. The large straw serves as a handle.

(g) Hold the prepared straws horizontally at room temperature until use.

2. Prepare the alcohol bath, controlled-rate freezer. Cool the chamber to a holding temperature of –7 °C.
3. Pipette the embryos into a 35 mm Petri dish of glycerol solution and equilibrate for 15 min at room temperature (20–25 °C).
4. Examine the embryos after 5 min of exposure to the glycerol solution and discard those remaining shrunken or with abnormal morphology.
5. Pipette the embryos into the glycerol column of each straw. Approximately 20–30 embryos can be frozen in each straw. Hold straws containing embryos horizontally until step 7.
6. Heat seal each straw using a plastic bag sealer. **Caution**: Do not warm the embryos suspension with the heat sealer or fingers. See Section 8.1.
7. Transfer the straw(s) into a straw rack submerged in the alcohol bath after the embryos have been in the glycerol solution for 20 min.
8. Seed ice formation in the sucrose column by partially withdrawing each straw from the alcohol and touching the midpoint of the sucrose column with a spatula precooled in liquid nitrogen. Return the straw to the straw rack. See Section 8.1
9. Check each straw approximately 5 min later for complete seeding of all columns. Repeat step 8 if straws do not seed completely.

Protocol 4. *Continued*

10. Cool the straws at 0.5 °C/min to –40 °C and hold at –40 °C for 10 min.
11. Plunge the straws directly into liquid nitrogen in a Dewar flask. Straws must be handled and stored at below –150 °C until thawing. Ideally, handle straws under the surface of liquid nitrogen at all times until thawing. A short period of exposure to room temperature air (<4 sec) is acceptable.
12. Place straws into precooled plastic goblets attached to labelled aluminium canes. Attach the handles to the cane with hollow cylindrical clip cut from a goblet.
13. Store the canes submerged in liquid nitrogen in a liquid nitrogen storage refrigerator.

6.5 Vitrification of embryos in vitrification solution VS3a

This is the vitrification method of Rall (16), routinely used for mouse and rat embryo banking at the Genetic Resource Section on the NIH campus in Bethesda, Maryland.

Protocol 5. Vitrification of eight-cell mouse embryos in vitrification solution VS3a

Equipment and reagents

- Small Dewar flask containing liquid nitrogen
- Polystyrene box with lid
- Straw rack (IMV)
- Plastic insemination straws, 0.25 ml, 133 mm long (Planer or IMV)
- Plastic insemination straws, 0.5 ml, 133 mm long (Planer or IMV), with plugs removed
- 1 ml syringe
- 27G needle
- Nylon (or steel) cannula rod (e.g. from IMV)
- Ruler
- Fine-tipped alcohol-resistant pen (Marker II Superfrost, Precision Dynamics Corp.)
- Impulse heat sealer for plastic bags and sterilization pouches
- Adhesive labels (vinyl cloth wire label or Datab markers, W. H. Brady Co.)
- PB1 medium containing 0.4% (w/v) BSA (6) (we use crystallized BSA (Bayer Pentex))
- Equilibration media: PB1 medium containing 6% (w/v) BSA; 1.6 M glycerol in PB1 medium containing 6% (w/v) BSA; 4.2 M glycerol in PB1 medium containing 6% (w/v) BSA; 1.0 M sucrose in PB1 medium containing 6% (w/v) BSA
- Vitrification solution VS3a: 6.5 M glycerol in PB1 medium containing 6% (w/v) BSA (this solution can be filtered and stored in the refrigerator for several years, providing sterile technique is used to withdraw aliquots)
- Goblets (IMV or Planer) and aluminium canes for storage in liquid nitrogen
- Eight-cell embryos held in PB1 at room temperature (20–25°C) or 4°C (see *Protocol 1*).

Method

1. Prepare straws:
   (a) Push the plug into the straw with a metal rod until the length from the open end to the plug is 11 cm (note the length of the plug may vary from batch to batch).

(b) Label the straws with an appropriate identification code for the model being frozen. Use wire labels or write on the straw with a very fine tipped marker pen.

(c) Mark each straw with guides for loading cryoprotectant solutions using a marker pen. Mark each straw at 7.5, 8 and 9 cm from the end of the plug.

(d) Place the sucrose and vitrification solutions into the straw using a 1 ml syringe fitted with a 1.5 inch, 27 gauge needle in a manner that results in an air bubble with a dry wall between columns of sucrose and vitrification solution.

(i) Hold the straw horizontally and insert the dry needle of the syringe containing the sucrose solution into the open end of the straw to the third mark.

(ii) Expel sucrose solution slowly until the plug is wetted.

(iii) Withdraw the needle, taking care not to wet the inside of the straw.

(iv) Place the needle of a syringe containing the vitrification solution in the straw at the second mark.

(v) Expel vitrification solution from the syringe until it reaches the first mark.

(vi) Withdraw the needle. Filled straws are shown in *Figure 3*.

(e) Heat seal the plugged end of straws.

(f) List strain, date and other information on a Datab label. Wrap the label around one end of a 0.5 ml plastic insemination straw and slip other end over the heat seal. The large straw serves as a handle.

(g) Hold the prepared straws horizontally at room temperature until use.

2. Place 3 ml of each equilibration solution in 35 mm Petri dishes: (a) PB1 medium containing 6% BSA, (b) 1.6 M glycerol in PB1 medium containing 6% BSA, and (c) 4.2 M glycerol in PB1 medium containing 6% BSA.

3. Pipette a group of 10–20 embryos into the dish of PB1 medium containing 6% BSA. Mix with the tip of the pipette and equilibrate at room temperature for 2–10 min (rinse step).

4. Pipette the embryos into the dish of 1.6 M glycerol in PB1 medium containing 6% BSA. Mix with the tip of the pipette and equilibrate at room temperature (rinse step).

5. Examine the embryos after 5 min of exposure to the glycerol solution and discard those remaining shrunken or with abnormal morphology (cryoprotectant permeation step).

Protocol 5. *Continued*

6. After 20 min exposure to 1.6 M glycerol, pipette the embryos into the dish of 4.2 M glycerol in PB1 medium containing 6% BSA. Mix with the tip of a pipette and equilibrate at room temperature (dehydration/ rinse step).
7. Pipette the embryos into the vitrification column of each straw after 1–1.5 min exposure to 4.2 M glycerol. Care must be taken to transfer embryos with as little 4.2 M glycerol solution as possible, ideally <5-10 times the total volume of the embryos. This can be readily accomplished using a capillary with an inner diameter just larger than that of the embryos.
8. Heat seal each straw using a plastic bag sealer. **Caution**: take care not to warm the embryo suspension with the heat sealer or fingers. See Section 8.1.
9. Place the straw on the straw rack in the polystyrene box in cold nitrogen vapour (below –150 °C) 1 min after transfer of embryos into the vitrification solution.
10. Plunge the straws directly into liquid nitrogen in the bottom of the polystyrene box after 2 min exposure to cold nitrogen vapour. Then transfer the straw to a Dewar flask. Straws must be handled and stored below –150 °C until thawing. Ideally, handle straws under the surface of liquid nitrogen at all times until thawing. A short period of exposure to room temperature air (<4 sec) is acceptable.
11. Place the straws into precooled plastic goblets attached to labelled aluminium canes. Attach the handles to the cane with a hollow cylindrical clip.
12. Store the canes submerged in liquid nitrogen in a liquid nitrogen storage refrigerator (see *Protocol 1*).

6.6 Recovery of embryos and live mice from embryos cryopreserved according to *Protocols 4* and *5*

This protocol, by Rall (16), describes the recovery of vitrified embryos and those frozen slowly in glycerol solutions.

Protocol 6. Recovery of embryos slowly frozen or vitrified in glycerol solutions

Equipment and reagents

- Small Dewar flask containing liquid nitrogen
- Two water baths at least 15 cm deep (room temperature and 37°C)

- Nylon (or steel) cannula rod (IMV)
- Pseudopregnant female mice (0.5 day) as embryo recipients (see Chapter 1, Section 6.2; Chapter 9, *Protocol 7*).
- PB1 medium containing 0.4% (w/v) BSA (or M2 medium containing 0.4% (w/v) BSA)
- Anaesthetic and instruments for embryo transfer (see Chapter 9, *Protocol 7*)

Method

1. Remove canes from liquid nitrogen refrigerator using long forceps and quickly transfer into small Dewar flask containing liquid nitrogen.
2. Select and transfer the straws for thawing into the Dewar flask and return the cane to the storage refrigerator.
3. Thaw the straw as follows:
   (a) Hold the straw vertically by the handle using forceps in room temperature air for 10 sec.
   (b) Immediately plunge the straw into room-temperature water bath for 10 sec.
   (c) Wipe the straw with a tissue and separate the handle from the straw. Then hold the straw firmly at the plug end and mix the contents of the straw with three or four sharp 'snaps' of the wrist, in a manner similar to that when shaking down a clinical thermometer. Properly mixed straws contain a single liquid column.
   (d) Place the straw in the 37 °C water bath, vertically, with the plug-end up for 3 min.
   (e) Transfer the straw into the room temperature water bath, vertically with the plug-end down for 1 min.
   (f) Wipe the straw with a tissue, then grasp the straw at the end opposite to the plug and shake the straw as described in step 3(c).
   (g) Place the straw in room temperature water vertically, with the plug-end up for at least 2 min, but no more than 5 min.
4. Recover the embryos from the straw as follows:
   (a) Wipe the straw with a tissue, then swab with sterile gauze moistened with 70% alcohol and allow the straw to dry.
   (b) Cut heat seals with scissors first at the middle of the adjacent air bubble and then at the middle of the plug.
   (c) Cannulate the plug end of the straw with a metal rod and expel the contents of the straw as a single drop into a sterile Petri dish by pushing the plug to the end of the straw. Do not expel the plug into the drop.
5. Rehydrate the embryos in fresh PB1 medium:
   (a) Collect shrunken embryos with a sterile mouth pipette.
   (b) Transfer the embryos into a fresh dish of PB1 medium at room temperature and equilibrate for 10 min.

Protocol 6. *Continued*

(c) Examine the embryos and discard those with abnormal morphology or more than two lysed blastomeres.

(d) Wash the embryos to dilute potentially contaminating pathogens by serial transfer through four or more dishes of fresh PB1 medium using a new sterile pipette for each transfer, and sterile technique.

6. Transfer embryos bilaterally into the oviducts (six to eight embryos/oviduct), of 0.5 day pseudopregnant recipient females (Chapter 9, *Protocol 7*). See *Protocol 3*, step 9, for additional suggestions.

7. Cryopreservation and recovery of mouse gametes

Here, we present protocols for the cryopreservation of mouse metaphase II oocytes and mouse sperm. The first method, for oocytes, was published by Carroll *et al.* in 1993 (29) and utilizes 1.5 M DMSO as cryoprotectant solution supplemented with fetal calf serum (FCS).

The second method is an adaptation of the sperm freezing protocol originally published by Okuyama *et al.* in 1990 (8). This method was subsequently refined and improved by Takeshima *et al.* (30), Nakagata and Takeshima (9) and Sztein *et al.* (10). Although mouse sperm freezing is still under development and cannot yet be applied to many inbred strains, recent progress merits inclusion of our current protocol in this chapter.

The final protocol describes IVF of mouse oocytes (fresh or frozen) using

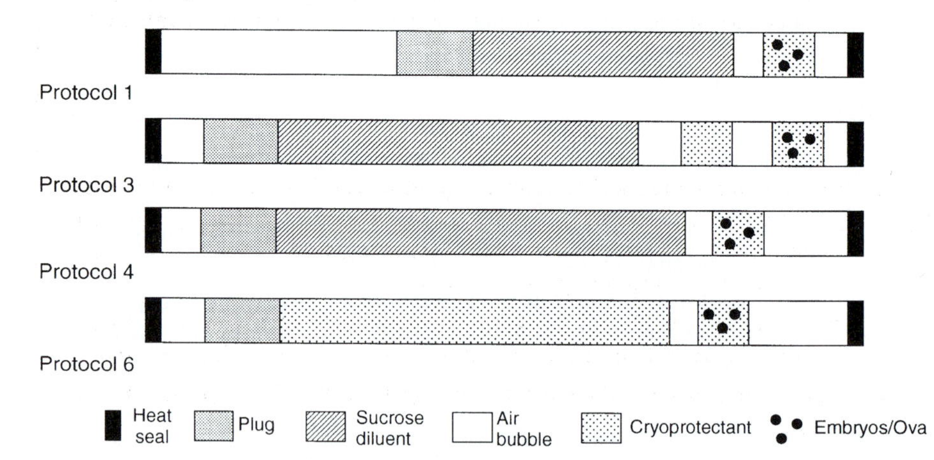

Figure 3. Appearance of plastic insemination straws immediately prior to cryopreservation showing the location of cryoprotectant solution, sucrose diluent and air spaces for each protocol.

fresh or frozen mouse sperm. This method is an evolution of several protocols and is presented as adapted at the MGU Harwell by Glenister and Thornton (13). There is an excellent section on IVF by Wood *et al.* in a previous volume of the 'Practical Approach' series (25).

7.1 Cryopreservation of mouse metaphase II oocytes

This protocol is that of Carroll *et al.* (29). The addition of FCS to the cryoprotectant prevents changes to the zona pellucida that occur during freezing and thawing. In the absence of FCS, the rate of fertilization of the frozen/thawed oocytes is much lower.

Protocol 7. Cryopreservation of mouse metaphase II oocytes using DMSO supplemented with fetal bovine serum (FBS) as cryoprotectant

Equipment and reagents

- Programmable freezing machine or home-made freezing apparatus, (see Section 5).
- Medium M2 (see *Protocol 1*), containing 10% fetal calf serum (FCS; ICN Flow); adjust the pH of M2 + FBS to 7.2 with 10 μl of 2 M HCl per 10 ml medium
- 1.5 M DMSO in M2 + FCS: add 1.07 ml DMSO (Analar grade, BDH) to 8.93 ml M2 + FCS; make up fresh on the day of use and do not filter sterilize
- Glass embryo dishes (BDH)
- Glass boiling tubes, or similar thick-walled glass tubes suitable for holding the straws (see Method).
- Plastic insemination straws, 0.25 ml, 133 mm long (Planer or IMV)
- 1 ml syringe
- Marker labels for identifying the straws (see *Protocol 1*)
- Equipment for sealing the straws (see *Protocol 1*)
- Plastic goblets (IMV or Planer) and aluminium canes for storage in liquid nitrogen
- Small Dewar flask containing liquid nitrogen
- Mouse oocytes collected ~15 h after hCG treatment and denuded of cumulus with hyaluronidase; hold the oocytes at 37 °C in M2 + FCS (it is important not to let the oocytes cool until they are placed in DMSO)

Method

1. Prepare the freezing apparatus. Cool and hold at 0 °C.
   (a) Prepare an ice slurry with crushed ice and water in a low-sided container. Cover the ice with aluminium foil. Lay two glass boiling tubes and a glass embryo dish on top of the foil. Place the container in the refrigerator until needed.
   (b) Place 3 ml of 1.5 M DMSO in M2 + FCS in the embryo dish and cool the dish on the ice slurry prepared in step 1(a) for at least 10 min.
2. Prepare the straws. Please refer to *Figure 3* for final appearance of straw.
   (a) Use a fine marker pen to make three guide marks on the straw to aid loading with solutions.
      (i) Mark 1: 17 mm from the end of the plug.

Protocol 7. *Continued*

(ii) Mark 2: 70 mm from mark 1.

(iii) Mark 3: 8 mm from mark 2.

3. Fill the straws as follows using a 1 ml syringe attached to the straw:
   (a) Aspirate a column of 1.5 M DMSO in M2 + FCS to mark 3.
   (b) Aspirate a column of air so that the meniscus reaches mark 2.
   (c) Aspirate a column of 1.5 M DMSO in M2 + FCS so that the meniscus reaches mark 1.
   (d) Aspirate air until the first column of 1.5 M DMSO reaches half way up the plug and forms a seal with the polyvinyl alcohol powder. The straws are now ready for loading with oocytes. Place the straws in one of the boiling tubes on ice for ~5 min while collecting the oocytes.
4. Transfer the prepared oocytes to the cold DMSO in the embryo dish. Equilibrate the dish for 2 min on ice. The oocytes will shrink and partially re-expand.
5. Using a mouth pipette, load the oocytes into the large column of DMSO in each straw. Up to 30 oocytes may be loaded per straw. For hints on loading and sealing straws, see *Protocol 1*. Seal each straw and replace in the second boiling tube on ice. Try to minimize the time for which oocytes are at room temperature when loading the straws. Return the embryo dish to the ice while loading the straws.
6. Quickly transfer the loaded straws to the cooling apparatus. *Total exposure to DMSO on ice is 12 min.*
7. Cool at 2°C/min to –8°C. Wait 5 min for the samples to equilibrate.
8. Seed the samples as in *Protocol 1* by touching the column of DMSO that does not contain the oocytes with pre-cooled forceps.
9. Wait for 5 min as in *Protocol 1* for the ice to migrate into the oocyte column.
10. Cool at 0.3°C/min to –40°C, then quickly remove the straws from the cooling apparatus and plunge them directly into a small Dewar flask of liquid nitrogen (see *Protocol 1*, the procedure is identical).
11. Store the straws in liquid nitrogen as in *Protocol 1.*

7.2 Recovery of oocytes frozen using *Protocol 7*

This procedure is essentially similar to the thawing of embryos as described in *Protocol 2*.

Protocol 8. Recovery of oocytes by rapid warming

Equipment and reagents

- Frozen straws of oocytes as required
- Glass embryo dishes (BDH)
- Water bath at 30°C
- Medium M2 + FCS as described in *Protocol 7*
- Prepared dishes of sperm for IVF (see *Protocol 10*)

Method

1. Transfer the appropriate straws from the storage refrigerator to a small Dewar flask of liquid nitrogen. Keep the straws submerged in liquid nitrogen.
2. Using forceps, hold the straw near the wire markers for 40 sec in room temperature air and then in water at 30°C until the ice disappears. Handle the straw gently and do not agitate it in the water bath.
3. Wipe the straw with a tissue, then cut off the bottom seal.
4. Expel the entire contents of the straw into a glass embryo dish either by pushing the cotton plug down the straw or by carefully cutting off the top seal with the straw positioned over the dish.
5. Keep the oocytes at room temperature for 5 min. Add 0.8 ml M2 + FCS at room temperature to the dish and swirl gently to mix.
6. Keep the dish at room temperature for 10 min.
7. Transfer the oocytes to a fresh dish of 2 ml M2 + FCS and maintain for a further 10 min.
8. Wash the oocytes in a fresh dish of M2 + FCS at 37°C and maintain at 37°C (for no more than 30 min) until ready to carry out the IVF.
9. Wash the oocytes through a dish of the chosen IVF medium at 37°C, then transfer the oocytes to the fertilization dishes for insemination.

7.3 Cryopreservation of mouse sperm

This is the method of Okuyama (8) as modified by Nakagata and Takeshima (9) and Sztein *et al.* (10). This protocol is relatively new to our laboratory and is still under development at the MGU Harwell by Peter Glenister and Claire Thornton (13). It is currently practised at the MGU Harwell and the Jackson laboratory, USA.

Protocol 9. Cryopreservation of mouse sperm

Equipment and reagents

- 1.8 ml Nunc cryotubes or 0.25 ml plastic insemination straws (see *Protocol 1*)
- Cryobox (Nalgene)
- Microcentrifuge (e.g. Eppendorf) and plastic, 1.5 ml microcentrifuge tubes
- 10 ml syringe

Protocol 9. *Continued*

- Cryoprotectant solution: 18% raffinose (Sigma) and 3% skimmed milk powder (Difco Betalab) in high quality water (prepared as described below)
- Gilson pipetters (or equivalent) suitable for pipetting volumes of 100–1000 μl.
- 35 mm Petri dishes (tissue culture quality, e.g. Falcon)
- Dissection tools
- Deep polystyrene box with lid suitable for holding liquid nitrogen
- Small Dewar flask of liquid nitrogen
- CO_2 incubator (37 °C, 5% CO_2)
- Heat block held at 37 °C
- Sexually mature male mice, at least 8 weeks old (preferably not previously mated)
- Screw-topped 15 ml tube (e.g. Falcon)

Method

1. Prepare the cryoprotectant solution.
   (a) Place 9 ml of Sigma H_2O or other high-quality water in a screw-topped 15 ml tube and equilibrate to 60 °C in a water bath.
   (b) Add 1.8 g raffinose and dissolve by gentle inversion. Add 0.3 g skimmed milk powder and dissolve by gentle inversion. Make volume up to 10 ml if necessary.
   (c) Aliquot into microcentrifuge tubes and centrifuge at 14,000 r.p.m. for 10 min. Alternatively, if available, use a centrifuge capable of holding the 15 ml tube.
   (d) Remove and pool the supernatant. Filter (0.45 μm, Millipore), into cryotubes or microcentrifuge tubes. Store at –20 °C in 1.1 ml aliquots.
2. Prepare the cooling apparatus. Place a platform (e.g. the insert from a Gilson yellow tip box) into the polystyrene box. This acts as a support for the cryobox. Carefully pour liquid nitrogen into the polystyrene box to just cover the platform. Place a cryobox (without the lid) on top of the platform so that it is suspended in liquid nitrogen vapour. Replace the lid on the polystyrene box and allow it to fill with liquid nitrogen vapour. Replenish the liquid nitrogen as necessary during the freezing session, but do not allow the level to rise above the platform.
3. Thaw one aliquot of cryoprotectant solution for each male mouse and bring to 37 °C in the incubator or heat block. Mix by inversion if there is any precipitation.
4. Pipette the cryoprotectant solution into a 35 mm Petri dish on the heat block at 37 °C.
5. Kill the male mouse.
6. Dissect the vasa deferentia and caudae epididymides from the mouse and carefully remove *all* fat and blood. This is best achieved by placing the organs on a tissue and examining them under a microscope. Work as quickly as possible, then place the trimmed tissue into the warm cryoprotectant solution.

7. Under the microscope, use two pairs of watchmaker's forceps to squeeze and mince the caudae and squeeze sperm gently out of the vas.
8. To disperse the sperm, tap and shake the dish gently for ~30 sec. Incubate for 10 min in a CO_2 incubator at 37 °C. Place the lid of the dish in the incubator and rest the sperm dish at an angle on the lid.
9. Keeping the dish at an angle, remove the epididymal and vas tissue from the suspension by scraping them to one side of the dish with the tip of a Gilson pipetter. Aliquot 100 μl into each of ten labelled cryotubes (or 10 × 100 μl in insemination straws). Replace the screw cap and tighten to seal the cryotube. If using insemination straws, use a Gilson pipetter to transfer 100 μl aliquots of suspension onto the lid of the Falcon dish. Aspirate the sperm suspension into the straws using a 1 ml syringe. Leave adequate space on each side of the suspension to allow for sealing the straw without heating the suspension. Carefully heat seal the straws on either side of the sperm aliquot.
10. Place the cryotubes into the cryobox contained in the liquid nitrogen vapour. If using straws, rest them horizontally on the cryobox (or use a straw rack, see *Protocol 3*). Replace the lid on the polystyrene box.
11. Leave undisturbed to cool for 10 min, then quickly plunge the containers directly into liquid nitrogen.
12. Store in liquid nitrogen refrigerators until required.

7.4 *In vitro* fertilization of mouse oocytes

Here we present a protocol for IVF using frozen sperm to fertilize freshly ovulated oocytes. This method is an evolution of several protocols adapted at the MGU Harwell by Peter Glenister and Claire Thornton (13). See introduction to Section 7. This protocol can be adapted for several purposes. For example, fresh sperm can be used to fertilize fresh or frozen oocytes. Similarly, frozen sperm could be used to fertilize frozen oocytes. There should be sufficient information in this section to enable individual researchers to vary the protocol as needed.

Many different media can be used for IVF. Some formulations work better for different strains of eggs and sperm. At Harwell, we are currently using a modified MEM medium (31) with considerable success (see *Protocol 10*). Alternative IVF media include HTF (32) and T6 (33).

Protocol 10. Recovery of frozen sperm and *in vitro* fertilization of mouse oocytes

Equipment and reagents

- Aliquot(s) of frozen sperm (*Protocol 8*)
- Water bath at 37 °C
- Microcentrifuge (e.g. Eppendorf) and plastic 1.5 ml microcentrifuge tubes
- Gilson pipetters (or equivalent), suitable for pipetting volumes of 10–1000 μl.
- CO_2 incubator (37 °C, 5% CO_2)
- Female mice primed with gonadotrophins; prepubertal females are preferred but mature mice can be used (24,25)
- Medium for IVF (see Section 7.4)
- Medium M2 containing 0.4% (w/v) BSA (see *Protocol 1*)
- Modified MEM medium prepared as follows: Sigma MEM (M-4655) plus (per 100 ml) 2.5 mg sodium pyruvate, 0.38 mg EDTA, 7 mg penicillin, 5 mg streptomycin and 300 mg BSA (all reagents from Sigma); dissolve the BSA without shaking. Filter sterilize (0.22 μm filter) and dispense ~10 ml aliquots into Falcon 2001 tubes. The medium can be frozen at −20 °C. If frozen medium is used, re-filter on thawing.
- Mineral oil, embryo-tested (Sigma)
- 35 and 60 mm Petri dishes

Method

1. The day before IVF, prepare an appropriate number of culture dishes and pre-incubate overnight to equilibrate. A set of dishes for IVF consists of:
   (a) a 35 mm Petri dish containing ~3 ml IVF medium for oocyte collection;
   (b) a 35 mm Petri dish containing either a 500 μl drop or a 200 μl drop of IVF medium overlaid with mineral oil (this is the IVF dish);
   (c) a 60 mm Petri dish containing four 100 μl drops of IVF medium overlaid with oil (this is the wash dish);
   (d) a 60 mm Petri dish containing four 100 μl drops of IVF medium overlaid with oil. This is the culture dish.
2. The number of sets of dishes and the size of the IVF drop depends entirely on the viability and fertilizing ability of the sperm sample in question. For example, if the sample is poor, then an entire aliquot of sperm would be used in one set of dishes utilizing a 500 μl IVF drop. Conversely, an excellent 100 μl aliquot could be distributed between up to five sets of dishes, each set utilizing a 200 μl IVF drop. For the rest of this protocol, we will assume we have a good 100 μl sperm sample, five sets of dishes and 200 μl IVF drops.
3. Transfer the sperm container (cryotube or straw) from the liquid nitrogen storage refrigerator to a small Dewar flask of liquid nitrogen.
4. Using forceps, hold the container in air for 30 sec and then thaw rapidly by placing in a 37 °C water bath. If using cryotubes, take special care that the tube has not filled with liquid nitrogen before plunging into the water bath (such tubes may explode). If liquid nitrogen is present in the tube, wait for it to evaporate first.

5. Using a Gilson pipetter set at 100 μl, transfer the thawed sperm from the cryotube to a microcentrifuge tube. If using straws, cut off the seals and expel the contents into a microcentrifuge tube. Centrifuge at 3000 r.p.m. for 4 min in a microcentrifuge.
6. Discard the supernatant and gently resuspend the pellet in 55 μl of IVF medium (preincubated at 37 °C). Do *not* flick the tube vigorously: use just enough force to lift the pellet into the medium.
7. Leave the lid of the microcentrifuge tube open and incubate for 10 min to allow the sperm to swim up into the medium.
8. Using a fresh tip each time, gently pipette 10 μl of sperm suspension into each fertilization drop. Do *not* pipette the sperm up and down in the drop. Return the fertilization dishes to the incubator as soon as possible.
9. Dissect oviducts from five (or more) superovulated unmated female mice (12–15 h after hCG treatment) per fertilization drop. Place the oviducts into the 3 ml dish of preincubated IVF medium. Using watchmaker's forceps, tear the ampullae and drag the cumulus masses into the medium. With a 1 ml Gilson pipetter set at 100 μl, transfer the cumulus masses from the medium into one of the fertilization drops, being careful to transfer as little medium as possible. Repeat for each fertilization dish. Incubate at 37°C for ~5 h (3–8 h).
10. Pick up the eggs from the fertilization drops and wash through four 100 μl drops of pre-incubated medium (one wash dish for each fertilization dish).
11. Transfer the washed eggs to the culture dish containing four 100 μl drops of pre-incubated medium. Distribute the eggs evenly between the drops.
12. Incubate overnight. On the next day (afternoon is preferable), transfer two-cell embryos to a holding dish of M2 medium at room temperature, and then to both oviducts of 0.5 day pseudopregnant foster mothers (five to eight per oviduct). See Chapter 9, *Protocol 7*.

8. Hints and tips

8.1 Embryo cryopreservation

Successful embryo cryopreservation requires careful attention to each of the steps listed in *Protocols 1–5*. Two steps require further discussion. First, hermetic sealing of straws is required to prevent potential microbial contamination of the frozen samples by leakage of contaminated liquid nitrogen into the straws (34). Leaky straws sometimes explode with the loss of embryos and potential injury to personnel when liquid nitrogen is rapidly converted into gas in the straw during warming. The most effective method for avoiding these problems is to hermetically seal both ends of the straw using an impulse heat sealer designed for plastic bags. Other sealing methods include the use of plastic sealing rods (IMV) and Cristaseal (Hawksley, UK), a clay-based putty used to seal capillary tubes in haematology laboratories. Some batches of Cristaseal can be softer than others and yield inadequate seals. Each batch must be tested before routine use.

Second, control of the formation of ice is required to prevent injury to embryo suspensions associated with supercooling. Supercooling refers to the ability of solutions to avoid freezing when cooled to temperatures below their thermodynamic freezing point. The presence of molar concentrations of cryoprotectants increases the likelihood that embryo suspensions will supercool. However, ice will eventually spontaneously form at low temperatures (usually between –10 and –20 °C) and when this happens, embryos are injured by growth of ice crystals into the blastomeres. The simplest method to prevent injury due to intracellular freezing is to manually induce ice to form in the suspension by 'seeding' the sucrose diluent solution in the straw with forceps or a spatula precooled in liquid nitrogen. Once ice has grown around the air spaces and into the column containing the embryos, the embryo suspension can be safely cooled to lower temperatures. Do not seed the column containing embryos, because this increases the extent of supercooling and likelihood of injury to the embryos.

8.2 Sperm cryopreservation and *in vitro* fertilization

The timing of the hCG hormone injection and collection of the ovulated eggs is important. If eggs are too 'old', many will not fertilize or may undergo parthenogenic activation. We find that hCG administered at ~18:00 h followed by collection and transfer of eggs into a sperm suspension at ~08:00 h the following morning is effective for some mouse strains and hybrids (our light cycle is 06:00 h–22:00 h). At the MGU Harwell most of our experience is with (C3H/HeH × 101/H) F1 eggs. Other strains may react differently.

If fresh sperm are used, a period in culture of 1–2 h is recommended prior to adding oocytes to allow capacitation, i.e. the processes that render sperm capable of fertilization. Cooling and freezing/thawing of mouse sperm results in capacitation-like changes, so an initial culture period is unnecessary (35). If fresh (or frozen) oocytes are fertilized with fresh sperm, the sperm dishes should be prepared before the oocytes are collected (or thawed). Fresh sperm suspensions are prepared by mincing the vas and cauda epidydimis in IVF medium and transferring aliquots of suspension to IVF dishes. Alternatively, an aliquot of a sperm suspension prepared for cryopreservation (i.e. vas and cauda minced in cryoprotectant solution) could be processed for IVF as an unfrozen control. The rest of the suspension could then be frozen for future use.

When large numbers of fertilized embryos are desired, a 100 μl frozen/thawed sperm sample can be used to prepare five fertilization dishes. This approach requires a lot of work, many oocyte donors and recipients, and may reduce the concentration of sperm below that needed for optimum fertilization. Furthermore, this should only be attempted with 'good' genotypes of mice. When fewer fertilized embryos are required or the quality of the thawed sperm is lower, a single fertilization drop (200–500 μl) also yields good results.

The concentration of motile spermatozoa in the insemination drop is a critical factor in the success of fertilization after cryopreservation and thawing. In the case of fresh mouse sperm, the desired concentration is about 1–2 $\times 10^6$ motile sperm/ml. In most cases, fresh sperm suspensions can be diluted to this range because most of the spermatozoa are motile. However, a sperm count alone is not sufficient to prepare the insemination drop using cryopreserved spermatozoa because many of the sperm are killed by the freezing–thawing process.

In the latter case, the sperm concentration must be adjusted by a visual estimate of the percentage of motile sperm in the thawed sample. This adjustment can be used to dilute or concentrate the thawed sperm suspension to the desired concentration. With experience, a single visual assessment can be used to adjust the concentration of thawed sperm. There is some evidence that the appropriate concentration of motile sperm varies for different strains. For example, fresh and thawed spermatozoa from C57BL/6J males must be concentrated to at least twice that of hybrid males to achieve acceptable rates of fertilization (J. M. Sztein, personal communication).

9. Long-term storage of embryos and gametes

Early research at Harwell indicated that mouse embryos frozen in liquid nitrogen could probably remain viable indefinitely and were unaffected by the cumulative dose of background radiation received during storage (36–38). Many years later, this theory shows no signs of being refuted. Several stocks have been recovered from the Harwell, NIH and Jackson embryo banks after 15–20 years of storage. In all cases, the viability of the embryos was similar to that when the stocks were initially frozen (*Figure 4*).

Similar long-term data are not available for oocytes and sperm but there is no reason to believe that prolonged storage would affect viability or genetic integrity. Recent studies on the fertilizing ability of cattle sperm after cryopreservation and storage for 37 years indicate normal rates of IVF and *in vitro* development to blastocysts (39).

9.1 Expected overall survival of cryopreserved embryos and gametes

Accurate information on the overall efficiency of embryo and gamete banking is essential for estimating the effort and resources needed to preserve a mouse

Figure 4. This litter of Rn^{Fkl} (*freckled*) animals was recovered from some of the oldest embryos stored in the Harwell bank. The embryos were frozen in February 1977 and thawed in August 1997. From one container (glass ampoule, 1.5 M DMSO) of 32 embryos, 20 looked morphologically normal and were transferred to pseudopregnant recipients. Six live young were obtained (19% of thawed, 30% of transferred). This rate of survival was similar to that of a viability test in 1977.

model. In the reproductive and cryopreservation literatures, disparities exist in the measures used to report success or efficiency of embryo cryopreservation. The most commonly reported measure is the proportion of normal offspring from only those embryos transferred to recipients establishing pregnancy. This does not account for the total number of embryos thawed nor recipients failing to establish pregnancy and therefore underestimates the number of thawed embryos required to establish a breeding colony.

We suggest that the most useful measure of efficiency is the overall percentage of thawed embryos yielding normal late-stage fetuses and pups. Recent research by Dinnyes *et al.* (40), indicates that the overall efficiency of rederiving an animal model from banked embryos varies depending on the genotype of the embryos. Overall, the efficiency of rederivation from banked embryos ranged from 14% for BALB/cAnN embryos to 51% for DBA/2N and a hybrid genotype (see *Table 1*).

The reason for these differences is poorly understood. These and other data suggest that the differences reflect either genotypic incompatibilities between the embryo and recipient female used for embryo transfer, inherent differences in the ability of embryos from these genotypes to develop *in vivo*, or both. Differences in the efficiency of rederiving future breeders from banked embryos means that some genotypes may require more banked embryos than others. For this reason, the efficiency of rederivation must be measured for

Table 1. Effect of genotype on the efficiency of rederivation of cryopreserved embryos[a]

Embryo genotype	Total number of embryos transferred to		*In vivo* development	
	recipients	recipients establishing pregnancy	pregnant recipients	overall
C57BL/6N	291 (26[b])	215 (19[c])	35%[d]	26%[e]
BALB/cAnN	303 (25)	229 (19)	19%	14%
C3H/HeN	294 (23)	225 (18)	26%	20%
DBA/2N	178 (16)	158 (14)	58%	51%
ICR X F1[f]	230 (22)	189 (18)	63%	51%

[a]Approximately equal numbers of embryos were cryopreserved using *Protocols 3* and *4*, and thawed using *Protocol 5*. Thawed embryos were transferred into the oviducts of (C57BL/6N x DBA/2N) F_1 females on day 1 of pseudopregnancy. For each genotype, no significant differences were found in the rate of development of slowly frozen and vitrified embryos ($P > 0.3$). Data from Dinnyes *et al.* (ref. 40).
[b]Total number of recipient females.
[c]Total number of recipients establishing pregnancy.
[d]Percentage of embryos transferred to recipients establishing pregnancy developing to normal day 18–19 fetuses.
[e]Percentage of all transferred embryos developing to normal day 18–19 fetuses.
[f]Embryos from cross of ICR (outbred) female and hybrid (C57BL/6N x DBA/2N) F_1 male.

each banked genotype to ensure that sufficient numbers of embryos are banked.

The efficiency of rederiving mouse models from cryopreserved gametes is more difficult to estimate. High rates of fertilization and development *in vivo* have been obtained with current cryopreservation procedures for mouse oocytes (29). No systematic studies have examined potential differences in the viability of thawed oocytes from different mouse genotypes.

Recent data indicate that the efficiency of rederivation from thawed mouse sperm varies markedly depending on the genotype of the sperm and the oocyte used for IVF. Songsasen and Leibo (14) report that sperm collected from three genotypes of mice (C57BL/6J, 129/J and hybrid (C57BL/6J × DBA/2J) F1 males) exhibit large differences in their ability to fertilize oocytes. The differences in the survival of thawed sperm from these genotypes correlated with their sensitivity to exposure to anisotonic solutions. The extremely low viability of thawed sperm from C57BL/6J and 129/J males suggests that caution is needed before mouse sperm cryopreservation is introduced as a general method for mouse model banking.

10. Genetic resource banks

Most of the effort to establish banks of cryopreserved embryos from laboratory animal models is focused at four international genetic resource programs, the Jackson Laboratory (Bar Harbor, Maine, USA), MRC Mammalian Genetics Unit (Harwell, UK), NIH Animal Genetic Resource (Bethesda, Maryland, USA), and Central Institute for Experimental Animals (Kawasaki,

Japan). A combined total of approximately two million embryos from 2700 mouse models have reportedly been cryopreserved at these centers. In addition, the NIH and Japanese banks have cryopreserved a total of approximately 75 000 embryos from 175 rat models. A number of research institutions, pharmaceutical/biotechnology companies and laboratory animal breeders have established in-house banking programmes for mouse models developed or used by their investigators.

Information on the availability of animal models from banked embryos from the three largest embryo banking programmes is available on the World Wide Web. Consult the URLs for the MRC Mammalian Genetics Unit (http://www.mgu.har.mrc.ac.uk/), NIH Animal Genetic Resource (http://www.nih.gov/od/ors/dirs/vrp/nihagr.htm) and Jackson Laboratory (http://www.jax.org) for further information.

Acknowledgements

We are indebted to Jorge Sztein and colleagues at the Jackson Laboratory, USA, Ulrike Huffstadt and Rudi Balling at the GS-Research Centre, Neuherberg, Germany, David Whittingham and Maureen Wood in St George's Hospital Medical School, London, and to Naomi Nakagata at Kumamoto University, Japan for much helpful advice.

References

1. Glenister, P. H., Whittingham, D. G. and Wood, M. J. (1990). *Genet. Res. (Camb.)*, **56**, 253.
2. Rall, W. F. (1992). *Animal Repro. Sci.*, **28**, 237.
3. Brown, S. D. and Peters, J. (1996). *Trends Genet.*, **12**, 433.
4. Whittingham, D. G., Leibo, S. P. and Mazur, P. (1972). *Science*, **178**, 411.
5. Wilmut, I. (1972). *Life Sci.*, **11**, 1071.
6. Whittingham, D. G. (1974). *Genetics*, **78**, 395.
7. Whittingham, D. G. (1977). *J. Repro. Fert.*, **49**, 89.
8. Okoyama, M., Isogai, S., Saga, M., Hamada, H. and Ogawa, S. (1990). *J. Fertil. Implant. (Tokyo)*, **7**, 116.
9. Nakagata, N. and Takeshima, T. (1992). *Theriogenology*, **37**, 1283.
10. Sztein, J. M., Farley, J. S., Young, A, F. and Mobraaten, L.E. (1997). *Cryobiology*, **35**, 46.
11. Wakayama, T., Whittingham, D.G. and Yanagimachi, R. (1998). *J. Reprod. Fertil.*, **112**, 11.
12. Wakayama, T. and Yanagimachi, R. (1998). *Nature Biotechnology*, **16**, 639.
13. Thornton, C. E., Brown, S. D. M. and Glenister, P. H. (1999). *Mammalian Genome* 10 (In press).
14. Songsasen, N. and Leibo, S. P. (1997). *Cryobiology*, **35**, 255.
15. Mazur, P. (1984). *Amer. J. Physiol.*, **247**, C125.
16. Rall, W. F. (1993). In *Genetic Conservation of Salmonid Fishes* (ed. Cloud, J. G. and Thorgaard, G. H.), pp. 137–58. Plenum Press, New York.

17. Rall, W. F. and Fahy, G. M. (1985). *Nature*, **313**, 573.
18. Mazur, P. and Schneider, U. (1986). *Cell Biophys.*, **8**, 259.
19. Mazur, P., Rall, W. F. and Leibo, S. P. (1984). *Cell Biophys.*, **6**, 197.
20. Willoughby, C. E., Mazur, P., Peter, A.T. and Critser, J.K. (1996). *Biol. Repro.*, **55**, 715.
21. Katkov, I. and Mazur, P. (1998). *J. Androl.*, **19**, 232.
22. Tada, N., Sato, M., Yamanoi, J., Mizorogi, T., Kasai, K. and Ogawa, S. (1990). *J. Repro. Fertil.*, **89**, 511.
23. Songsasen, N., Betteridge, K. J. and Leibo, S. P. (1997). *Biol. Repro.*, **56**, 143.
24. Hogan, B., Beddington, R., Costantini, F. and Lacy, E. (eds) (1994). *Manipulating the Mouse Embryo*, 2nd edn. Cold Spring Harbor Laboratory Press, Cold Spring Harbor, NY.
25. Monk, M. (ed.) (1987). *Mammalian Development: a Practical Approach.* IRL Press, Oxford.
26. Renard, J.-P. and Babinet, C. (1984). *J. Exp. Zool.*, **230**, 443.
27. Rall, W. F. and Polge, C. (1984). *J. Repro. Fertil.*, **70**, 285.
28. Leibo, S. P. (1984). *Theriogenology*, **21**, 767.
29. Carroll, J., Wood, M. J. and Whittingham, D. G. (1993). *Biol. Repro.*, **48**, 606.
30. Takeshima, T., Nakagata, N. and Ogawa, S. (1991) *Exp. Anim.*, **40**, 495.
31. Ho, Y., Wigglesworth, K., Eppig, J. and Schultz, R. (1995). *Mol. Repro. Dev.*, **41**, 232.
32. Quinn, P., Kerin, J. F. and Warnes, G. M. (1985). *Fertil. Steril.*, **44**, 493.
33. Quinn, P., Barros, C. and Whittingham, D. G. (1982). *J. Repro. Fertil.*, **66**, 161.
34. Schaffer, T. W., Everett, J., Silver, G. H. and Came, P. E. (1976). *Health Lab. Sci.*, **13**, 23.
35. Fuller, S. J. and Whittingham, D. G. (1996). *J. Repro. Fertil.*, **108**, 139.
36. Whittingham, D. G., Lyon, M. F. and Glenister, P. H. (1977). *Genet. Res. (Camb.)*, **29**, 171.
37. Glenister, P. H., Whittingham, D. G. and Lyon, M. F. (1984). *J. Repro. Fertil.*, **70**, 229.
38. Glenister, P. H. and Lyon, M. F. (1986). *J. In Vitro Fertil. Embryo Transfer*, **3**, 20
39. Leibo, S. P., Semple, M. E. and Kroetsch, T. G. (1994). *Theriogenology*, **42**, 1257.
40. Dinnyes, A., Wallace, G. A. and Rall, W. F. (1995). *Mol. Repro. Dev.*, **40**, 429.
41. Rall, W. F. (1991). Prospects for the cryopreservation of mammalian spermatozoa by vitrification. In: *Reproduction in Domestic Animals, Supplement 1, Proceedings of the 2nd International Conference on Boar Semen Cryopreservation,* (L. A. Johnson and D. Rath, eds.), Paul Parey Scientific Publishers, Berlin, pp. 65–80. Reproduction in Domestic Animals, Supplement 1.

3

Spatial analysis of gene expression

STEFAN G. NONCHEV and MARK K. MACONOCHIE

1. Introduction

The generation of mRNA is, in most cases, a prerequisite for gene function. Thus the study of mRNA or transcript generation is important for understanding the potential functions of genes, and many molecular techniques have been used to analyse gene expression. Routine molecular technologies involve either the immobilization of total RNA or poly(A)$^{+}$ RNA on membranes followed by hybridization of DNA or RNA probes, or the use of oligonucleotide primers and reverse transcription–polymerase chain reaction (RT-PCR). However, further molecular methods are being developed, and it is likely that DNA chip and microarray technology will permit more global assays of gene expression. All the above methods are two-dimensional, since spatial information is lost by removal of the cells and tissues to provide the material for assay. It has become clear that the restriction of gene expression to discrete domains in higher eukaryotes is one important method of regulating gene function, and thus gene expression assays which are able to reflect the three dimensions of embryos, tissues and organs are likely to remain important components of genetic analyses for some time to come.

This chapter aims to present some of the techniques available for analysing gene expression in three dimensions. Gene expression can be assayed either directly by analysing the presence of transcripts using *in situ* hybridization or indirectly by using reporter genes to provide a read-out of gene expression. Combinations of these two types of system can be employed, thus permitting simultaneous detection of expression from different genes. It is also possible to use multiple detection systems with different labels in *in situ* hybridization studies.

The ability to analyse transcription from different gene family members presents an important predictor of potential functional redundancy which may be revealed on phenotype examination of single gene mutations. In addition, compensatory mechanisms involving a functional rescue by other gene family members can be examined by expression analysis in the particular mutant background(s). Finally, of course, it is important to establish the expression of newly identified genes in relation to established molecular markers in spatial terms. Thus combinatorial gene expression analysis is an important tool for mutant analysis and gene characterization and represents one of the more active areas of current research.

The authors are particularly interested in the spatial and temporal analysis of gene expression during mouse development. Thus, much of this chapter describes the use of protocols for analysing gene expression in developing mouse embryos. However, the techniques can be easily adjusted for analysis in a variety of contexts.

2. Dissection of post-implantation embryos

The appropriate legal requirements must be fulfilled prior to harvesting embryos. In the UK, project and personal licences may be required (depending on the stage of the embryo) and, in order to obtain these, the appropriate training is necessary. In other countries the national and local legal and ethical regulations must of course be adhered to—a useful first port of call for further details is to consult your animal unit.

Timed matings are established to provide embryos (see Chapter 1, Section 6.1). On average, mice tend to mate halfway through the dark cycle; noon on the day on which the copulation plug is found is referred to as 0.5 days *post coitum* (dpc).

Protocol 1. Harvest of post-implantation embryos

Equipment and reagents

- PBS: 8.0 g/l NaCl, 0.2 g/l KCl, 1.44 g/l Na_2HPO_4, 0.24 g/l KH_2PO_4; adjust to pH 7.2 with HCl
- Watchmakers #5 forceps
- Fine-point iris scissors

Method

1. Kill a pregnant mouse containing embryos at the appropriate stage by cervical dislocation or CO_2 asphyxiation. Thorough training in this procedure should be available from your animal house or other project/ personal licence holders.
2. Lay the mouse on absorbent tissue on its back and saturate the abdominal area with 70% ethanol to help prevent hairs becoming entangled with dissection.
3. Make an incision in the skin using sharp scissors midway between the rib cage and the anus. Grasp the skin on either side of the incision and pull in opposite directions, towards the head and tail. This exposes the body wall covering the abdominal cavity.
4. Open the abdominal cavity with scissors and forceps by making an incision just anterior (approximately 1.0 cm) to the base of the tail, and

cutting outwards to both sides of the mouse. Push the fat and gut to one side to display the reproductive tract.

5. The mouse uterus has two horns. Remove the uterus by cutting at the cervix and at both utero-tubual junctions and transfer uterus to PBS to rinse. In order to separate individual embryo implants (decidua), blot the uterus on tissue and grasp at one end with forceps. Carefully cut between the decidua and transfer each to PBS in a bacteriological Petri dish. If conceptuses are packed closely together, perform this separation in the PBS dish itself, since embryos will respond to release of the pressure following the cut and may well pop out uncontrollably.
6. Remove the decidua from within the uterus using both forceps and applying small tearing motions to the uterine wall all the way round decidua until released. The final stages of dissection are somewhat dictated by the embryonic stages being recovered: for early post-implantation embryos up to 7.5 dpc, make an incision midway through the decidua and roll embryos out carefully using the tips of closed forceps. For later stages, remove the embryo and associated membranes by a lateral incision using forceps midway through the decidual wall, and follow the incision equatorially. The embryo should gradually be revealed in its membranes. Finally, pull away the membranes to reveal the embryo.
7. Transfer the embryos to the appropriate fixative for the subsequent procedure (refer to the appropriate protocols).

3. Whole-mount non-radioactive *in situ* hybridization

3.1 Precautions for RNA work

There is a clear danger of target RNA degradation during *in situ* studies due to the prevalence of ribonucleases (RNases) in the laboratory environment. However, simple routines can be adopted which can considerably reduce this risk.

Solutions should be treated with diethyl pyrocarbonate (DEPC) when possible. DEPC acts by covalent modification of the amine groups of proteins and cannot, of course, be used to treat protein solutions. In addition it is itself inactivated by Tris. Tris-based buffers should be made up in DEPC-treated water. DEPC treatment is performed as follows:

(a) Add DEPC to 0.1%, shake vigorously to mix and leave to stand overnight.

(b) Destroy the DEPC by thorough autoclaving at 15 lb/sq. in. (1 bar) at 121 °C for 15-20 min.

It is also wise to keep all biochemicals for RNA use separate from general laboratory biochemicals. Disposable gloves must be worn at all times during the procedure. Disposable gamma-irradiated plasticware is particularly suitable for reactions. If glassware is to be used it should be thoroughly baked prior to use at a temperature able to destroy RNases; use 180 °C overnight. Tip boxes and tubes should all be set aside for RNA use only.

3.2 Theory of *in situ* hybridization

The principle of *in situ* hybridization relies on the hybridization of a labelled single-stranded probe against the target mRNA. There is no need to denature the target, since mRNA is single stranded, and the resultant hybrid formed following *in situ* hybridization is detected using an antibody-conjugate directed against the particular label incorporated into the probe. The antibody is conjugated with a variety of different enzymes (such as alkaline phosphatase or horseradish peroxidase) and this permits a colour reaction to be used for detecting the presence and location of the hybrid formed during *in situ* hybridization. The practical problems are to permit probe penetration into cells without destroying cellular integrity, and to inhibit endogenous enzymatic activities which may potentially compete with the antibody-linked activity. RNA–RNA hybrids are particularly strong and thus permit the use of high stringency hybridization and washing regimes, which aids in probe specificity, although the appropriate controls should nevertheless be routinely carried out.

Non-radioactive methods are particularly suitable since not only can resolution to the level of single cells be obtained, but isotope usage in the laboratory is also reduced. However, the sensitivity is not as great as that of radioactive methods. Radioactive methods do have additional drawbacks in addition to the safety issue in the length of time required to perform the experiment and also a loss of cellular resolution. Since hybridization using radioactive methods currently is performed on embryo sections, the spatial information obtained is initially two-dimensional, and reconstruction is required to put sections into a three-dimensional context. Since this chapter focuses mainly on techniques allowing immediate three-dimensional imaging of transcript localization, the reader requiring radioactive techniques is referred to ref. 1. Highly expressed and moderately expressed transcripts are readily identified by non-radioactive *in situ* hybridization using the protocols outlined below, and these have been widely used in a number of laboratories.

Detailed protocols covering the whole range of different *in situ* hybridization techniques are available in other volumes of the 'Practical Approach' series (1–3). Here we are concerned with combining different detection systems for gene expression analysis in normal, mutant and transgenic mice, and therefore we outline protocols for non-radioactive whole mount *in situ* hybridization.

3.3 Generation of riboprobe

Although probes can be RNA, DNA or oligonucleotides, single-stranded RNA probes (riboprobes) are particularly sensitive for whole mount applications and do not require denaturing. Furthermore, riboprobes will not renature (notwithstanding repeated elements) and so form strong hybrids with the target. The probe to be tested should be sub-cloned in a vector with a cloning site flanked by RNA polymerase initiation sites such as the pBluescript series (Stratagene) or pGEM series (Promega) of vectors. *In vitro* transcription (*Protocol 2*) is used to generate riboprobe from 'cold' nucleotide precursors and incorporating a hapten-conjugated nucleotide label. A choice of three labels is currently available: digoxigenin-UTP (DIG-UTP), fluorescein isothiocyanate-UTP (FITC-UTP) and biotin-UTP. The DIG detection system is currently the most sensitive, and initial experiments are best attempted using this label.

Clean template, free of RNA and protein, is essential for efficient *in vitro* transcription. Plasmid prepared by banding in CsCl gradients is ideal, although many of the commercial kits incorporating resins or columns are also probably able to deliver plasmid DNA of sufficient quality. Prior to transcription, the plasmid must be linearized using a restriction enzyme unique to the polylinker, but separated from the RNA polymerase transcription site to be used for *in vitro* transcription by the insert. This ensures that the transcription reaction generates probe consisting of insert only. The antisense strand is generated as a riboprobe, but it is also useful to independently transcribe the sense strand as a control for measuring non-specific binding in the *in situ* hybridization.

Protocol 2. *In vitro* transcription of riboprobe

Reagents

- Riboprobe systems (Promega) or DIG RNA labelling kit (Boehringer)
- DIG labelling mix (Boehringer)[a]
- Appropriate RNA polymerases (T3, T7 and/or SP6), 20 units/μl
- Linearized plasmid template
- 0.1 M dithiothreitol (DTT)
- RNase-free water
- Lambda DNA digested with *Hin*dIII
- TE buffer: 10 mM Tris–HCl, 1 mM EDTA, pH 8.0
- 4 M LiCl

Method

1. Set up the following transcription reaction in a microcentrifuge tube:

linearized plasmid	1.0 μg
0.1 M DTT	2.0 μl
5 × transcription buffer (as supplied in the kit)	2.0 μl
DIG labelling mix[a]	2.0 μl
RNase inhibitor (as supplied in the kit)	0.5 μl

Protocol 2. *Continued*

RNA polymerase	20 units
RNase-free water	
(total volume	20.0 μl)

Incubate the reaction mixture at 37 °C for 2 h.

2. Run 1 μl of the reaction on a minigel at 100 V, along with a DNA concentration standard such as a lambda DNA digested with *Hin*dIII. The riboprobe should migrate further than the linearized plasmid, and for an efficient transcription reaction should fluoresce after ethidium bromide staining up to ten-fold more strongly than the plasmid band, as judged by comparison with the concentration standard.
3. Remove plasmid DNA from the rest of the reaction mixture by adding 2.0 μl of RNase-free DNase I (as supplied in the kit) and incubating at 37 °C for 15 min.
4. Adjust the volume to 100 μl with TE pH 8.0. Precipitate the riboprobe by adding 10 μl of 4 M LiCl and 300 μl ethanol and incubate at –20 °C for a period of 2 h to overnight.
5. Pellet the riboprobe by centrifugation for 20 min at 4 °C.
6. Wash the pellet with 70% ethanol and centrifuge for a further 5 min.
7. Allow to air dry inverted on a clean tissue.
8. Resuspend the pellet in 50 ml of TE buffer
9. Run 1.0 μl on a minigel together with a concentration standard. Estimate the riboprobe concentration by comparison with the concentration standard. DNA concentration standards will give an underestimate of the true RNA concentration since single-stranded RNA has a weaker fluorescence. Nevertheless for *in situ* hybridization purposes such estimates are adequate.
10. Riboprobe can be stored at –20 °C for prolonged periods. Some laboratories add RNase inhibitors, but if RNase-free conditions have been used throughout, the riboprobe is stable in TE buffer.

[a] FITC and biotin RNA labelling mixes are also available.

Riboprobes are ideally generated from cDNAs and need to be specific to the gene under study. Care should be taken when making probes from members of gene families that the part of the cDNA used as template is not so similar to other family members that it may cross-hybridize. The 3′-untranslated region is often suitable. Generally templates of 0.5–2.0 kb are of a suitable size for generating riboprobes which are able to penetrate tissue in whole mount *in situ* hybridization; there is little to be gained using much larger probes since probe penetration will become increasingly difficult.

3.4 Fixation and pre-treatment of embryos

Following dissection, embryos are fixed and bleached. A gentle proteinase treatment serves to permeabilize the embryo to permit probe penetration in the hybridization step. The embryo incubations in this section are carried out in plastic 30 ml 'Universal' tubes. Solutions are aspirated using a 1.0 ml pipette tip; it is better to leave a little residual solution from each wash as this will reduce the risk of mechanical embryo damage. Incubations are at room temperature unless otherwise stated, and are performed with shaking on either a linear shaker or a rocking platform.

Protocol 3. Embryo fixation and permeabilization

Reagents

- 4% paraformaldehyde: dissolve 2 g of paraformaldehyde in 50 ml PBS; incubate at 65°C for a few hours to aid dissolution, inverting every half hour until dissolved. The solution should be freshly made.
- 6% hydrogen peroxide: 30% (w/w) H_2O_2 diluted 4:1 in PBT
- PBT: PBS + 0.1% Tween-20 (Sigma)
- 10 μg/ml proteinase K (Boehringer Mannheim) in water
- 2 mg/ml glycine in PBT
- 4% paraformaldehyde + 0.2% glutaraldehyde

Method

1. Following dissection (*Protocol 1*), fix the embryos in 5–10 ml of PFA at 4°C for 2 h (for embryos at 8.5 dpc or younger) or overnight (for embryos at 8.5–10.5 dpc).
2. Remove the paraformaldehyde by three 5 min washes in PBT on a shaking platform.
3. Dehydrate the embryos by passing through a graded methanol series of incubations: (a) 5 min in 25% methanol in PBT; (b) 5 min in 50% methanol in PBT; (c) 5 min in 75% methanol in PBT; (d) 5 min in 100% methanol. Embryos can be stored in 100% methanol for short periods (days to weeks) at –20 °C
4. Rehydrate the embryos by passing through the graded methanol series in reverse order: (a) 5 min in 75% methanol in PBT; (b) 5 min in 50% methanol in PBT; (c) 5 min in 25% methanol in PBT; (d) 5 min in PBT, twice each.
5. Bleach the embryos by incubation in 6% hydrogen peroxide for 1 h at room temperature. Equilibrate in PBT with three 5 min PBT incubations.
6. Permeabilize the embryos by incubation in 10 mg/ml proteinase K for 10 min.
7. Inactivate the proteinase K by two 5 min incubations in freshly made 2 mg/ml glycine in PBT. Equilibrate back to PBT by three 5 min washes in PBT.

Protocol 4. *Continued*

8. Re-fix embryos in 4% paraformaldehyde + 0.2% glutaraldehyde for 20 min. Equilibrate to PBT by three 5 min washes in PBT. The embryos are now ready for hybridization, and can be stored in hybridization buffer at –20 °C.

The incubations above are for embryos at 10.5 dpc and younger. If older embryos are to be hybridized, then incubation times should be increased; for embryos up to 12.5 dpc the 5 min incubations are replaced with 10 min incubations. For adult tissues and organs in *in situ* studies, the appropriate incubation times should be determined empirically to permit adequate penetration of solutions.

3.5 Hybridization and removal of non-specific hybrids

Embryos are conveniently hybridized in 2.0 ml round-bottomed tubes. All incubations are performed with gentle shaking or agitation. Solutions are aspirated using 1.0 ml tips and pipette, taking care not to damage the embryos.

Protocol 4. Hybridization and post-hybridization washes

Equipment and reagents

- Hybridization buffer: 50% formamide, 5 × SSC (0.75 M NaCl, 75 mM sodium citrate pH 5.0), 50 μg/ml yeast tRNA (type III, Sigma), 1% SDS, 100 mg/ml heparin (H9399, Sigma), 0.2% Tween-20, 0.5% CHAPS (3-[(3-cholamidopropyl)dimethylammonio]-1-propane-sulphonate) (Sigma), 5 mM EDTA; store at –20°C
- 20 μg/ml RNase A (Sigma) in RNase buffer
- Wash I: 50% formamide, 5 × SSC pH 5.0, 1% SDS
- Heating block placed sideways on rocker or hybridization oven with rocking platform
- RNase buffer: 0.5 M NaCl, 10 mM Tris pH 7.5, 0.1% Tween
- Wash II: 50% formamide, 2 × SSC pH 5.0, 1% SDS

Method

1. Transfer the embryos to be hybridized to 2.0 ml microcentrifuge tubes. Rinse twice in hybridization buffer at room temperature.
2. Pretreat in hybridization buffer at 65°C with gentle rocking for a period of 1 h to overnight.
3. Remove pretreatment buffer and add hybridization buffer containing riboprobe from *Protocol 2* at 1 μg/ml. A volume of 0.3–1.0 ml of hybridization buffer is ideal for three mid-gestation embryos. Hybridize overnight at 65°C with gentle rocking.
4. Remove the unbound probe by two room-temperature rinses with hybridization buffer.

5. Rinse the embryos in wash I and then perform two 30 min washes by incubating the embryos in wash I at 65 °C.
6. Equilibrate the embryos in RNase buffer. First replace wash I with a 1:1 mix of wash I and RNase buffer. Incubate the embryos at 65 °C for 10 min. Then perform three 5 min incubations in RNase buffer at room temperature.
7. Remove non-specific hybrids by RNase treatment. Incubate embryos in 20 μg/ml RNase A at 37 °C for 5 min, twice.
8. Rinse with wash II and perform high stringency washes by incubating in wash II at 65 °C for 30 min, twice.

As with conventional membrane hybridization assays, individual probes behave differently and although most probes will work under the standard conditions outlined above, some parameter optimization may be required for the best results. *Protocol 4* represents the ideal stage at which to vary the stringency of hybridization by altering the temperature of hybridization and wash I/II incubations (steps 3, 5 and 8).

3.6 Immunodetection of signals

The riboprobe includes a labelled nucleotide, and thus the hybrid generated can be detected with an antibody directed against this label. This 'label-specific' antibody is conjugated with a detection system such as alkaline phosphatase (AP), horseradish peroxidase or β-galactosidase. There are a variety of combinations available for the three labels, DIG, FITC and biotin, with different conjugated activities for detection. Here we only describe the use of a DIG–AP detection system as this appears to be the most sensitive under standard conditions. Only minor changes are required when using the anti-FITC–AP and streptavidin–AP combinations.

Incubations are at room temperature unless otherwise stated, and all incubations are with gentle agitation to ensure circulation of solution around the embryo.

Protocol 5. Immunodetection and development of signal

Reagents

- 10 × TBST: 1.4 M NaCl, 27 mM KCl, 0.25 M Tris–HCl pH 7.5, 1% Tween-20
- NTMT: 100 mM Tris–HCl pH 9.5, 100 mM NaCl, 50 mM $MgCl_2$, 0.1% Tween-20
- Blocking reagent (Boehringer)
- NBT: 75 mg/ml nitroblue tetrazolium salt in 70% dimethyl formamide; store at –20 °C in the dark
- PBT (see *Protocol 3*)
- Embryo/tissue powder: homogenize embryos or tissue, of the same stage at which the probe is to be assayed, in PBS and incubate in ice-cold acetone for 30 min. Pellet the embryo fragments by centrifugation in a microcentrifuge for 10 min at 4°C and remove the supernatant. Grind the pellet to a fine powder, divide it into aliquots and store at –20°C.
- DIG–AP antibody conjugate (Boehringer)

Protocol 5. *Continued*

- BCIP: 50 mg/ml 5-bromo-4-chloro-3-indolyl phosphate, tolusinium salt in 100% dimethylformamide; store at –20 °C in the dark
- Lamb serum (Sigma) heated at 56°C for 30 min
- 2 mM levamisole hydrochloride (L[–]-2,3,5,6-tetrahydro-6-phenylimidazo[2,1-b]thiazole hydrochloride) (Sigma) in TBST, made fresh each time

Method

1. Allow post-hybridization washed embryos from *Protocol 4* to cool to room temperature, and then equilibrate with three 5 min incubations of freshly made 2 mM levamisole in TBST.
2. 'Pre-block' the embryos in 2 mM levamisole in TBST containing 2% blocking reagent and 10% lamb serum at room temperature for 1–2 h.
3. During the pre-blocking incubation it is convenient to pre-absorb the DIG antibody. Add a few grains of embryo or tissue powder to 0.5 ml TBST and incubate it at 70 °C for 30 min. Allow it to cool on ice, add lamb serum to 1%, blocking reagent to 2% and 1 μl of DIG antibody. Shake gently for 1 h at room temperature, pellet in a microcentrifuge at 4°C for 10 min and add the supernatant to 1.5 ml TBST, 1% lamb serum and 2% blocking reagent. The antibody is now at 1:2000 dilution and is ready for incubation.
4. Remove the pre-block solution and replace it with diluted antibody. Incubate at 4 °C overnight with gentle rocking.
5. Wash off the unbound antibody by repeated washes of TBST at room temperature. Three 5 min washes in 2 mM levamisole in TBST are followed by five 1 h washes in 2 mM levamisole in TBST.
6. Equilibrate the embryos in staining buffer. Remove the TBST and replace with freshly made NTMT + 2 mM levamisole. Perform three 15 min incubations in NTMT at room temperature.
7. Stain the embryos by incubating them in NTMT + 2 mM levamisole to which has been added with 4.5 μl/ml NBT and 3.5 μl/ml BCIP at room temperature in the dark. Check the colour development occasionally; highly expressed probes will begin to stain after 30 min, others may take several hours.
8. When sufficient colour has developed, stop the reaction by washing with PBT pH 5.5 (see *Protocol 3*) for three 5 min incubations and then fix the embryos in 4% paraformaldehyde + 0.1% glutaraldehyde (see *Protocol 3*) for 1 h at room temperature. Embryos can be kept at 4 °C in PBT.
9. Embryos can be cleared for better visualization of signal by incubation at room temperature in 50% glycerol in PBT for 1 h followed by 80% glycerol in PBT for 1 h.

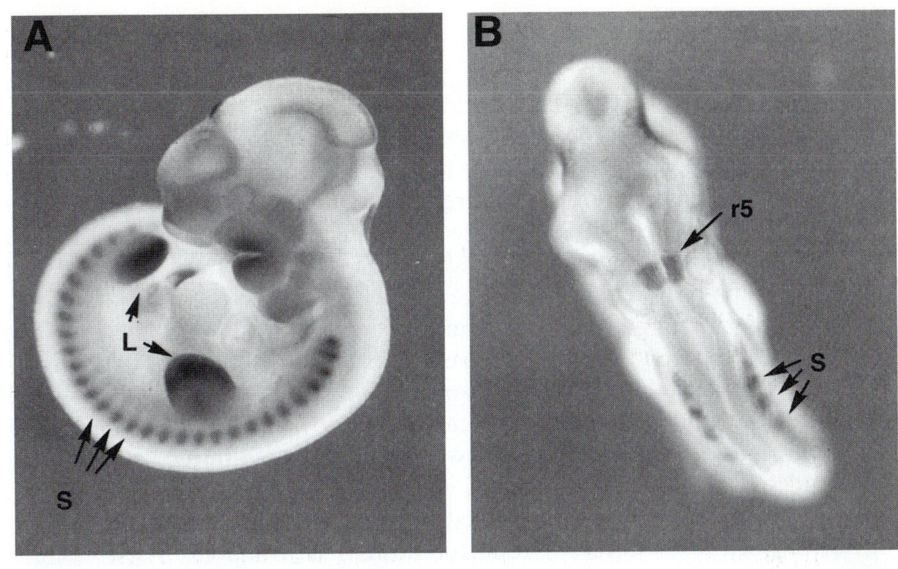

Figure 1. Direct visualization of gene expression patterns by non-radioactive whole-mount *in situ* hybridization. Using DIG-labelled riboprobes with detection by an anti-DIG antibody conjugated with alkaline phosphatase, the precise localization of transcripts in the developing embryo can be investigated. (A) Lateral and (B) dorsal views of a 10.5 dpc embryo showing discrete localisation of transcripts to the developing somites, limbs and in the hindbrain in rhombomere 5 (r5). S, somites; L, limbs.

3.7 Troubleshooting

The development of signal from the alkaline phosphatase reaction catalysed by the DIG–AP antibody conjugate competes with endogenous AP activity which produces background staining. An important first step in analysing the transcription of a gene is to have some evidence from either Northern blot analyses or RT-PCR that the gene under study is being transcribed at the stage or tissue being assayed. The following controls should be routinely included: (i) hybridization with sense riboprobe; and (ii) hybridization with no riboprobe. These controls demonstrate specificity of signal and development of background staining respectively. If the test (antisense) riboprobe generates a non-specific signal, then the hybridization and post-hybridization washing stringency should be increased. If there is background due to the antibody incubation, then further pre-absorption is required. If no signal is generated, then the stringency should be reduced. In addition it is also useful to obtain a probe which has been documented for the particular stage under assay and hybridize this as a positive control.

4. Reporter transgenes

4.1 General considerations

The principle underlying the use of reporter transgenes is to tag a gene or its regulatory region with a gene for which a convenient expression assay is available. The assay is therefore indirect but can be particularly sensitive. A number of reporter genes are available commercially, but assays for luciferase and chloramphenicol acetyl transferase (CAT) are not consistent with the spatial and temporal localization of gene expression. On the other hand, enzymatic assays which generate a colour reaction with β-galactosidase (4) and alkaline phosphatase (5) make these gene products ideal reporters and protocols are presented for their assay. More recently, fluorescent reporter genes have been developed for use in mice (6); these are ideal for applications where embryos, tissue or organs are to be kept alive in different culture regimens and a non-invasive read-out of gene expression is required. However, more specialized equipment is required.

A wide variety of different reporter designs is available. These include reporter vectors which also carry different basal promoter elements, with modified translational sequences such as polyadenylation sites and internal ribosome entry site (IRES) sequences, retroviral-based vectors and a fusion between the antibiotic resistance gene, *neo* and the reporter *lacZ*, known as βgeo. The protocols presented here are assays suitable for detection of the reporters which rely on detection of the activity of β-galactosidase or alkaline phosphatase, or the presence of green fluorescent protein (GFP).

4.2 Introduction and uses of reporter genes

Although reporter genes have been used widely in tissue culture assays for some time (7), the benefits of using reporters *in vivo* have only found widespread use in the last decade. For spatial analyses of gene expression, the goal is to place the particular reporter in the mouse genome in a heritable fashion, and this is achieved through either embryonic stem (ES) cell manipulation or pronuclear injection, depending on the particular biological questions to be addressed. The generation of transgenic mice is dealt with in detail elsewhere in this volume (Chapters 9 and 10). Some of the more common applications for reporter genes are outlined below:

(a) Enhancer mapping studies: in order to identify the minimal *cis*-acting DNA sequences required for normal spatial and temporal expression of a gene, candidate regulatory sequences from the gene can be fused to a reporter which has a minimal promoter, or the reporter can be fused in-frame within the coding region of a gene. A series of constructs will be required to progressively delimit the sequence elements critical for regulation of the reporter (8,9).

(b) Gene targeting strategies can benefit by inserting a reporter gene within the coding region of a gene whose function is to be tested through replacement via homologous recombination in ES cells (10). The reporter is therefore used as a mutagen, as well as serving to tag gene expression from the mutant allele (11). The βgeo reporter combines the enzymatic activity of *lacZ* with the antibiotic resistance conferred by the *neo* gene.

(c) Randomly tagging the genome: gene trap, promoter trap and enhancer trap strategies rely on the random integration of a reporter in the genome, and this is commonly carried out in ES cells (12,13). Tagging genes with a reporter in ES cells additionally allows for many *in vitro* screening assays of expression to be performed prior to analysis of expression of the integration *in vivo*.

Transgenic mouse lines carrying reporters are also particularly useful for investigating potential genetic relationships by appropriate breeding schemes with mutant alleles. Combinatorial gene expression analysis allows the spatial contribution of individual genes to be compared. Reporter genes have already been shown to be of great value and versatility at all levels of analysis of genetic pathways, from addressing the issue of gene function, to isolating upstream regulators and the identification of downstream targets.

4.3 Detection of β-galactosidase activity

The bacterial *lacZ* reporter gene has the advantage of permitting a highly sensitive *in situ* visualization of β-galactosidase activity in all tissues expressing the transgene. The histochemical staining procedure is based on the β-galactosidic cleavage of the chromogenic substrate X-Gal (5-bromo-4-chloro-3-indolyl-β-D-galactosidase), leading to a blue reaction product in the cell, which can diffuse extracellularly before precipitation, but once precipitated is sufficiently stable to survive a variety of further treatments (14,15). With transgenic constructs harbouring a nuclear localization signal, most of the blue product will display intra- or peri-nuclear staining.

Using the procedure outlined below for embryos up to 12.5–13.5 dpc, background staining is usually negligible (*Figure 3B*, *D* and *F*). However, after this stage the developing skin acts as a barrier to reagent penetration for fixation and staining. To avoid these problems, embryos at 13.5 dpc and older should be partly dissected to permit full penetration of reagents (4).

Protocol 6. LacZ staining

Reagents

- PBS (see *Protocol 1*)
- Washing solution: 2 mM $MgCl_2$, 0.01% sodium deoxycholate, 0.02% NP40, 5 mM EGTA, all in PBS
- LacZ fixative: 1% formaldehyde, 0.2% glutaraldehyde, 2 mM $MgCl_2$, 5 mM EGTA (from stock of 100 mM, pH 8.0), 0.02% Nonidet P40 (NP40) (Sigma), all in PBS

Protocol 6. *Continued*

- X-Gal (5-bromo-4-chloro-3-indolyl-β-D-galactosidase) (Melford Laboratories); store in the dark at −20 °C. Stock solution: 40 mg/ml X-Gal, dissolved in dimethylformamide (Sigma) store in dark at 4 °C
- Staining solution: 5 mM potassium ferricyanide, $K_3Fe(CN)_6$, 5 mM potassium ferrocyanide, $K_4Fe(CN)_6 \cdot 3H_2O$, 2 mM $MgCl_2$, 0.01% sodium deoxycholate, 0.02% NP-40, 1 mg/ml X-Gal from stock, all in PBS (store in dark at 4°C)

Method

1. Harvest and wash the embryos in PBS (*Protocol 1*).
2. Fix the embryos in LacZ fixative by incubation at 4°C for 30–60 min. This and subsequent steps are conveniently carried out in plastic scew-capped tubes or in 24-well tissue culture plates.
3. Wash the embryos three times, for 20 min each, at room temperature in washing solution.
4. Stain the embryos in the dark, in staining solution.[a]
5. Wash the embryos in PBS and store in PBS in the fridge for a few days to intensify the blue stain.
6. Store the embryos indefinitely in 70% ethanol at 4 °C.

[a] The duration of this step can vary and is dependent on the level of expression of the transgenic construct. In cases where there are high levels of expression, signals can be detected in a few hours. If expression levels are low, staining can be done at 37 °C overnight. With later stage embryos (>12.5dpc) the staining reaction should only be performed at room temperature, using longer incubations times if necessary, in order to avoid background staining problems.

4.4 Detection of alkaline phosphatase activity

A second reporter gene which can be detected histochemically is human placental alkaline phosphatase (PLAP); this has been shown to be extremely versatile in studies that require marking of cells and tissues of transgenic embryos (5,16). The highly sensitive detection of alkaline phosphatase is based on the dephosphorylation of the substrate X-phosphate (BCIP) to give a dark blue dye as a further oxidation product. The oxidant in the reaction is nitroblue tetrazolium chloride (NBT). Reduction of NBT intensifies the colour and makes the detection more sensitive; see *Figure 3A* and *E*. When using the assay presented in *Protocol 7* for PLAP staining, take particular care to ensure that endogenous AP activity, which is abundantly present in all living tissues, is fully quenched.

Protocol 7. Staining procedure for human placental alkaline phosphatase

Reagents

- Alkaline phosphatase (AP) staining buffer: 100 mM Tris–HCl pH 8.5, 100 mM NaCl, 50 mM $MgCl_2$
- NBT (Boehringer Mannheim), stock solution at 50 mg/ml in 70% dimethylformamide
- BCIP (Boehringer Mannheim), stock solution at 10 mg/ml in 70% dimethylformamide
- 50× levamisole stock solution (12 mg/ml in water or staining buffer)
- Staining solution: 1 mg/ml NBT, 100 μg/ml BCIP, 240 μg/ml levamisole in AP staining buffer
- Stop solution: 50 mM EDTA, pH 5.0

Method

1. Harvest embryos and wash thoroughly in PBS (see *Protocol 1*).
2. Fix in paraformaldehyde (see *Protocol 3*) at 4 °C for 20–40 min.
3. Equilibrate the embryos in PBS by three 5 min incubations in PBS at room temperature. Inactivate the endogenous AP activity by heating the embryos at 65–70 °C for at least 1 h (heat in PBS, not in AP staining buffer).
4. Cool the embryos for 30 min at room temperature.
5. Incubate the embryos in 1× levamisole solution (600 μg/ml) for at least 1 h at room temperature.
6. Transfer the embryos to the staining solution and incubate in the dark at room temperature.
7. Briefly check the staining reaction after 1 h (the colour reaction is light sensitive and is generally complete within 1–2 h).
8. When the colour reaction has developed to the desired extent, stop the reaction by replacing the staining solution with stop solution for at least 1 h. Refix in paraformaldehyde (see *Protocol 3*) overnight at 4 °C and store in PBS for a limited period since background staining will slowly increase.

4.5 Use of the green fluorescent protein (GFP) in transgenesis

The green fluorescent protein (GFP) is responsible for the green bioluminescence from the jellyfish *Aequorea victoria.* GFP is a single peptide, of 238 amino acids, that absorbs blue light and emits green fluorescence without the need for exogenous substrates or cofactors. It is rapidly becoming the reporter molecule of choice for the *in vivo* analysis of gene expression. Importantly, assays using GFP do not require any prior embryo preparation or substrate loading. Therefore the main advantages of this unusually stable reporter protein are the easy detection using standard long-wave UV light sources and the possibility of real-time *in vivo* staining of early embryos. In addition, variants of GFP have been generated with different spectral properties, permitting their use in combinatorial gene expression analyses. This latter feature allows the simultaneous labelling of multiple proteins, obviating the use of additional substrates which are commonly used by other reporter systems (17). A variety of constructs have been designed using the cDNA encoding GFP to generate transgenic mice in order to monitor enhancer specificity, fusion protein distribution and lineage analysis (6,18–20). A major problem associated with GFP expression in transgenic embryos is the correct *in vivo* generation of the chromophore by native folding of the polypeptide so that the final product displays fluorescent properties. Indeed, a number of communications indicate that certain constructs do not express the GFP reporter in mice, and the reporter also appears to be particularly vulnerable to the effects of integration site.

Protocol 8. Observation and imaging of GFP-transgenic embryos

Equipment and reagents

- Epifluorescence microscope
- Confocal laser scanning microscope system
- FITC filter set: Band Pass, 450-490 nm, Dichroic Mirror, 512 nm, Barrier filter, 515 nm
- Perfusion chamber
- Vibratome
- M2 embryo culture medium (Sigma)
- Ringer's physiological solution: 9 g NaCl, 0.42 g KCl, 0.25 g $CaCl_2$ per litre of water

A. *Pre-implantation embryos*

1. Harvest pre-implantation stage embryos (Chapter 1, *Protocol 6*).
2. Place the embryos in microdrops of medium on coverslips under paraffin oil (see Chapter 9, *Protocol 5*, step 11).
3. Observe GFP expression using epifluorescence microscope equipped with a filter set for fluorescein (excitation at 488 nm and detecting emission at wavelengths >515 nm).

B. *Post-implantation stage and transgenic tissues*

1. Dissect the embryos and remove organs from later stages fetuses in PBS (see *Protocol 1*).
2. Cut vibratome slices of 100–300 μm or prepare whole mounts of organs or tissues.[a]
3. Place in a perfusion chamber with Ringer's physiological solution for live observation.
4. Observe immediately as in A, and analyse GFP images by a confocal laser scanning microscope system.

[a] If better and more detailed morphological integrity is required, then fix tissues in 4% paraformaldehyde prior to further treatments.

5. Multiple and combined detection systems

We have presented detailed protocols for direct and indirect assays to reveal the spatial pattern(s) of gene expression. However, combining these approaches permits us to explore the dynamic relationships between genes and integrate this information with morphological aspects of the developing embryo or in a tissue or organ context. In addition, differential labelling of either genes and/or regulatory elements permits us to precisely localize expression domains with reference to specific molecular markers. Although a large variety of multiple staining assays are possible, we suggest the following choice of combinations which have proven effective in detecting simultaneous gene expression and reporter activity in whole-mount preparations and sectioned material. These protocols have been used to investigate embryonic patterning (*Figure 2C*), trace gene co-expression (*Figure 2G*) and follow cell migratory pathways (*Figure 2H*).

5.1 Whole-mount staining of early (8.0–12.5 dpc) embryos

For whole-mount applications using transgenic mice, double reporter staining for AP and β-galactosidase with consecutive *in situ* hybridization analysis is suitable for combined visualization of gene expression. The assays are usually performed with whole litters from the appropriate crosses of transgenic lines, and this is also important for correlating genotype with specific patterns of reporter expression. As outlined earlier, whole-mount techniques display negligible background staining prior to 12.5 dpc, but minor modifications can help reduce background development (see Section 4.3). Double marking experiments necessitate slight modifications of the single-detection protocols outlined above (*Protocols 6* and *7*), and additional steps of thorough washing between staining procedures are required.

Protocol 9. Combined LacZ staining and *in situ* hybridization

Reagents

See *Protocols 2–6*

Method

1. Dissect out the embryos in cold PBS.
2. Incubate the embryos in cold LacZ fixative (*Protocol 6*) for 10 min at 4 °C.
3. Wash the embryos in cold PBS plus 0.02% NP40 (*Protocol 6*) for 10 min at 4 °C.
4. Stain the embryos for the minimal period of time in LacZ staining solution (*Protocol 6*) at 37 °C until sufficient colour has developed.

Protocol 10. *Continued*

5. Wash the embryos three times for 20 min in PBS plus 0.02% NP40 at room temperature.
6. Process the embryos for non-radioactive whole-mount *in situ* hybridization according to *Protocols 2–5*.

Protocol 10. Double staining for β-galactosidase and alkaline phosphatase

Reagents

- See *Protocols 6* and *7*
- 4% paraformaldehyde (see *Protocol 3*)

Method

1. Harvest the embryos in PBS (see *Protocol 1*).
2. Fix the embryos in paraformaldehyde at 4 °C for 15–30 min.
3. Wash the embryos three times in PBS for 20–30 min.
4. Stain the embryos for β-galactosidase (*Protocol 6*) at room temperature, overnight.
5. Wash the embryos twice in PBS for 20–30 min.
6. Heat the embryos (in PBS) for 40 min at 65°C to inactivate endogenous AP.
7. Incubate the embryos in levamisole solution (*Protocol 7*) for 1–2 h at room temperature.
8. Place the embryos in AP staining solution (*Protocol 7*) for a few hours until colour develops.
9. Incubate the embryos in stop buffer for 30 min and refix in paraformaldehyde at 4 °C overnight.

5.2 Multiple detection on cryostat and paraffin sections

A serious limitation to the use of whole-mount staining protocols on later stage embryos is the poor penetration of fixative, stains and probes to the internal sites of gene expression. For these late stages, improved access of the reagents to the fetus is achieved by prior sectioning of the material.

Figure 2. Detection of reporters in the central nervous system and the mesoderm. (A–C) Differential staining of hindbrain rhombomeres by detection of (A) placental alkaline phosphatase (AP), (B) β-galactosidase activity (βgal) and (C) combined detection. (D–F) Reporter staining from different enhancers and genes illustrates that the spatial pattern of gene expression can be tightly controlled, and careful examination of these patterns reveals overlapping and non-overlapping domains in tissues and organs that may have functional significance. (G) Analysis of the cellular distribution of combined AP (yellow)

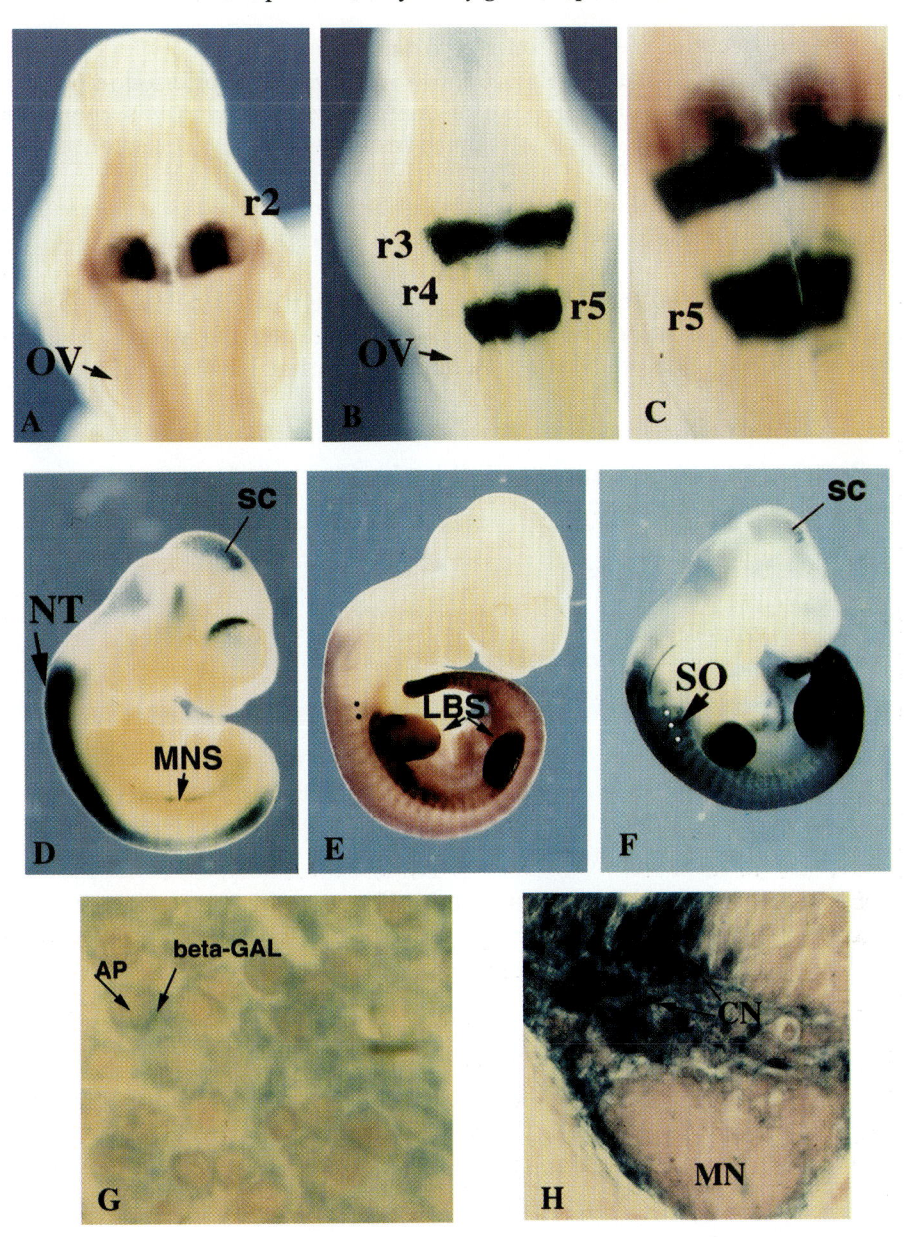

and βgal (blue-green) staining showing co-expression in the same cells of the developing neural tube, although restricted to different areas within these cells. (H) βgal-eosin stained paraffin transverse sections through the developing spinal cord illustrating the commissural pathways in blue (arrows) and the ventral motor neurons (pink) which are negative for the reporter. OV, otic vesicle; r, rhombomere; NT, neural tube; MNS, mesonephros; LBS, limbs; SO, somites; SC, superior colliculus; MN, motor neurons; CN, commissural neurons. This figure was sponsored by Leica Microsystems.

Adjacent sections prepared on a cryostat can be processed with different histochemical stains, immunocytochemistry and *in situ* hybridization (*Figure 2G*). Cryostat sections require shorter incubations than paraffin sections for tissue penetration and display a higher sensitivity of signal detection. However, sections obtained from embryos embedded in paraffin wax give excellent morphological resolution as cellular integrity is maintained (see *Figure 2H*) and also permit the opportunity to process a large number of thin serial sections.

Sectioned material should be transferred to pre-treated slides as this provides suitable conditions for section retention during the various staining procedures, and slide preparation ('subbing') is outlined in *Protocol 11*.

Protocol 11. Pre-treatment of slides for frozen and wax sections

Equipment and reagents

- Metal slide trays
- Ovens at 150 °C and 42 °C
- TESPA (3-aminopropyl-triethoxysilane) (Sigma), 2% solution in acetone
- Acetone
- 10% HCl
- 70% and 90% ethanol

Method

1. Immerse slide trays containing glass slides in 10% HCl followed by 70% ethanol, distilled water and finally in 95% ethanol. Allow them to dry at room temperature.
2. Bake the washed slides in an oven for 10 min at 150 °C and allow them to cool to room temperature.
3. Dip the slides for 10 sec in 2% TESPA in acetone.
4. Wash the slides twice in acetone and once in distilled water for approximately 10 sec each.
5. Dry the slides overnight at 42 °C in a dust-free oven.
6. Store the pre-treated slides at room temperature in a slide box or metal trays wrapped in aluminium foil to protect from dust.

Minimal sample preparation is required for cryostat sectioning of embryonic material, tissues and organs. Collecting adjacent sections on different slides allows the detection of different probes and application of different stains to neighbouring sections. The limited nature of tissue pre-treatment results in minimal signal loss, and also permits straightforward application of the protocols for detecting gene expression.

Protocol 12. Multiple staining on cryosections

Equipment and reagents

- Cryostat
- Plastic moulds
- Narrow slide-containers
- Cryoprotectant: 20–30% phosphate-buffered sucrose
- Superfrost slides (BDH)
- Embedding material: OCT Tissue Tek embedding medium (Miles Scientific)
- Liquid nitrogen or dry ice
- 4% paraformaldehyde (see *Protocol 3*)

Method

1. Dissect out the embryos in cold PBS and fix in paraformaldehyde at 4°C overnight.
2. Equilibrate the embryos in cryoprotectant overnight at 4°C.
3. Place the embryos in thin plastic moulds and embed in OCT compound.
4. Freeze the moulds by placing on dry ice or on the surface of liquid nitrogen.
5. Mount the mould on a cryostat chuck.
6. Cut frozen sections at about –26°C with a thickness of 8–15 μm.
7. Thaw-mount the sections on Superfrost pre-treated slides, dry at room temperature and store at –70°C.
8. Staining procedures: Place slides with adjacent sections in narrow slide-containers, and fix and stain separately for β-galactosidase (*Protocol 6*) or AP (*Protocol 7*) or treat for non-radioactive *in situ* hybridization (*Protocols 2–5*).

The obvious benefits of generating reporter transgenic lines is the ability to analyse detailed cellular expression of the marked gene throughout the embryo. This can be combined with a wide range of tissue-specific, cytoplasmic or nuclear histochemical stains on paraffin sections and this contributes to the spatial interpretation of gene activity. *Protocol 13* outlines the procedures suitable for embedding and staining *lacZ* transgenic embryos at 12.5 dpc. Incubation times should be adjusted for later gestational stages accordingly.

Protocol 13. Embedding, sectioning and counterstaining of *lacZ* embryos

Equipment and reagents

- Microtome
- Paraffin dispenser
- Paraffin section-mounting water bath (Raymond Lamb)
- Plastic moulds
- Pre-treated slides (*Protocol 11*)
- Paraffin wax: pastillated Fibrowax (BDH)
- Saline (0.9% NaCl)

Protocol 12. *Continued*

- PBS
- Toluene[a]
- 4% paraformaldehyde (see *Protocol 3*)
- DPX mounting medium (Raymond Lamb)
- Eosin B (Sigma), 0.25% in 30% ethanol (dissolve the solid and leave the solution for several days at room temperature with occasional stirring)

Method

1. Place stained, X-gal-positive embryos in 5–10 ml ice-cold paraformaldehyde and leave overnight at 4 °C.
2. Incubate with 5 ml PBS at 4 °C for 30 min, then with 5 ml PBS at 4 °C for 30 min.
3. Dehydrate the embryos by treating them with a series of 50%, 70%, 90% and 100% ethanols, each for 30 min at room temperature.
4. Clear the embryos by incubating in 10 ml of toluene at least twice for 30 min at room temperature.
5. Equilibrate the embryos in a mixture of 1:1 toluene:paraffin wax for 20 min at 60 °C.
6. Replace the solution with pure molten wax, by incubating the embryos with three fresh changes of wax, each for 20 min at 60 °C (do not exceed these times).
7. Transfer the embryos to fresh wax in a plastic mould, orientate them as required with a warmed needle and allow the wax to set.
8. Cut ribbons of 6 μm sections on the microtome.
9. With a brush or needle lower the sections and float them on a section-mounting water bath at 50 °C.
10. Allow any creases to unfold and disappear for 2–5 min, then lift and collect sections onto pretreated slides (*Protocol 11*).
11. Drain excess water onto a tissue and place the slides on a slide drier for 30 min at 37°C.
12. Allow the slides to dry at room temperature and store them at 4 °C in a box with desiccant.
13. Remove the wax by immersing twice in toluene[a] for 10 min each at room temperature.
14. Immerse the slides for a few seconds each in a graded ethanol series: successively in 100%, 80%, 70%, 50% and 30% ethanols.
15. Wash the slides in PBS or saline for 10 min.
16. Counterstain the sections in eosin or other histochemical stains for appropriate lengths of time determined by desired intensity required for the specific counterstain. For eosin, start with 5–10 min and examine the intensity.

17. Dehydrate the sections by successive immersion through the ethanol series (30–100%); a few seconds in each is sufficient.
18. Rinse the slides in toluene for 5 sec and mount in DPX mounting medium under coverslip.

[a] Toluene can be replaced by xylene or Histoclear.

6. Whole-mount skeletal analysis

Mutational analyses in the mouse have demonstrated that many genes are required for patterning the developing embryo and for the processes of chondrification and ossification to generate the complete axial and appendicular skeleton, and the skull vault. Skeletal preparations are critical in

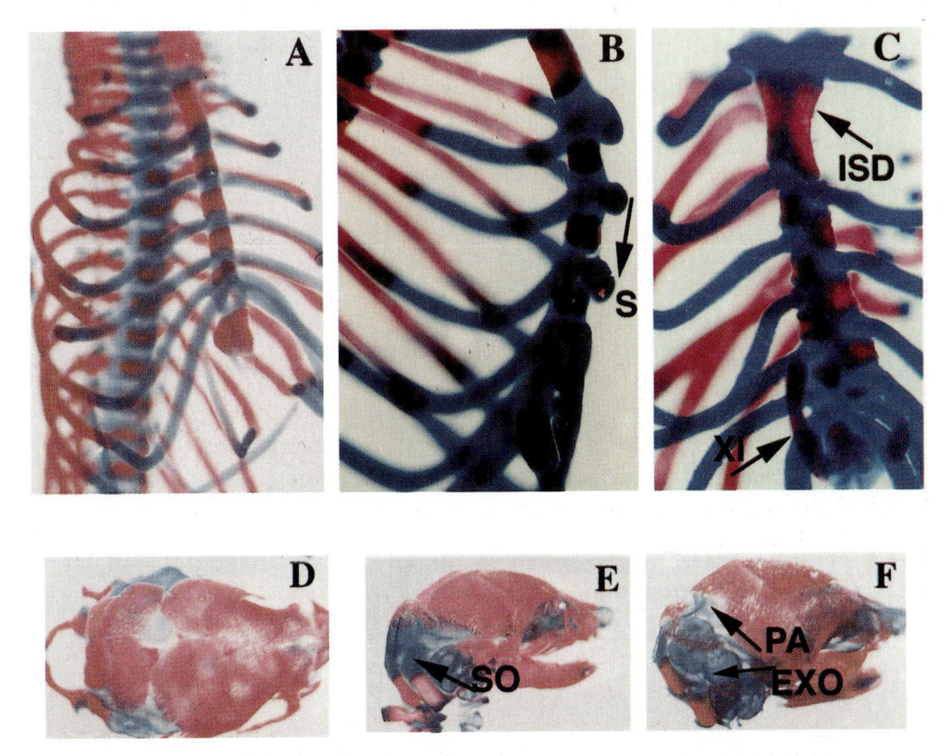

Figure 3. Skeletal preparations of normal and mutant mouse fetuses. (A) Wild-type sternum and rib-cage; (B,C) 18.5 dpc skeletons displaying abnormalities of the sternebrae, the inter-sternebral discs and the xyphoid process; (D) wild-type skull; (E,F) skull phenotypes demonstrating poor ossification, severe reduction or complete absence of the supra-occipital, parietal and exoccipital bones. S, sternebrae; ISD, inter-sternebral discs; XI, xiphoid process; SO, supraoccipital bone; PA, parietal bone; EXO, exoccipital bone. This figure was sponsored by Leica Microsystems.

defining any affected skeletal structures in phenotype analysis. Whether the mutation is spontaneous or induced, is the result of targeted mutation or ectopic mis-expression, skeletal preparations can delimit the precise functional domains and target sites in the developing skeleton. Furthermore, skeletal preparations are required in mutant breeding programmes which seek to reveal the genetic components underlying skeletal morphogenesis.

Chondrification in the mouse embryo starts at about 12.5 dpc (43–48 somites) and at about 14.5 dpc (61–63 somites) the first ossification centres begin to form (21). At earlier stages, when cartilage prevails, the embryos are stained only with Alcian Blue; as ossification proceeds later in gestation and post-partum, Alizarin Red is used to examine the completion of osteogenesis.

We present here a version of a protocol currently used in several laboratories (22–25), with illustrations from our skeletal analysis of gene ablation and mis-expression experiments affecting the development of the sternum and skull (*Figure 3*).

Protocol 14. Staining of cartilage and bones

Reagents

- Glycerol series: 20% and 50% glycerols in ethanol and 100% glycerol
- Alcian Blue 8GS (Sigma). 15–30 mg in 100 ml of 5% acid ethanol (5% glacial acetic acid in ethanol)
- Alizarin Red S (Sigma): stock is 15 mg/ml in ethanol; dilute 1 in 200 in 1% KOH for working solution (75 μg/ml)
- Destaining solution: 20% glycerol, 1% KOH

Method

1. Deliver and dissect the late stage fetuses from the uterus by caesarian section into PBS (see *Protocol 1*).
2. Kill the fetus by chloroform or CO_2 asphyxiation and remove as much of the skin and muscles as possible without damaging the skeleton.
3. Eviscerate the fetus through a midline abdominal incision.
4. Fix the carcasses in an ample volume of 96% ethanol (50–100 ml per embryo) for a minimum of 24 h.
5. Stain the cartilage in Alcian Blue overnight for 15.5–18.5 dpc fetuses. For post-partum skeletons longer fixation periods (several days) are necessary.
6. Rinse the preparations at least twice in 96% ethanol to remove debris.
7. Clear the skeletons in 2% KOH for approximately 12 h.
8. Stain the bones in Alizarin Red overnight.
9. If necessary, skulls and limbs can be partially restained in a weaker Alcian Blue staining solution (0.005% Alcian Blue in 70% ethanol) for a few more hours.

10. Place the skeletons in destaining solution for a few days and dissect out any remaining loose bits of undissolved tissues (mainly muscles). Change the destaining solution daily.
11. Pass the carcasses through the glycerol/ethanol series for 1–2 days in each of the 20% and 50% glycerols, and finally store them in 100% glycerol.
12. Photgraph the skeletal preparations in 100% glycerol with a stereomicroscope using the maximum incident illumination possible.

Acknowledgements

We would like to thank Jack Price for introducing us to the LacZ staining techniques, James Sharpe, Linda McNaughton and David Wilkinson for valuable advice, and members of Robb Krumlauf's laboratory for helpful discussions. S.N. was supported by a European Commission Biotechnology grant (BIO CT 930060) and all work was supported by the Medical Research Council. *Figures 2* and *3* were sponsored by Leica Microsystems.

References

1. Wilkinson, D. G. (1992) In *In Situ Hybridization: A Practical Approach* (ed. Wilkinson, D. G.), pp. 75–83. Oxford University Press.
2. Wilkinson, D. & Green, J. (1990) In *Postimplantation Mouse Embryos: A Practical Approach* (ed. Copp, A. J., Cockcroft, D. L. & Hames, B. D.), pp. 155–71. IRL Press, Oxford.
3. Qiling Xu & Wilkinson, D. (1998) In *In Situ Hybridization: A Practical Approach*, 2nd edn. (ed. Wilkinson, D. G.), p. 87. Oxford University Press.
4. Whiting, J., Marshall, H., Cook, M., Krumlauf, R., Rigby, P. W. J., Stott, D. and Allemann, R. K. (1991) *Genes Dev.* **5**, 2048.
5. Sharpe, J., Nonchev, S., Gould, A., Whiting, J. and Krumlauf, R. (1998) *EMBO J.* **17**, 1788.
6. ZernikaGoetz, M., Pines, J., Hunter, S. M., Dixon, J., Siemering, K., Haseloff, J. and Evans, M. J. (1997) *Development* **124**, 1133.
7. Sanes, J. R., Rubenstein, J. L. R. and Nicolas, J.-F. (1986) *EMBO J.* **5**, 3133.
8. Maconochie, M., Nonchev, S., Studer, M., Chan, S.-K., Pöpperl, H., Sham, M.-H., Mann, R. and Krumlauf, R. (1997) *Genes Dev.* **11**, 1885.
9. Nonchev, S., Vesque, C., Maconochie, M., Seitanidou, T., Ariza-McNaughton, L., Frain, M., Marshall, H., Sham, M. H., Krumlauf, R. and Charnay, P. (1996) *Development* **122**, 543.
10. Schneider-Maunoury, S., Topilko, P., Seitanidou, T., Levi, G., Cohen-Tannoudji, M., Pournin, S., Babinet, C. and Charnay, P. (1993) *Cell* **75**, 1199.
11. Maconochie, M. K., Simpkins, A. H., Damien, E., Coulton, G., Greenfield, A. J. and Brown, S. D. M. (1996) *Transgenic Res.* **5**, 123.
12. Skarnes, W. C., Auerbach, A. B. and Joyner, A. L. (1992) *Genes Dev.* **6**, 903.

13. Friedrich, G. and Soriano, P. (1991) *Genes Dev.* **5,** 1513.
14. Lojda, Z. (1970) *Histochemie* **23**, 266.
15. Shapira, S. K., Chou, J., Richaud, F. V. and Casadaban, M. G. (1983) *Gene* **25**, 71.
16. Kam, W., Clauser, E., Kim, Y. S., Kan, Y. W. and Rutter, W. J. (1985) *Proc. Natl Acad. Sci. USA* **82**, 8715.
17. Chalfie, M., Tu, Y., Euskirchen, G., Ward, W. W. and Prasher, D. C. (1994) *Science* **263**, 802.
18. Brenner, M. and Messing, A. (1996) *Methods* **10**, 351.
19. Zhuo, L., Sun, B., Zhang, C. L., Fine, A., Chiu, S.-Y. and Messing, A. (1997) *Dev. Biol.* **137**, 36.
20. Okabe, M., Ikawa, M., Kominami, K., Nakanishi, T. and Nishimune, Y. (1997) *FEBS Lett.* **407**, 313.
21. Kaufman, M. H. (1992) *Atlas of Mouse Development.* Academic Press, London.
22. Lufkin, T., Mark, M., Hart, C., Dolle, P., LeMeur, M. and Chambon, P. (1992) *Nature* **359**, 835.
23. Mansour, S. L., Goddard, J. M. and Capecchi, M. R. (1993) *Development* **117**, 13.
24. Nonchev, S., Maconochie, M., Gould, A., Morrison, A. and Krumlauf, R. (1997) *Cold Spring Harbor Symp. Quant. Biol.* **62**, 313.
25. Gavalas, A., Studer, M., Lumsden, A., Rijli, F., Krumlauf, R. and Chambon, P. (1998) *Development* **125,** 1123.

4

Mapping phenotypic trait loci

BENJAMIN A. TAYLOR

1. Introduction

When a genetic locus is known only by a phenotype, mapping generally requires analysis of segregating crosses (i.e. meiotic mapping). Phenotypic trait loci vary from major mutations affecting simple qualitative trait differences such as coat colour, to quantitative trait loci (QTLs) that exert small statistical effects on complex traits such as body weight. If the phenotypic trait variant is a new mutation, a new cross must be generated to detect linkage. On the other hand, if the new locus represents a variant that distinguishes inbred strains, special genetic systems, such as recombinant inbred strains, may provide an attractive alternative to a traditional linkage cross.

2. Rationale for mapping

2.1 Genetic definition of trait locus

Assigning the responsible locus to a specific chromosomal region gives it genetic definition, distinguishing it from similar loci that map elsewhere in the genome. For example, by mapping a new neurological mutation of the mouse to a specific chromosomal region, allelism with phenotypically similar mutations may be excluded (or not). In the case of QTLs, linkage data usually constitute the only evidence for the existence of a locus.

2.2 Identification of candidate genes

Mapping a trait locus will often suggest or exclude potential candidate genes. As the number of mapped and cloned genes increases in the mouse, this is an increasingly powerful justification for mapping. For example, mapping the skeletal mutation *chondrodysplasia* (*cho*) to the distal part of chromosome 3 led to the discovery of a frameshift mutation in the procollagen, type XI, alpha 1 gene (1).

2.3 Genetic manipulation

A third motivation for mapping a phenotypic trait locus is to facilitate genetic manipulation. For example, mapping information can accelerate the transfer of a recessive mutation onto a new genetic background, and enable the identification of mutant mice prior to the appearance of an overt phenotype. Likewise, mapping a QTL allows the construction of a congenic strain, which may be useful in further defining the QTL. Genetic mapping of a trait locus is the first step toward positional cloning.

3. Background information for mapping phenotypic trait loci of the mouse

3.1 Definitions

Phenotypic traits may be either monogenic (under the control of a single gene locus) or multigenic (under the control of more than one locus). The number of loci contributing to variation in a trait is always specific to a particular genetic cross. Meiotic mapping depends on the independent reassortment of genes located on non-homologous chromosomes as well as the genetic exchanges (recombinations) between different genetic loci on the same chromosome as the result of crossing over between homologous chromosomes. When alleles at a pair of heterozygous loci are transmitted to offspring more frequently in parental combinations (both alleles of the two loci from the same parent) than in recombinant combinations (one allele of the pair from each parent), the loci are said to show linkage. The ratio of the number of recombinant progeny (A) to the total (N, i.e. the number of parental and recombinant progeny) is referred to as the recombination frequency

$$r = A/N \tag{1}$$

The genetic distance between two loci on the same chromosome is measured in morgans. A morgan is defined as a length of chromosome in which the expected number of crossovers per gamete is 1.0. A centimorgan (cM) is one hundredth of a morgan and, therefore, a segment within which the expected number of crossovers per gamete is 0.01. For small chromosomal segments (e.g. $<$10 cM), a given percentage recombination can be said to estimate the same number of cM (e.g. 1% recombination is equivalent to 1 cM). For larger segments, this relationship may not hold because of the occurrence of multiple crossovers (chiefly doubles) in the same interval. Because of the phenomenon of genetic interference, the numbers of double and higher order crossovers in short segments are typically far fewer than would be predicted if the occurrence of different crossovers were completely independent. The locus to be mapped will be referred to here as the target locus, and any other previously mapped locus scored in a mapping experiment is referred to as a genetic marker. The distance on either side of a marker locus at which linkage with the target locus can be detected with a specified probability is called the swept radius. The swept radius depends on such factors as the type of cross and number of progeny.

3.2 Mouse genome

The standard laboratory mouse karyotype consists of 19 pairs of autosomes and the X–Y pair. The estimated genetic lengths (based on 1998 Chromosome Committee reports; Mouse Genome Informatics (MGI) Resource, Mouse Genome Informatics, The Jackson Laboratory, Bar Harbor, Maine; see http://www.informatics.jax.org/, 8 October, 1998) range from 55 cM for chromosome 19 to 120 cM for chromosome 2, with a total length of 1575 cM. However, these numbers should not be regarded as absolutes, as genetic distances may vary between different crosses depending on the gender and genotype of the hybrid parent(s). Generally, it has been thought that recombination is somewhat more frequent in oogenesis than in spermatogenesis (2). (In humans, overall map length is 60% greater in females than males (3).) However, a recent mouse study suggests that recombination in spermatogenesis is equally frequent as in oogenesis, but more concentrated near the ends of chromosomes (4). Since there are $\sim 3 \times 10^9$ base pairs of DNA per haploid mouse genome, 1 cM is roughly equivalent to 2 megabases (3000 megabases/1575 cM), although this may vary considerably from one region to another due to non-randomness of recombination.

3.3 Mapping basics

Unlike human geneticists, who must rely on available families, mouse geneticists can plan and select matings to facilitate linkage analysis. For a mating to be informative, at least one parent must be heterozygous for both the target locus and marker loci. By crossing mice of two inbred strains, an investigator can assure that hybrid parents are uniformly heterozygous at the same set of loci. Investigators will generally choose between one of two kinds of crosses: backcrosses and intercrosses. In a backcross, F1 hybrid mice are mated to mice of one of the parental strains to produce a segregating population. In an intercross, F1 hybrid mice are mated *inter se* to produce the segregating population. *Figure 1* illustrates the genetic consequences of backcrossing and intercrossing. If the target is an inbred strain variant rather than a new mutation, the investigator may be able take advantage of special recombinant strains, which have been inbred following an initial strain cross (see Section 7).

4. A backcross mapping experiment: a model for simple mapping experiments

4.1 Identifying the phenotypic variant

The first step is to identify a phenotypic variant, such as a new mutation or a polymorphic variant in an inbred strain. Preliminary studies are usually done to establish that the phenotypic variant is heritable, i.e. that it is genetically

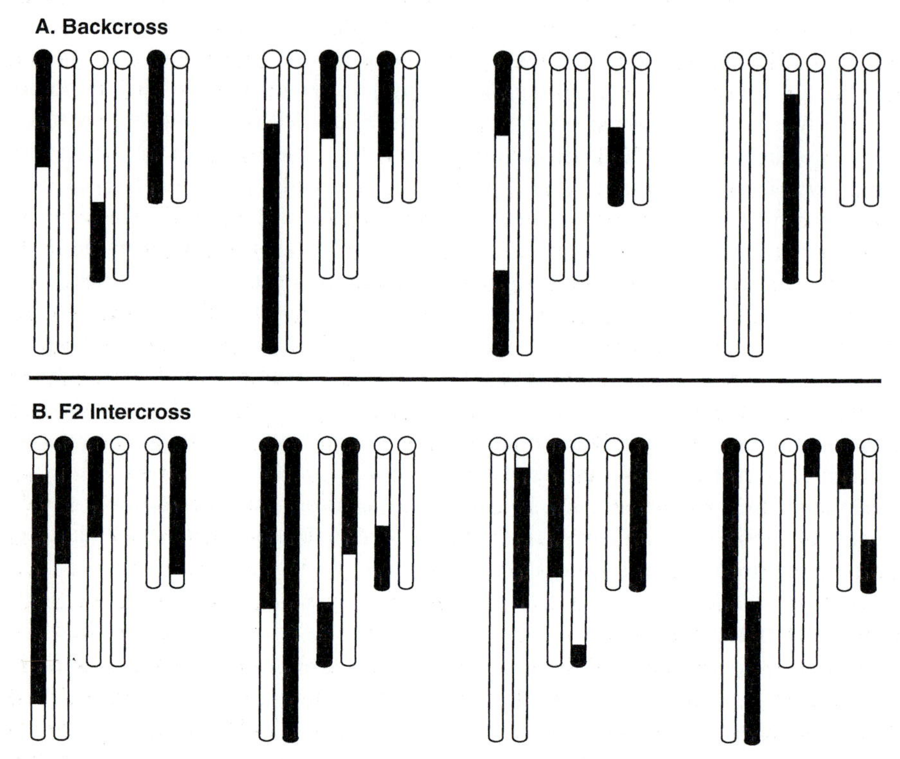

Figure 1. Genetic consequences of recombination and segregation in backcross and intercross progeny derived from crossing two inbred parental strains, A and B. Chromatin derived from the A and B parents are shown as white and black, respectively. Three pairs of autosomes representing large (100 cM), medium (75 cM) and small (50 cM) chromosomes are shown for four progeny of each type. Note that backcross progeny receive one complete set of non-recombinant chromosomes from their inbred parent, while in intercross progeny, chromosomes inherited from both parents are subject to recombination. Positions of crossovers were assigned arbitrarily, but are intended to reflect the frequency of recombination on chromosomes of different lengths and also the

transmitted. These studies will usually indicate whether the variant is inherited as a recessive, dominant or semi-dominant trait, and whether its inheritance conforms to predicted Mendelian segregation ratios. We will assume that the mutation or variant is found in an inbred strain.

4.2 Choosing a tester strain

A tester strain is one chosen to cross with the stock bearing the variant. The choice of a tester strain can be very important. It must contrast with the mutant strain at the target locus, something that can be assumed unless the mutation or variant is common. A tester stock also must carry different alleles from the variant-bearing stock at multiple, readily genotyped, marker loci. Initially, geneticists had to rely on phenotypic mutant genes as genetic markers. Linkage testing stocks, each carrying a combination of useful marker genes (usually several semi-dominant, or several recessive, visible mutations), were constructed which allowed linkage testing of several chromosomes in a single cross (5). Visible markers have been replaced by polymorphic molecular markers, first by protein electrophoretic variants, and more recently by DNA variants (chiefly restriction fragment length polymorphisms and microsatellite variants). If both the target-bearing and tester stocks are inbred, one can be assured that the F1 hybrid will be heterozygous for all loci at which the two strains carry distinct alleles, a major simplifying condition. The tester strain should be a good, or at least an adequate, breeder. While any inbred strain, unrelated to the mutant- or variant-bearing stock, can be expected to exhibit multiple DNA polymorphisms on all chromosomes, the frequency and distribution of these variants in certain strain combinations may limit fine mapping (6). Consequently, geneticists usually choose an evolutionarily divergent stock as the tester strain. The gene pool of most laboratory mice is derived principally from *Mus musculus domesticus*, the common house-mouse of western Europe, with a smaller contribution from related subspecies, probably *M. m. molossinus*, the Japanese house-mouse (7,8). Inbred strains have been derived from several *Mus* subspecies, including *M. m. castaneus* from Thailand (e.g. strain CAST), *M. m. molossinus* from Japan (e.g. strain MOLD) and *M. m. musculus* from eastern Europe (e.g. strain SKIVE) (8). The karyotypes of these strains are very similar to those of standard laboratory strains. These strains have proved to be excellent linkage testing strains when crossed with common laboratory stocks because of their marked genetic divergence, and also because both male and female F1 hybrids are fertile. The CAST/Ei strain is commonly used for linkage testing at The Jackson Laboratory. Crosses with *Mus spretus*, a distinct *Mus* species found in southern Spain and northern Africa, afford even greater genetic divergence (9). F1 hybrid females between common mouse stocks and *M. spretus* are fertile, although F1 males are sterile. Investigators who have not handled derivatives of wild mice before should be prepared to use special precautions to avoid escapees and bites!

A major concern is that classification of genotypes at the target locus be unambiguous. Modifier genes contributed by the tester strain may either enhance or reduce expression of a mutant gene in segregants of a linkage cross. While modifier genes can be legitimate objects of study (10–12), their presence may obscure differences between target locus genotypes (13,14). Unfortunately, such epistasis is unpredictable in advance of actually making a cross. Generally, the greater the genetic divergence between the variant bearing stock and the tester stock, the greater the likelihood that modifier genes will be important.

Choice of the tester stock also can affect the map distances obtained. Crossing over appears to be inhibited in certain regions of specific crosses (and enhanced elsewhere). Different crosses may yield strikingly different distributions of crossovers in some regions (15). Although inversion polymorphisms that distinguish common laboratory strains from wild-derived strains may be the source of variability in some cases (16), few such differences have been documented, and other factors such as recombination hotspots may be involved (17). In the Whitehead Institute (C57BL/6J-*Lepr*^{ob} × CAST) intersubspecific intercross used to map thousands of microsatellite markers, the entire genetic map is estimated to be 1361 cM, while analysis of a 46-member subset of the Frederick Cancer Center (C57BL/6J × SPRET)F1 × C57BL/6J interspecific backcross, which has been extensively typed for both microsatellite and restriction fragment length variants, yielded a 1385 cM map (18). Note that the first map is an average for both spermatogenesis and oogenesis, while the latter measures recombination in oogenesis alone. The fact that these maps, which must encompass nearly all of the genome, are less than 1400 cM suggests that the consensus map (1575 cM) might be inflated. Since no very complete map has been generated from crosses between two common laboratory strains, it is unclear whether the larger consensus map size pertains to such crosses.

4.3 Mating mutant and tester stock to produce F1 progeny

In general, matings between the mutant stock and the tester stock may be made in either direction (i.e. mutant female × tester male or vice versa). However, matings between *M. spretus* males and laboratory strain females are more productive than the reciprocal cross. In most cases, whether to breed from tester males or tester females is a practical choice that depends on availability and reproductive ability of males and females of both stocks. If the mutation-bearing parental stock is not homozygous for the mutant allele, and the target mutation is recessive, progeny testing will be necessary to identify F1 carriers. By checking target locus phenotypes of F1 males and females, it may be possible to test for X-linkage at this stage (if not already excluded). (It is wise to pedigree all F1 and subsequent matings in case genetic heterogeneity is encountered.) Ovarian transplantation can be used to circumvent the problem posed by premature lethality or infertility of mutant mice in establishing a linkage cross (19).

4.4 Producing backcross progeny

Mate selected F1 progeny to the recessive parent strain to produce backcross progeny. If the target locus could be X-linked, choose heterozygous F1 females for mating. When the target mutation is dominant, one can mate F1 carriers to wild-type individuals of either parental stock. F1 females tend to be very prolific, so it may be advantageous to use hybrid females for backcrossing to inbred males. As noted above, the gender of the hybrid parent may influence recombination frequency in many regions of the genome. To be able to evaluate the effect of gender on recombination frequency, matings of both types are required. If the goal is simply to map an autosomal mutation, the direction of the cross can be chosen arbitrarily. Record the gender of the F1 parent for all backcross progeny, and report the resulting data accordingly. If F1 parents of both genders are used, report the resulting linkage data separately by mating type, as well as combined. The size of the cross will depend on the goals of the study and is discussed further below. (It is a good precaution to extract a DNA sample from each parent of the backcross progeny.)

4.5 Classifying progeny for the target locus

Classify backcross progeny at the target locus (heterozygous versus homozygous). Accurate classification, including data recording, is one of the most important steps in a mapping experiment. Examine the first few litters carefully to see whether any ambiguous phenotypes are found. Deviations from standard phenotypes should be noted. If a range of phenotypes is encountered, it may be desirable to adopt a grading system for scoring progeny. Gender and any segregating visible marker genotypes (e.g. coat colour genotypes) should also be recorded, as such information can be useful for detecting sample mix-ups. The target locus segregation data should be tested (using the chi-square goodness-of-fit test) for conformity to a 1:1 ratio in the case of a backcross (or a 3:1 ratio in the case of a dominant/recessive gene segregating in an intercross). If significant deviations are found, one should suspect incomplete penetrance or reduced viability, phenocopies, transmission distortion or some other disturbance.

4.6 Extracting a DNA sample

Extract a DNA sample from each backcross mouse, determine its concentration and purity, and store an aliquot at standard concentration, in a microtitre array. (Deep-well microtitre trays are available which will accommodate 0.8 ml samples (Marsh Biomedical Products).) Freeze-preserve a back-up sample of tissue from each mouse in case a sample is lost or a mix-up is suspected. Tissues from several litters may be frozen, and processed as a batch. Kits for isolating DNA using DNA-binding resins in a 96-well format (Qiagen) can be used for mouse tails and other tissue [Schwarz, 1997; http://www.elsevier.com/locate/tto; T01146] saving time and supplies, while reducing opportunities for sample mix-ups.

4.7 A linkage testing plan

Developing a strategy for linkage detection requires an estimate of the swept radius, the distance between two loci such that linkage can be detected with a specified probability. The swept radius depends on the number of gametes

Table 1. Critical number of recombinants (A_m) and swept radius for detecting linkage in backcross of size N

No. of backcross progeny (N)	Critical number[a] (A_m)	Swept radius[b] (cM) at a power of			
		0.5	0.75	0.9	0.95
22	5	28	20	15	12
34	10	36	28	23	20
46	15	41	33	27	25
58	21	47	39	33	29
70	26	49	41	35	32
82	31	50	43	37	34
94	37	54	46	40	37

[a]Critical number: if there are A_m or fewer recombinants among N backcross progeny, the null hypothesis of no linkage is rejected ($\alpha \leq 0.025$).
[b]The swept radius is the distance between a marker and the target locus at which the power to detect linkage attains a given level. Power levels, defined as the probability of rejecting the null hypothesis of no linkage when it is false, are specified in the column headings: 0.5, 0.75, 0.9 and 0.95, respectively. For example, if linkage is declared when five or fewer recombinants are obtained among 22 progeny, the probability of detecting linkage for a marker–target distance of 28 cM is 0.5. To improve the detection probability to 0.95 requires reducing the swept radius to 12 cM, and therefore the distance between adjacent markers would need to be reduced from 56 cM to 24 cM. Recombination frequency is assumed to relate to map distance according to the Kosambi mapping function (see *Equation 4*). These different definitions of swept radius can be used to decide how many markers are needed to cover chromosomes of different lengths for a given linkage testing plan.

tested (N), the type I error rate (α), the specified minimum probability of linkage detection (i.e., power), and the distribution of multiple crossovers. *Table 1* shows the swept radius for different combinations of N, associated with four different probabilities of linkage detection (0.5, 0.75, 0.9 and 0.95), and at α level of <0.025. Also shown is the critical number, A_m, the maximum number of recombinants consistent with the declaration of linkage at this significance level. Thus, a simple plan is as follows. Select 22 backcross progeny for genotyping, and reject the null hypothesis if the number of recombinants with any marker (A) is ≤ 5. This provides a swept radius of 20 cM. Thus the entire genome can be scanned by selecting 46 well-placed markers: three on chromosomes 1–5 and 8, and two on each of the remaining chromosomes. (X-chromosome markers may be omitted if the mutation is known to be autosomal.) Where two markers cover a chromosome, these ideally would be placed at slightly less than one-quarter, and slightly more than three-quarters, of the chromosome length. The number of markers per chromosome recommended here is based on Chromosome Committee map lengths.

This specific plan provides a type I error rate of 0.008, a minimal probability of linkage detection of 0.75, and an estimated swept radius of 20 cM. Thus, the autosomes may be screened with 44 markers, typed in 22 backcross progeny (plus controls), for a total of 1056 tests. The sample size (N), 22, was chosen to fit in two rows of a microtitre tray, with space for two controls. While a power level of 0.75 may not seem sufficiently high, it should be recognized that this is the level in the worst case, i.e. when the target locus happens to fall a full swept radius from the closest marker. Over most of the genome, the target to marker distance will be <20 cM, enhancing the probability of linkage detection. Also, if the target falls half-way between two markers, linkage may be detected with either. Thus, the effective swept radius is substantially underestimated in regions between two markers. Finally, some consensus chromosome map lengths are probably inflated.

Alternative screening plans can be developed using *Table 1* as a guide. For example, with $N = 34$ (accomodated in three rows of a microtitre tray) and $A_m = 10$, complete autosomal coverage can be achieved with only 38 markers, for a total of 1368 tests. (This test criterion has a type I error probability of 0.012, a minimal probability of linkage detection of 0.75 and an estimated swept radius of 28 cM.). Select two markers on all chromosomes, except chromosomes 2 (three markers) and 19 (one marker). Thus, the number of markers can be reduced at the expense of needing to test more progeny. Under either of these screening plans, the expected number of false positives will be fewer than one per complete genome screening. Individuals to be included in the screen may be chosen arbitrarily (e.g. the first progeny weaned) or randomly from a larger population. It should be evident that testing large numbers of progeny is unnecessary and counterproductive for detecting linkage of a monogenic, fully penetrant, trait locus.

Note that the significance level recommended here (<0.025) is insufficient for establishing linkage beyond reasonable doubt. Firm proof requires a P value of ≤0.0001. This is easily attained by typing additional progeny and/or typing markers closer to the target locus.

4.8 Genotyping a subset of backcross progeny for selected markers

Score the selected subset of backcross progeny sequentially for markers distributed on different chromosomes. If there are obvious candidate genes for the mutation, test markers in the vicinity of the candidate genes first. Since linkage may be detected with any given marker, it makes sense to carry out the genotyping in stages, a few markers at a time, looking for evidence of linkage before proceeding to test other markers. Test one marker per chromosome until all chromosomes have been tested with one marker. But if a strong hint of linkage is found with some marker, test other markers on the same chromosome before proceeding. For example, if six recombinants were scored among 22 progeny in the first testing plan, it would make sense to test a second marker on the same chromosome. If no linkage is detected in this first pass (one marker per chromosome), test additional markers to fill in gaps. In a worst-case scenario, one might screen the entire genome (44 markers) without detecting significant linkage. One would then review the data to see what regions are most consistent with linkage, and test additional markers in those regions. Usually, linkage will be detected before a single marker has been typed on each chromosome.

4.9 Recording data

Record data in a spreadsheet or other computer file for storage and analysis by Map Manager (20), Mapmaker (21) or similar computer program. Genotypes should be coded generically, e.g. heterozygous (H) or homozygous (A) to facilitate the recognition of linkage. Rescore the original data (e.g. gel photographs or autoradiograms) against a printout of the computer file to detect scoring and data entry errors. It is a good idea for different individuals to do the original scoring and the error checking.

4.10 Scanning data for linkage

Scan the genotype data for clear linkage (deficiency of recombinant progeny) either electronically or manually. Tabulate the number of recombinants between each marker and the target locus. If this is done manually, once the number of recombinants exceeds the critical number by two, the locus may be considered unlinked. A more general way to scan the data is to compute the estimated recombination frequency (see *Equation 1*) and its standard error,

$$s_r = [r(1 - r)/N]^{0.5}, \tag{2}$$

for each marker with the target locus. The statistic $[(0.5 - r)/s_r]^2$ is distributed approximately as chi-square with one degree of freedom, and tests for significant departures from free recombination. A simple 2 × 2 contingency chi-square test provides a more exact test to judge whether there is a significant deviation from independent segregation (see *Table 2*). P values of <0.025 should be pursued further. However, linkage should not be considered firmly

Table 2. Testing linkage and estimating recombination frequency from an [(*AA BB*) × (*aa bb*)]F1 × (*aa bb*) backcross

Allele transmitted	***A***	***a***	**Totals**
B	d	e	$d + e$
b	f	g	$f + g$
Totals	$d + f$	$e + g$	$N = d + e + f + g$

Recombination frequency $(r) = (f + e)/N$
Chi-squared test for independence (one degree of freedom):
$\chi^2 = [N(dg - ef)^2]/[N/(d + e)(f + g)(d + f)(e + g)]$
A and *a* (and *B* and *b*) represent the alternative alleles transmitted by the hybrid parents to their backcross progeny. The combinations *AB* and *ab* constitute the parental (non-recombinant) gametes, while *aB* and *Ab* constitute the non-parental (recombinant) gametes. In the 2 × 2 contingency table, the number of individuals inheriting the allelic combination *AB*, *aB*, *Ab* and *ab* are denoted by the letters *d*, *e*, *f* and *g*, respectively. The null hypothesis of no linkage is tentatively rejected ($\alpha \leq 0.025$) if $r < 0.5$ and the computed value of χ^2 is >3.84.

established until the probability of obtaining by chance the observed number, or fewer, recombinants is <0.0001.

4.11 Finding markers that flank the target

Once clear evidence of linkage is obtained, type additional markers that are proximal and distal to the first linked marker. For example, if in the initial screen, a central marker exhibits a recombination frequency of 0.15 with the target locus, try scoring markers 20 cM proximal and distal to the initial marker. If the suspected linkage is real, then one of the flanking markers should show much stronger evidence of linkage. Inspect or analyse the data to judge the likely three-point order. This involves trying the three possible locus orders and picking the one which requires the fewest double-crossovers. Computer programs such as Mapmaker and Map Manager have the capability of comparing multiple different orders, and of ranking these orders by their relative likelihood, expressed as lod scores. An order with an associated lod score which is 2.0 lod units smaller (100-fold less likely) than the preferred order may be firmly excluded. With backcross data, involving correctly scored and well spaced loci, gene order is usually unambiguous, even when there is only a single crossover between adjacent loci. Search for flanking markers that exhibit clear linkage with the target locus.

4.12 Finding closer markers

Once closely linked (<10 cM) markers, both proximal and distal to the target locus, are identified, try other markers to identify flanking markers even closer (those which show one or two crossovers in the screening set). Then type the remainder of the backcross with these two flanking markers. Cross-over mice can then be typed with respect to other markers that fall between these flanking markers. (Non-crossover mice do not require further analysis.) Based on the results of typing these markers in all recombinant progeny, additional markers can be selected. This process of genotyping selected recombinants with intervening markers is continued until (a) markers are identified that exhibit zero or single crossovers with the target locus in the entire cross, (b) the desired level of resolution is achieved, or (c) the supply of useful markers is exhausted. Selective genotyping based on flanking markers is an efficient way to place the target locus relative to closely linked markers, but does not provide full information about the distances among partially typed markers. One caution: it is important to type each marker in enough progeny to verify that it is linked to the region. Markers that do not map to their originally assigned region are occasionally encountered.

4.13 Analysing the complete data set to determine gene order and distance

Analyse the entire data set with a computer program such as Mapmaker to check for best locus order and estimated distances among linked loci. Check any apparent double crossovers for possible errors. If one must postulate double crossovers, regardless of where the target locus is placed among the markers, suspect misclassification. Recall that for markers typed only on the 'recombinant' subpanel, untyped progeny are assumed to be non-recombinant, and these need to be accounted for in computing recombination frequencies. It is legitimate to 'infer' genotypes based on the assumption that double crossovers do not occur within a short region—so long as this assumption is kept in mind and the data are reported accordingly. The program Mapmaker uses maximum likelihood estimation to account for such missing data in an optimal way, making it unnecessary to provide inferred genotypes. Keep in mind that, due to interference, progeny selected for recombination in one interval will tend not to have crossovers in adjacent intervals. Thus, in such selected progeny, recombination between markers that define the selected interval and markers just outside the interval will be strongly reduced.

4.14 Reporting linkage data

Traditionally, linkage data have been reported as the estimated recombination frequency with its standard error of estimate, $r \pm s_r$. The estimate of the standard error (see Section 4.10) is based on a normal approximation of the binomial distribution, which is good when r is 0.2–0.5 and N is large. Unfortunately, this approximation is very bad when r is small—the most relevant linkages. A better way to report linkage results is to give the estimated recombination frequency along with the 95% confidence limits (CL). The latter can be obtained from the terms of the cumulative binomial distribution. To obtain the lower limit, find the value of r such that the probability of obtaining the observed (A) or greater number of recombinants among N progeny is 0.975. This is equivalent to finding the value of r such that 1 minus the probability of obtaining $A - 1$ or fewer recombinants among N progeny equals 0.025. For the upper limit, find the value of r such that the probability of obtaining the observed (A) or fewer recombinants is 0.025. The terms of the cumulative binomial distribution have been published in tabular form (22). The spreadsheet program Excel (Microsoft) has a statistical function for computing the terms of the cumulative binomial distribution. Thus, the lower CL is obtained by finding (by successive approximation) the value of r such that, 1 – BINOMDIST(A–1,N,r,TRUE) = 0.025, $A \neq 0$. The upper CL is obtained by finding the value of r such that BINOMDIST(A,N,r,TRUE) = 0.025. For example, if there were one recombinant among 25 progeny, the estimated recombination frequency is 0.04, the lower CL would be ~0.001 and the upper CL would be ~0.2035. In the present example, it will be seen that Excel will return the value 0.02494... when directed to compute the function, 1 – BINOMDIST(0,25,0.00101,TRUE). Similarly, the value 0.0251... is returned by the function, BINOMDIST(1,25,0.2035,TRUE).

For a pair of loci that fail to recombine, the 95% upper confidence limit

($UCL_{95\%}$) of the recombination frequency may be computed more simply (based on the Poisson approximation of the binomial distribution) as

$$UCL_{95\%} = 1 - e^{(\ln 0.05)/N}, \qquad (3)$$

an approximation which is good when N is large.

Thus, the $UCL_{95\%}$ of the recombination frequency beween non-recombining loci is ~$3/N$. In cases where no recombinants are found, a one-sided UCL ($\alpha = 0.05$) is computed since it is impossible to set a non-zero lower limit.

Results may be reported as recombination frequencies directly or converted to cM (e.g. equating a recombination frequency of 0.04 to 4.0 cM). Because of the possibility of undetected double crossovers, recombination frequencies >0.2 should not be converted into cM in this linear fashion, but may be converted using an appropriate mapping function. A good, commonly applied mapping function is Kosambi's,

$$m = 25[\log_e(1 + 2r) - \log_e(1 - 2r)] \qquad (4)$$

where m is map distance in cM, and r is recombination frequency (23).

A mapping function that should not be used is Haldane's (24). It is based on an assumption of zero interference, so recombination frequencies are over-corrected, and short distances are systematically overestimated. Because of its computational simplicity and the fame of its originator, the Haldane function is frequently used. Haldane recognized that this mapping function was a poor choice in the paper in which he first proposed it, but it keeps recurring in genetic literature (25)!

Computer files containing linkage data may be submitted to Mouse Genome Informatics electronically (e-mail: submissions@informatics.jax.org).

4.15 Size of cross

The decision as to how many backcross progeny to analyse depends on how the data are to be used, and availability of resources. Mapping based on a 100-mouse backcross is generally considered adequate for reporting a new linkage. Such a cross permits one to look at the consensus map for potential candidate genes. Of course, to carry out a positional cloning effort would require a much larger cross (1000–3000 progeny).

The size of the cross determines the potential average mapping resolution. A backcross of N progeny can be used to generate a map in which crossovers occur on average every $100/N$ cM. From Section 4.14, we see that the $UCL_{95\%}$ of the distance (in cM) beween nonrecombining loci is ~$300/N$. Thus, nonrecombining loci in a 100-mouse backcross are 95% certain to lie within 3 cM of each other. In this case, the target locus would be mapped within a 6 cM region. However, in most cases, the target locus will show recombinants with closely linked markers whose positions may be well established, thus serving to define specific proximal and distal bounds. This depends on prior knowledge about the position of the nearest markers in a consensus map. If such information is lacking or inadequate, consider mapping selected markers in one or more standard crosses, such as the BSS backcross panel available from the Jackson Laboratory (26).

4.16 Markers for fine mapping

To obtain the greatest precision from a given cross, one should try to identify markers that exhibit no crossovers with the target locus as well as flanking markers that exhibit a single crossover, but this is certainly not obligatory, especially if the cross is large. If the cross involves an evolutionarily divergent strain, one can generally find MIT microsatellite markers (see Section 8) that are closely linked to the target. Ultimately, how many markers one scores depends on the goal of mapping.

5. Incomplete penetrance, phenocopies and reduced viability

Some variants and mutations do not exhibit a consistent phenotype, a phenomenon referred to as incomplete or variable penetrance. A simple example would be a recessive mutation for which 90% of homozygous mutant (*m*/*m*) individuals exhibit a specific defect, but the remaining 10% are indistinguishable from normal (+/+ or +/*m*) mice. Incomplete penetrance can seriously inflate estimates of recombination since non-recombinant progeny are apt to be classified as recombinants. Often the degree of penetrance depends on the genetic background, in which case one presumes that modifier loci play a role in the functional consequences of the mutant genotype. A low level of incomplete penetrance should not seriously impede the detection of linkage. However, it does result in inflated estimates of recombination frequencies and can obscure gene order. Generally, if incomplete penetrance is suspected, restrict analysis to mice that manifest the mutant phenotype for estimating genetic distances and determining gene order. However, it is important that some wild-type progeny be genotyped. Otherwise, one cannot discriminate between linkage and abnormal marker gene segregation due to causes unrelated to the mutation. Heterozygotes for the semi-dominant, *loop tail* mutation (*Lp*), a model for neural tube defects, show incomplete penetrance, with some carriers being indistinguishable from wild-type. Accurate mapping required restricting analysis to backcross progeny manifesting the mutation (27). In human genetics, where incomplete penetrance is commonly encountered, special linkage analysis programs have been devised which provide maximum likelihood estimates of recombination frequency in the presence of incomplete penetrance. These programs provide estimates of the degree of penetrance from the data.

Phenocopies are the opposite of incomplete penetrance. Here, some mice which do not carry the mutant genotype nonetheless exhibit the phenotype associated with the mutant gene. The cause may be unknown environmental factors, or other genetic loci segregating in the cross. For example, a mouse runted because of environmental factors may be difficult to distinguish from an endocrine dwarf mutation. If phenocopies are suspected in the data, one can try to restrict mapping analysis to wild-type mice. In general, it is useful to make note of any deviation from the standard phenotypes. However, if both incomplete penetrance and phenocopies are suspected, one cannot assume that either phenotype is exclusively associated with any specific genetic class. In this case, one needs to adopt a model that allows for the fact that phenotype is not strictly tied to genotype, and utilize one of the computer programs specifically designed for mapping complex traits. One should then think of loci as affecting the risk of a particular variant phenotype, rather than determining the phenotype directly (see below).

In any case, if misclassification is suspected, the number of progeny screened for linkage should be increased.

Reduced viability of mutant classes is not a major problem for linkage analysis unless recombinant and non-recombinant classes have different viabilities. Inviability of certain classes was a greater problem when visible markers were used, and certain double mutant genotypes might be under-represented.

6. Mapping by intercross

Although the backcross design is straightforward and workable, consideration should be given to an intercross design. An intercross results from mating males and females that are heterozygous for both the target locus and linkage markers. Generally, intercrosses are produced by mating F1 hybrids between two inbred strains. The genotypes of the resulting F2 progeny reflect recombination events in both parental gametes, but it is not possible to discriminate between crossovers in paternal and maternal gametes. Consequently, intercross data provides an estimate of the average recombination frequency in male and female gametes, but does not afford separate estimates of male and female recombination.

A major reason for choosing an intercross rather than a backcross is that an intercross is potentially twice as efficient as a backcross of the same size. This is simply a function of being able to score twice as many meioses in intercross progeny. However, if the target mutation is dominant or recessive (as opposed to semi-dominant or co-dominant), this full efficiency is not realized, and the relative efficiency varies between equality with the backcross for values of r close to zero, and two-thirds as efficient, when the value of r is close to 0.5 (28). Thus, for a dominant or recessive trait, the two-fold gain in efficiency applies only to the recessive class. Nonetheless, the potential efficiency of the intercross can be recovered by progeny testing. First, the target locus is mapped approximately to identify closely linked flanking markers. The remainder of the cross is then typed with these flanking markers to identify crossovers near the target. Then recombinant mice, whose genotype at the target locus is uncertain, are progeny tested. This requires keeping progeny with dominant phenotypes alive until flanking markers can be scored. Once the target gene has been mapped to a small region, the number of such progeny tests required is modest. This strategy can be applied to advantage in positional cloning projects (29), but may not be practical when only moderate mapping resolution is required. However, for the purpose of linkage detection, one can selectively analyse only mutant homozygotes (in the case of a recessive), thus taking advantage of the two-fold gain in efficiency.

If homozygotes for the target mutation are lethal or sterile, it may be necessary to choose an intercross rather than a backcross. Of course, in these cases it is necessary to progeny test the F1 hybrids to identify carriers of the mutant gene. Alternatively, mate multiple F1 pairs, and retain and analyse only the expected one in four matings that produce affected progeny. Since most F1 hybrids are excellent breeders, intercross matings tend to produce large and frequent litters.

One problem with intercross matings is that recombination cannot be calculated from an algebraic equation as is possible for backcross data. Instead, the method of maximum likelihood estimation is required. This involves a somewhat cumbersome, iterative estimation procedure, but one that can be carried out manually with the aid of published tables of scores which are applied to different genotypic classes (30). These computational difficulties are circumvented entirely by the use of computer programs such as Mapmaker and Map Manager that automate the estimation process.

One disadvantage of the intercross design is that mistyped progeny are not always apparent. Because interference is partially masked, mis-scored individuals are not as easily recognized as apparent 'double crossovers'. Since intercross progeny may by chance, inherit closely linked crossovers, one on each parental chromosome (i.e. ABc/aBC), identification of mistyped individuals is not simple. As crosses become larger, and markers become more dense, this limitation tends to disappear, but for small crosses and sparse markers, it is a distinct disadvantage.

7. QTL mapping, including the use of derivitive strains

7.1 General nature of QTL mapping

Many traits of biomedical interest exhibit quantitative variation and are under the control of multiple genes as well as undefined non-genetic factors. Complex traits of the mouse, such as epilepsy (31) and obesity (32), have been analysed by the method known as QTL mapping. QTL analysis is generally applied when the phenotypic differences between trait locus genotypes become statistical rather than deterministic. The approach is basically an attempt to detect linkage between genetic markers and the loci that contribute to variation in a phenotypic trait. Linkage is inferred by a statistical association between the phenotypic trait and the marker genotype. Association tests may be done one marker at a time, or by a method known as interval mapping (33,34). The latter involves calculating the relative likelihood of the observed results assuming that a QTL exists at various positions between linked markers, versus the likelihood of the same results, assuming no linkage. Results obtained by this method and the older, single locus association test methods are similar (35). Since a putative QTL may exert any non-zero effect on the trait, power increases with the size of the cross. Thus, very large crosses could potentially detect quite small gene effects, but such minor factors might be of little biological interest. Cross size also affects the confidence limits about detected QTLs. While a general treatment of the complex subject of QTL mapping is beyond the scope of this chapter, and excellent sources are available, we will discuss some general QTL mapping issues, particularly special resources that are available, and point to the original sources for details.

QTL analysis usually involves inbred strain differences, although it is possible to do QTL analysis when only one of the parental stocks is inbred (36). In strain crosses, the source of variability is primarily polymorphic variation that accumulates in natural populations under the influence of mutation, selection and genetic drift. Detection of such variability involves strain crosses, usually strains that differ markedly with respect to the trait of interest. Recombinant inbred strains, congenic strains, recombinant congenic strains, and consomic strains are special genetic tools useful for dissecting interstrain differences. In some circumstances, these special strains provide alternative or auxiliary approaches for QTL detection and analysis (see Section 7.4).

7.2 QTL mapping using strain crosses

7.2.1 Strain selection

One will generally want to cross strains that differ markedly with respect to the phenotype of interest. If the parental strain difference in mean phenotype between parental strains is large relative to the standard deviation of the trait within strains, one can have greater confidence that QTLs can be identified in a moderately large cross. But QTLs also may be detected in crosses between strains that are phenotypically similar (37). In order to assure adequate marker polymorphism, avoid crossing strains that are known to be related.

7.2.2 Backcross versus intercross

As for single gene linkage detection, whether to use a backcross or an intercross is a complex choice. If there is evidence of directional dominance, one might prefer to backcross to the 'recessive' parental strain. This would be particularly true if one suspects that a particular phenotype seen in one strain is due to a combination of two or more recessive factors. However, the intercross design allows the expression of recessive factors from both parent strains. It is approximately twice as powerful for detecting QTLs that lack dominance (i.e. when heterozygotes are phenotypically intermediate between the two homozygotes). Of course the intercross design also provides evidence as to the degree of dominance of QTLs, while the backcross design does not. There are circumstances where a backcross is more powerful for detecting a QTL, but an intercross tends to provide more precise mapping (38). Overall, the intercross is preferred in the typical case where little can be assumed about the genetic control of the trait.

7.2.3 Number of progeny

Unless a prior survey of inbred strains with respect to the quantitative trait indicates a clustering of strains into distinct phenotypic groups (suggesting a major gene), one should probably plan to phenotype several hundred progeny. Even if a major gene is involved, a larger cross will yield more exact mapping. The choice of cross size depends on how small a QTL effect one wishes to be able to detect. Darvasi (38) has found by computer simulation that the number of progeny required to provide a 0.5 probability of attaining a significant lod score (see Section 7.2.6) are $65.5/(2d^2 + h^2)$ and $60.5/(d + h)^2$, for an intercross and backcross, respectively, where d is the standardized allele effect (standardized by dividing by the within-genotype standard deviation), and h is the standardized dominance effect. For example, when $d = 0.5$, and $h = 0$, the requisite numbers of intercross and backcross progeny for 50% power are 131 and 242, respectively. These expressions assume a dense marker map. Because of the many opportunities for chance association in a genome wide scan, one is obliged to use a stringent statistical significance threshold (33,39), which means that small crosses have little power. In planning an experiment, because of the length of time required to replicate a two-generation study, it is better to err on the side of analysing extra progeny, rather than being obliged to replicate the entire cross in order to obtain statistically significant results. One should also recognize that while small experiments with low power have a reduced chance of detecting real QTLs, they are equally likely to generate false positives. Even if a real QTL is detected in a small experiment, the investigator may be disappointed to find that, upon replication, the effect proves to be smaller than initially suggested, a statistical phenomenon known as regression toward the mean.

7.2.4 Marker spacing

QTL mapping generally requires more markers for full genome coverage than does the mapping of Mendelian traits. The expected value of a lod score declines as $(1 - 2r)^2$ as a function of the recombination frequency (r) between the QTL and a marker locus. Thus, a marker 10 cM from a QTL is expected to exhibit a lod score only 0.64 as large as a marker which never recombines with the QTL. Thus, to detect marginally significant QTLs, markers should probably be spaced little further than 20 cM apart, and no more than 10 cM from chromosome ends. Of course, if a sizable lod score (e.g. >2.0) is obtained with markers spaced at 20 cM, one should try to narrow gaps in the region to test for the presence of a significant QTL. Once a significant QTL has been identified, typing additional markers may help to define the confidence interval about the QTL as well as other parameters concerning the QTL. However, unless the QTL has an unusually large effect, or the cross is very large, markers more closely spaced than 10 cM provide little additional information (35).

7.2.5 Selective genotyping

Since individuals in the two tails of a trait distribution contribute disproportionately to any genetic associations, considerable savings can be achieved, without significant loss of information, by restricting genotyping to extreme individuals (33). Thus, by genotyping only the 15% of progeny in each tail of a normal phenotypic distribution (30% of total) one can extract ~80% of the information that could be obtained by typing all progeny. Such selective genotyping is certainly adequate to determine whether additional genotyping in the region of a suspected QTL is worthwhile. In general, one would expect lod scores for a QTL identified to increase by only 20% as the result of typing the intermediate progeny (70% of total), so typing intermediates is worthwhile only for QTLs that are either significant or close to significant. If no association is found with a marker genotyped in the tails of the phenotypic distribution, it is very unlikely that a statistically significant QTL is nearby. If several traits are measured, the amount of information obtained from selective genotyping about traits other than the one chosen as the basis for selection will vary in relation to their correlation with the primary trait. (For uncorrelated traits, the amount of information gained about the secondary trait is equivalent to genotyping the same number of progeny picked at random.) If there are multiple traits of equal interest, consider genotyping mice in the 10% extremes of each trait. Computer programs such as Mapmaker QTL take into account the effects of selective genotyping in estimating genetic parameters, e.g. gene effect, but it is essential that phenotypic data on non-genotyped progeny be included in the analysis. Otherwise, exaggerated estimates of gene effect and percentage variation controlled by a QTL will result. For obtaining best estimates of QTL parameters and for evaluating the role of epistasis between different QTLs, it is important that genotyping be complete at one or more markers close to each QTL. Complete genotyping at markers flanking the QTL lod peak can also reduce the confidence interval about the QTL (see Section 7.2.7).

7.2.6 Significance thresholds

Unlike linkage testing of monogenic traits, with QTL mapping, there is no prior knowledge of how many (if any) potentially detectable QTLs exist. Thus, a given experiment may provide evidence of several, or no, QTLs, depending on the interpretation of statistical evidence. The magnitude of specific effects of QTLs is inferred from the data. In intercrosses, the degree of dominance is also estimated from the data. Finally, the investigator typically searches most of the genome for evidence of QTLs, and if suggestive evidence is obtained, additional markers are often typed in the implicated region. These circumstances present a complex multiple-comparisons problem, requiring a more stringent significance threshold than that appropriate for testing trait association with a single marker. Lander and colleagues have addressed all these issues, and have proposed a set of lod score thresholds for achieving a valid 0.05 significance level in both human and mouse studies (39). They recommend lod score thresholds of 3.3 and 4.3 for backcrosses and intercrosses, respectively. These thresholds are applicable to a saturated genetic map, i.e. one in which each pair of adjacent markers is separated by no more than one crossover. Lander and Kruglyak have argued that these stringent thresholds should be applied even when genotyping is far from saturated, lest the literature be swamped with spurious QTL reports (39). Others have criticized these thresholds as too stringent, and emphasized the need for more flexible standards (40). Permutation tests have been proposed as a means of adjusting thresholds for experiments which differ in marker density (41). However, these tests can be invalidated by additional genotyping in regions where an association is suggested by an initial screen (a common and reasonable practice). Overall, the author finds the Lander and Kruglyak arguments more persuasive than those of their critics. If data are partitioned (e.g. analysed separately by gender), the applicable lod score threshold is increased by 0.5. If lod scores are not computed, equivalent P-value thresholds should be applied.

Another multiple-comparison issue, that of multiple traits, is rarely acknowledged. Frequently investigators measure a number of related traits in the same experiment, but apply the same significance threshold to each of the traits as would be applied to a single trait. For example, in a study of growth, body weights might be taken at several different ages, and data for each age analysed separately. To the extent that the different measurements are incompletely correlated, this creates a multiple-comparisons issue. (A worst-case scenario is one in which the investigator measures numerous traits, but reports only the analysis of a single 'significant' trait.) One solution is to identify (prior to data collection) a single trait of primary interest to which the standard threshold is applied without adjustment. P values for secondary traits should then be considered as only rough measures of actual significance levels. If there are two traits of interest for essentially different reasons, one can argue that the investigator should not be penalized for studying them both in the same, rather than separate, experiments. Such philosophical issues are difficult to resolve, but one should be aware of them in interpreting data.

7.2.7 Confidence intervals

The map positions associated with the points at which a QTL lod plot falls 1.0 and 2.0 lod units below the peak are traditionally used to indicate regions within which the QTL is expected to lie with increasing confidence (33). Recent analysis suggests that the 1.5 lod support interval is roughly equivalent to a 95% confidence interval (CI) (42). These CIs are sensitive to the assumption that phenotypes are distributed normally, an assumption that is usually not strictly correct. Thus, such CIs should only be taken as approximate. Darvasi has obtained expressions for the expected 95% CI (in cM), by computer simulation. For a backcross,

$$\mathrm{CI} = 3000/[N(d + h)^2] \quad (5)$$

and for an intercross

$$\mathrm{CI} = 1500/(Nd^2)] \quad (6)$$

where d and h have the same meaning as in Section 7.2.3 (38). Note that in the absence of dominance (i.e. when $h = 0$), the expected QTL CI for an intercross is one-half that for a backcross. The actual CI will depend in part on whether or not phenotypically extreme progeny happen to have recombinants near the QTL peak. These equations are based on an assumption that the marker map is dense.

As discussed below (see Section 9), the amount of genotyping required in a QTL screen can be greatly reduced by DNA pooling.

7.3 Fine mapping QTLs

Unless a QTL controls a high percentage of phenotypic variance, merely increasing sample size in a backcross or F2 will not efficiently reduce the associated CI. The CI declines inversely with sample size, so to halve the CI requires doubling sample size. Several approaches (and combinations of approaches) have been used or proposed for more efficiently reducing sample size.

7.3.1 Selective phenotyping

Since only progeny with recombinations in the vicinity of the QTL contribute additional mapping information, an effective strategy for reducing a CI is to genotype segregating progeny for markers flanking the QTL prior to phenotype determination, and then phenotype only recombinant mice. One or more additional markers within the interval will be needed at this stage. If phenotyping is expensive relative to genotyping, this is a good strategy. However, once the CI has been reduced substantially by such selective phenotyping, further progress is limited by the rarity of crossover progeny in the now reduced interval.

7.3.2 Progeny testing

To make further progress efficiently requires some form of progeny testing. The purpose of progeny testing is to reduce or remove uncertainty about the QTL genotype of recombinant mice. Thus, one identifies progeny (usually males) with crossovers in the critical region, and then mates them to homozygous tester animals (usually one of the parental strains) to produce offspring for phenotyping. The offspring are then typed for a segregating marker near the crossover to determine whether the marker is associated with the trait of interest. A positive result indicates that the QTL is on the same side of the crossover as the marker, while a negative result (no association) indicates that the QTL is on the opposite side of the crossover.

In progeny testing, one should balance the number of recombinants tested simultaneously with the number of mice analysed per progeny test. It is more efficient to spread effort over several progeny tests at the expense of greater certainty about the genotype of each recombinant tested. Thus, it is better to test several crossovers partially than to test a single crossover exhaustively. However, as the QTL region is reduced, and relevant recombinants become more difficult to find, a larger number of progeny in each test is justified.

7.3.3 Congenic strain construction

A third strategy is to begin transferring a QTL allele from one strain to another (using marker-assisted selection), with the idea that eliminating variability due to other loci affecting the trait will allow the transferred QTL to be mapped more efficiently. Once the QTL segment has been transferred to an inbred background, it can be further reduced by recombination and tested to see if the transferred QTL allele has been retained or lost (38,43,44). The introduced segment (or portions of it) may be made homozygous prior to phenotypic testing, or testing may be carried out in a segregating population. If comparisons are made between congenic and background strains, it is important to be cognisant of environmental influences shared by littermates. Because of epistasis, the effect of a QTL allele in a specific inbred background is unpredictable. QTL effects may increase, decrease or disappear entirely in response to the new genetic background.

7.3.4 Advanced intercross populations

Development of an advanced intercross population by random mating for several generations beginning with the F2 is another approach that has been proposed for increasing QTL mapping precision (45). Each two generations of random mating adds to the frequency of crossovers in the population an amount equivalent to that found in the F2 generation. Thus, analysis of a population that is eight generations beyond the F2 would be expected to result in a CI that is approximately one-fifth that obtained from analysis of a similar number of F2 progeny. This approach has the advantage that precision is improved throughout the genome, a consideration if one is interested in fine-mapping several QTLs. However, once the CI has been reduced substantially, this approach is inefficient for obtaining further reductions. It is important that a minimum population size be maintained during the generations of random mating to avoid excessive genetic drift.

7.4 Derivative inbred strains for QTL mapping

Special inbred strains (recombinant inbred, congenic, recombinant congenic, and consomic strains), derived from pre-existing strains provide an alternative to backcrosses and intercrosses for QTL analysis. The genetic compositions of each of these types of strains are illustrated in *Figure 2*. The pros and cons of these alternatives to strain crosses for analysis of complex traits are discussed below.

7.4.1 Recombinant inbred strains

Recombinant inbred (RI) strains, derived by brother–sister mating from the F2 generation of an initial strain cross (46,47), provide a potentially efficient

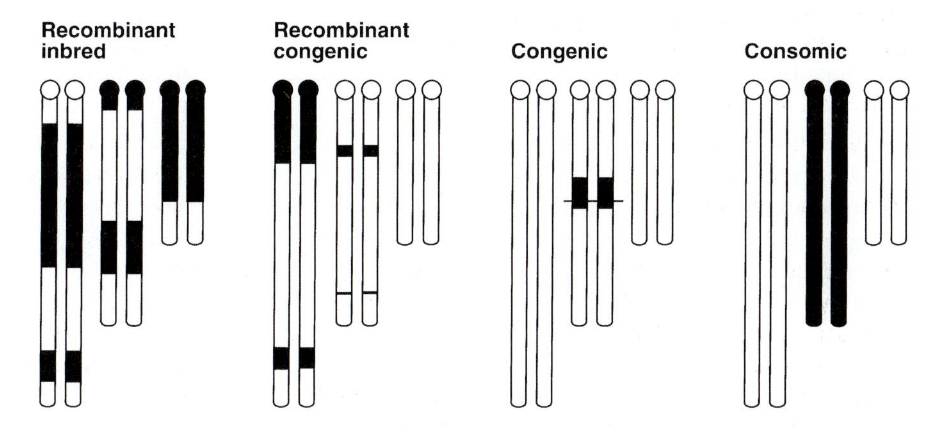

Figure 2. Genetic consequences of recombinant inbred, recombinant congenic, congenic and consomic strain formation. Hypothetical example of segregation and recombination in four kinds of recombinant strains derived from two pre-existing inbred strains designated A and B. Chromatin derived from progenitor strains A and B is shown as white and black, respectively. Three pairs of autosomes are shown representing large (100 cM), medium (75 cM) and small (50 cM) chromosomes. Note that for recombinant inbred strains, each parent contributes equally. For recombinant congenic strains derived by brother–sister inbreeding, beginning with second backcross progeny, the donor strain contributes 12.5% of the genome. For congenic strains, the average length of a donor segment (from the B progenitor) containing a single selected locus is approximately 200/*G*, where *G* is the number of backcross generations. For $G = 10$, the average introduced segment is 20 cM, or ~1.25% of the genome. The transverse line within the donor segment indicates the site of a hypothetical selected locus. In consomic strains, a single complete chromosome of the donor strain B is transferred to the background strain A.

means of detecting associations between quantitative traits and genetic markers. Advantages of RI strains are:

(a) For established RI strain sets, there may already be a dense marker base which the investigator does not need to duplicate. Consequently, little or no genotyping is required.
(b) RI strain analysis allows replication of fixed genotypes for control of environmental noise.
(c) RI strain mice are analysed directly, thus the two-generation delay associated with a standard cross is avoided.
(d) Recombinants accumulate during the inbreeding process, resulting in a four-fold expansion of the map (autosomal loci) and surprisingly good mapping resolution.
(e) Since RI strains are inbred, heterozygotes are eliminated and genetic differences tend to be magnified.
(f) It is feasible to measure genetic correlations among different traits that cannot be measured in a single individual (e.g. adiposity in males and females).

Disadvantages and limitations of RI strains are:

(a) Since the number of RI strains in a given set is generally small, power to detect QTLs is weak, limiting linkage detection to QTLs that account for a large proportion of the genetic variance.
(b) RI strains are available for only specific progenitor strain combinations, which may not be most suitable for the analysis of the trait of interest.
(c) Some RI strains are poor breeders, a factor which may limit the number of animals available for experimentation.
(d) Mutations slowly accumulate during RI strain propagation and may influence the trait of interest.

The first two limitations are serious ones. However, analysis of a set of RI strains as a first attempt at QTL detection (and for estimating the genetic complexity of the trait) is an efficient strategy if one is interested primarily in QTLs with large effects.

Some investigators use RI strains for a primary QTL screen with a relaxed significance threshold, and then attempt to confirm suspected linkage in a backcross or intercross (48). Some of these candidate regions are confirmed, others are not. Since marker genotyping in the segregating generation can be limited to candidate regions suggested by the RI data, this approach can be relatively efficient. However, because of low power in the RI phase of the study, QTLs may be missed. RI strains also can be a source of specific recombinant genotypes which may be progeny tested by intercrosses or backcrosses involving the progenitor strains as a means of fine mapping a QTL (38).

7.4.2 Congenic strains

Existing congenic strains also have been used to map (49,50), or further analyse, QTLs (51). Here, one simply surveys a set of congenic strains for phenotypes that differ significantly from the background strain. When differences are found, one presumes that a QTL affecting the trait is within the donor chromosomal segment of the congenic strain. However, because of the possibility of unlinked contaminant loci and mutations, linkages identified this way should be confirmed. This approach has the significant advantage that a positive result means that one already has the QTL difference separated from other loci affecting the trait, and the QTL is limited to a discrete region. The major limitation of this approach is that, since the introduced segment is typically ~20 cM in length, many congenic strains are required to cover the entire genome. No currently available set is nearly that large. The largest set of congenic strains, the C.B/By strains (52), are thought to cover one-third to one-half of the genome. Although dozens of congenic strains exist for the major histocompatibility region of mouse chromosome 17, coverage of the rest of the genome is quite spotty. However, once a QTL has been mapped to a particular region, it makes sense to compare relevant congenic strains with their background strains. Unless the combination of donor and background strains is the same as that used to identify the QTL initially, it is only a presumption that the same locus is involved.

7.4.3 Recombinant congenic strains

This strategy is intermediate between RI and congenic strain strategies. In recombinant congenic (RC) strain formation, two backcrosses (usually) are made to one of the progenitors (the recipient strain) before initiating continuous brother-sister mating (53). For two backcrosses, the resulting RC strains are each expected to have inherited 87.5% of genes of the recipient parent. As detailed by Demant and colleagues, RC strains provide a powerful tool for genetically dissecting multigenic traits. Thus, if a trait is controlled principally by three or four genes, each RC strain is likely to differ from the backcross parent strain with respect to, at most, one factor affecting the trait. When the RC strains are typed for numerous marker loci, it is possible to associate markers with RC strain phenotypes.

The RC strain strategy is similar to that for RI strains in many respects and shares many of the same advantages and disadvantages. It is especially useful if a particular phenotype found in the backcross parent (e.g. susceptibility to a particular kind of cancer) requires a combination of three or four alleles lacking in the donor strain. Particular limitations of this approach are the need for a dense marker base (to identify especially small segments of donor strain chromosome), and the fact that phenotypes found in the donor parent may not be recovered in any RC strain. Demant has created several sets of RC strains for use in analysis of tumour susceptibility (54–56).

7.4.4 Consomic strains

A consomic strain is one in which a single entire chromosome of the background strain has been replaced with the corresponding chromosome of a donor strain. Although such strains now exist only for the Y chromosome (57), other such strains are being developed. With a set of consomic strains, it would be fairly easy to rapidly determine which chromosomes harbour QTLs by comparing each consomic strain with the background strain. Once it is determined that a QTL is located on a particular chromosome, a congenic strain for the QTL-containing segment could be easily derived. It remains to be seen whether whole autosome or X-chromosome consomic strains will be fertile as homogeneous stocks. Potentially, unfavourable epistatic interactions between the introduced chromosome and the background genome might prevent the establishment of homozygous consomic stocks, especially those made from genetically divergent strains.

8. Markers

Microsatellite markers are now used almost exclusively for general linkage analysis. Many thousands of microsatellite sequences, consisting of dinucleotide repeats (e.g. $(CA)_n$, where n is between 10 and 60), are widely dispersed in mammalian genomes (58). The number of dinucleotide repeats at any site is subject to mutation due to errors in DNA replication, resulting in abundant length variation in populations. Length variation is readily detected by amplifying microsatellite sequences by the polymerase chain reaction (PCR) using site-specific primers, and separating the amplified DNA fragments electrophoretically. Microsatellite loci are highly polymorphic, usually co-dominant, and inexpensive and rapid to use. The same assay system can be used for thousands of loci. A total of 6331 microsatellite markers have been identified and mapped in the mouse genome by the the Center for Genome Research at the Whitehead Institute (18). (see MIT website at: http://www.genome.wi.mit.edu/.) These markers are denoted by symbols *D_Mit#*, where the *D* indicates an anonymous DNA marker, _ identifies the chromosome, *Mit* is the Laboratory Registration Code for MIT, and # is a serial number given to loci assigned to the particular chromosome (for example, *D2Mit3* is the third MIT microsatellite marker assigned to chromosome 2). Their initial mapping, based on a 46-member set of (C57BL/6J-*ob* × CAST)F2 progeny, was not very exact (6). Many markers also have now been mapped in 46 interspecific backcross progeny (18), and most are assigned to YAC contigs. Also, many of these have been used by various investigators in diverse mapping projects. A panel of inbred strains have been characterized with respect to fragment sizes, and this information is on the MIT/Whitehead website. Primer pairs for amplifying these loci are available commercially (Research

Genetics). It is a good idea to test the primers on the parental strain and F1 hybrid DNAs prior to using them on the backcross DNAs.

One difficulty with microsatellite markers is that the DNA of some strains does not amplify well in competition with DNA from other strains (i.e. in heterozygotes), presumably due to sequence variation in the primer sites. Occasionally, primers will not amplify the DNA of a given strain at all. In some cases, reducing the annealing temperature of the amplification reaction may equalize amplification. Since primers were designed to be complementary to strain C57BL/6J, amplification of the DNA of this strain is expected. The strong sequence divergence between the various *Mus* subspecies from laboratory mice means that microsatellite amplication from such strains is frequently a problem. This is especially true in the case of *M. spretus*.

Microsatellite fragment length polymorphism is detected using either agarose or polyacrylamide gel electrophoresis. For agarose gels, ethidium bromide is used to visualize DNA fragments, while silver staining, radioisotopes or fluorescent label is used to detect DNA fragments in acrylamide gels (6). Agarose gels are generally faster to set up and analyse, but limit polymorphism detection to allelic fragment size differences of ~8% or greater. Several methods for improving the efficiency of microsatellite analysis using agarose gels has been described (59). Multiplexing, the amplification and/or analysis of multiple markers in a single assay, is possible. Access to an automated sequencing facility can greatly increase genotyping throughput. Fluorescently labelled primers increase the number of loci that can be scored in a single run (60). Single-strand conformation polymorphism (SSCP) has been used to detect polymorphism in the 3′-untranslated region of specific genes (61) (see also Chapter 5).

Restriction fragment length polymorphisms provide another means of mapping a candidate or reference locus relative to a phenotypic variant. Dispersed repetitive sequences, such as endogenous retrovirus genomes (62–64), minisatellite sequences (65) and sequences flanked by B1 repeat elements (66) can provide multiple markers involving single assays. However, visualization of these DNA variants requires hybridization. Also, these approaches do not offer the same degree of choice regarding location as do microsatellite markers.

In the future, more rapid means of discriminating genetic differences will undoubtedly become available. These will probably incorporate automated genotypic scoring and recording as well as efficient genotypic discrimination. A system is needed that would allow the scoring of 20–100 loci per mouse in a reliable, inexpensive and automatable way. A system that does not require elaborate instrumentation would be desirable. Current research focuses on the development of oligonucleotide arrays synthesized on silicon chips as a means of detecting single nucleotide polymorphisms (SNPs) by hybridization (67).

9. DNA pooling

The time-consuming and expensive task of genotyping linkage crosses can be drastically reduced by adopting the strategy of phenotypic DNA pooling. The rationale for pooling is that DNA pools made from phenotypically distinct backcross or intercross progeny will differ in the representation of marker alleles at loci linked to the genes that determine phenotype. Linkage detection depends on recognition of this differential representation of allelic DNA fragments in pools. DNA pooling will generally shorten the search for linkage of major variants (68,69), and can also be efficient in QTL detection (70–73). *Protocol 1* describes the DNA pooling method for detecting linkage of a recessive mutation in an intercross population using microsatellite markers.

Protocol 1. Mapping recessive visible mutation by analysis of DNA pools

Reagents

- TE buffer: 10 mM Tris–HCl pH 8.0, 1 mM EDTA
- MIT microsatellite primers (Research Genetics)
- 10× PCR buffer: 17.5 mM $MgCl_2$, 500 mM KCl, 100 mM Tris–HCl pH 9.0, 1% Triton X-100, 60% sucrose, 1 mM Cresol Red (Fluka)
- dNTPs (e.g. Promega)
- *Taq* DNA polymerase (e.g. Promega)
- 3× dye mix: 15% Ficoll (mol. wt 400 000; Sigma), 0.125% xylene cynanol (Bio-Rad), 0.125% bromophenol blue (Sigma), 25 mM EDTA. Dilute the mixture three-fold with distilled water before use.

Method

1. Determine concentration of each intercross DNA sample contributing to pools by UV absorption spectrometry and dilute, if necessary, to 0.5 mg/ml in TE buffer.
2. Pool a 3´μg aliquot of DNA from each of 25 intercross mutant (*m*/*m*) homozygotes and (at least) 25 intercross wild-type progeny and dilute ten-fold in distilled water, for a final concentration of 50 μg/ml.
3. Mix pools thoroughly by repeated pipetting and gentle vortexing. Pools may be frozen, then thawed and vortexed again prior to use.
4. Select MIT microsatellite primers which yield parental strain variants which are separable on agarose gels, and give a co-dominant F1 hybrid pattern. A list of markers suitable for a cross between laboratory inbred strains and CAST/Ei can be found in ref. 68.
5. Amplify pools, parental strains and F1 DNAs with MIT microsatellite primers. PCR reaction mixture is as follows: 125 ng (2.5 μl) genomic DNA, 0.2 μM primers, 0.8 mM dNTPs (Promega), 0.25 U *Taq* polymerase, 1× PCR buffer, for a final volume of 14 μl. Program the thermal cycler for the following PCR cycling times: 94 °C for 2 min, followed by

35 cycles of 94°C for 20 sec, 55°C for 30 sec, 72°C for 30 sec, followed by cooling for 5 min at 4°C.

6. Add 1× dye mix (5 μl) to each reaction, and separate the dye–PCR product mixture (9 μl) in 3% agarose gels (2% Metaphor, 1% SeaKem GTG (FMC); 0.09 M Tris–borate, 0.002 M EDTA) containing ethidium bromide (2 ng/ml) for approximately 3 h at 105 V.
7. Photograph gel under UV light, and look for increased representation of the amplified allelic fragment from the mutant-bearing strain in the mutant pool and a decrease in the representation of this same fragment in the wild-type pool.

For detecting linkage of monogenic traits by the DNA pooling strategy, two pools are made representing the two phenotypes for recessive/dominant traits segregating in an intercross or backcross (*Figure 3*), or the two homozygotes for semi-dominant or co-dominant trait loci segregating in an intercross. In general, the larger the number of individuals contributing to the two pools, the better the statistical efficiency of the test. Good statistical power is provided by representing at least 50 meioses in each DNA pool (e.g. 50 heterozygotes and 50 homozygotes in a backcross, or 25 homozygous (recessive) mutants and 25 wild-type progeny in an intercross). Relatively small improvement is achieved with larger numbers. However, one should not hesitate to use smaller numbers in pools (e.g. if the entire cross is small); one merely has to accept that there will be more regions showing a hint of linkage, and that a larger number of markers will be needed to cover the genome. With 50 gametes represented in each pool, markers spaced at 40 cM (and within 20 cM of chromosome ends) are adequate for detecting linkage of monogenic loci. For an intercross, it is useful to include both parental strains and the F1 hybrid as controls. For a backcross, appropriate DNA controls are an F1 hybrid, the backcross parental strain, and a 1:1 mixture of these two samples.

For pooling to work, it is essential that allelic DNA fragments (e.g. allelic microsatellite PCR products) are clearly resolvable in heterozygotes. Thus, larger fragment size differences are required for pooling than for individual genotyping. It is also important that both parental fragments amplify well in any pooled-PCR sample in the absence of linkage, although equal amplification is not essential. Intercrosses are more amenable to pooling than backcrosses for this reason. Pool results may be evaluated subjectively. One can discern by simple inspection differences of 15–20% in allelic representation between intercross pools (71). Densitometry can be used as an aid for judging differences, but this is not essential. A caution should be added: since linkage of 'pool' markers is not confirmed, an incorrectly mapped or misidentified marker does not test its nominal segment of chromosome for the presence of the target locus. Instead, it gives a false test, one representing some other region. Thus, if two markers on a chromosome suggest possible

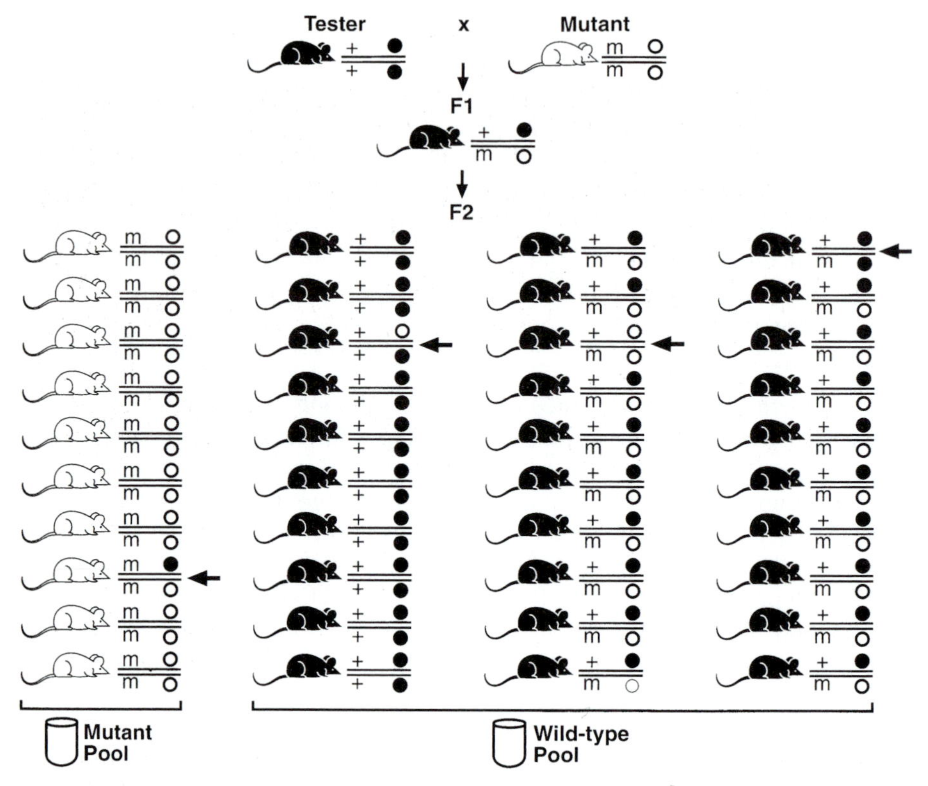

Figure 3. Linkage detection of recessive mutant gene in an intercross experiment by DNA pooling. The consequences of pooling DNA from mutant (white; *m/m*) and wild-type (black; +/+ and +/*m*) mice on a linked locus. White and black balls represent marker locus alleles from the mutant-bearing and tester strains, respectively. Arrows point to individual mice with a single recombinant chromosome. In this example, there are four crossovers among 80 chromosomes, for a recombination frequency of 0.05. The mutant and wild-type pools consist of 5% (one of 20) and 66.7% (40 of 60) tester strain alleles at the marker locus, respectively.

association and an intervening marker does not, suspect that the latter result is invalid.

The efficacy of DNA pooling to detect linkage is illustrated by pools constructed based on the segregation of the recessive *dilute* coat colour mutation ($Myo5a^{d}$) in a large intercross between CAST/Ei (+/+) and MEV (*d/d*) (*Figure 4*). *Dilute* is located centrally on chromosome 9. From this intercross, three pairs of DNA pools were made to simulate three separate linkage tests. These pools were amplified with the chromosome 9 primer pair corresponding to *D9Mit2*, a marker which shows 23% recombination with *d* in this sample material. A distinct diminution of the CAST allelic fragment is seen in each of the *d/d* pools (a, c and e) relative to both the F1 hybrid and the

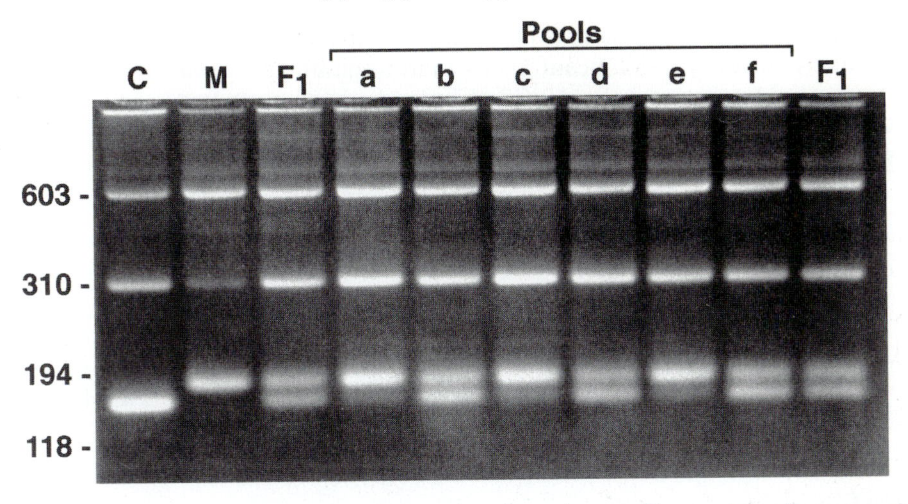

Figure 4. Analysis of DNA pools and controls using microsatellite markers. Agarose gel electrophoresis of microsatellite products from DNA pools and controls: The results of PCR amplifying pooled DNAs from three independent pairs of dilute homozygote (*d/d*) and wild-type (+/+ and +/*d*) DNA pools from (CAST/Ei × MEV)F2. Lanes: C (CAST/Ei); M (MEV); F1 [(CAST/Ei × MEV)F1 hybrid]; a (23 *d/d* F2 pool); b (75 +/? F2 pool); c (25 *d/d* F2 pool); d (75 +/? F2 pool); e (25 *d/d* F2 pool); f (65 +/? F2 pool); and F1 [(CAST/Ei × MEV)F1 hybrid] (see text). Each DNA sample or pool was amplified with the *D9Mit2* primer pair.

wild-type (+/+ and +/*d*) pools (b, d and f) in each of the three independent comparisons. The wild-type pools appear to be slightly enriched for the CAST allele at the *D9Mit2* locus. Little variation in the degree of differential enrichment is observed between the three pairs of pools, illustrating the reproducibility of the method.

As noted above, pooling also can be quite effective for QTL mapping. For a quantitative trait, select individuals from the two tails of the phenotypic distribution to create a 'high' and a 'low' DNA pool. About 20–30% of the total progeny should contribute to the two pools (10–15% from each tail). A variation of the technique is to weight each individual's contribution by its relative squared deviation from the entire population mean of the phenotypic trait (71). The squared deviation of each individual contributing to a particular pool is computed; deviations are then summed over all individuals contributing to the pool, and individual weights are assigned according to the ratio of the individual squared deviation to this sum. This procedure is expected to better reflect the strength of any statistical association between a marker and the phenotypic trait than unweighted pooling. If there are two or more traits of primary interest, separate pools need to be made for each trait. Each extra trait requires two additional assays for each marker. Markers spaced at 20 cM (and within 10 cM of chromosome ends) should be adequate for detecting statistically significicant QTLs.

The goal of DNA pooling is to rapidly identify regions of the genome that are likely to contain significant QTLs (and to eliminate regions that do not). Once candidate regions are identified, individual genotyping of markers in the region is used to rigorously test statistical significance and to estimate genetic parameters. It is strongly recommended that DNA pooling be given serious consideration prior to embarking on any linkage search.

Acknowledgements

I wish to thank Sandra Phillips and Jennifer Smith for help with figures, and Muriel Davisson, Wayne Frankel and Janan Eppig for comments on the manuscript. This work was supported by NIH Research Grants GM18684, CA33093 and DK50692.

References

1. Li, Y. *et al.* (1995). *Cell*, **80**, 423.
2. Roderick, T. H., Hillyard, A. L., Maltais, L. J. and Blake, C. S. (1996). In *Genetic Variants and Strains of the Laboratory Mouse* (ed. Lyon, M. F., Rastan, M. F. and Brown, S. D. M.), p. 929. Oxford University Press.
3. Dib, C. *et al.* (1996). *Nature*, **380**, 152.
4. Lawrie, N. M., Tease, C. and Hulten, M. A. (1995). *Chromosoma*, **140**, 308.
5. Carter, T. C. and Falconer, D. S. (1951). *J. Genet.*, **50**, 307.
6. Dietrich, W., Katz, H., Lincoln, S. E., Shin, H.-S., Friedman, J., Dracopoli, N. and Lander, E. S. (1992). *Genetics*, **131**, 423.
7. Blank, R. D., Campbell, G. R. and D'Eustachio, P. (1986). *Genetics*, **114**, 1257.
8. Bonhomme, F. and Guenet, J.-L. (1996). In *Genetic Variants and Strains of the Laboratory Mouse* (ed. Lyon, M. F., Rastan, M. F. and Brown, S. D. M.), p. 1577. Oxford University Press.
9. Robert, B., Barton, P., Minty, A., Daubas, P., Weydert, A., Bonhomme, F., Catalan, J., Chazottes, D., Guenet, J.-L. and Buckingham, M. (1985). *Nature*, **314**, 181.
10. Moser, A. R., Dove, W. F., Roth, K. A. and Gordon, J. I. (1992). *J. Cell Biol.*, **116**, 1517.
11. MacPhee, M., Chepenik, K. P., Liddell, R. A., Nelson, K. K., Siracusa, L. D. and Buchberg, A. M. (1995). *Cell*, **81**, 957.
12. Cormier, R. T., Hong, K. H., Halberg, R. B., Hawkins, T. L., Richardson, P., Mulherkar, R., Dove, W. F. and Lander, E. S. (1997). *Nature Genet.*, **17**, 88.
13. Friedman, J. M., Leibel, R. L. and Bahary, N. (1991). *Mammal. Genome*, **1**, 130.
14. DeJager, P. L., Harvey, D., Polydorides, A. D., Zuo, J. and Heintz, N. (1998). *Genomics*, **48**, 346.
15. Reeves, R. H., Crowley, M. R., Moseley, W. S. and Seldin, M. F. (1991). *Mammal. Genome*, **1**, 158.
16. Himmelbauer, H. and Silver, L. M. (1993). *Genomics*, **17**, 110.
17. Shiroishi, T., Sagai, T. and Moriwaki, K. (1993). *Genetica*, **88**, 187.
18. Dietrich, W. F. *et al.* (1996). *Nature*, **380**, 149.

19. Davisson, M. T., Guay-Woodford, L. M., Harris, H. W. and D'Eustachio, P. (1991). *Genomics*, **9**, 778.
20. Manly, K. F. (1993). *Mammal. Genome*, **4**, 303.
21. Lander, E. S., Green, P., Abrahamson, J., Barlow, A., Daly, M. J., Lincoln, S. E. and Newburg, L. (1987). *Genomics*, **1**, 174.
22. Diem, K. and Lentner, C. (1975). *Scientific Tables*, p. 85. Geigy Pharmaceuticals, Ardsley, NY.
23. Kosambi, D. D. (1944). *Ann. Eugen.*, **12**, 172.
24. Haldane, J. B. S. (1919). *J. Genet.*, **8**, 299.
25. Crow, J. F. (1990). *Genetics*, **125**, 669.
26. Rowe, L. B., Nadeau, J. H., Turner, R., Frankel, W. N., Letts, V. A., Eppig, J. T., Ko, M. S. H., Thurston, S. J. and Birkenmeier, E. H. (1994). *Mammal. Genome*, **5**, 253.
27. Mullick, A., Groulx, N., Trasler, D. and Gros, P. (1995). *Mammal. Genome*, **6**, 700.
28. Green, E. L. (1981). *Genetics and Probability in Animal Breeding Experiments*, p. 271. Oxford University Press, New York.
29. Segre, J. A., Nemhauser, J. L., Taylor, B. A., Nadeau, J. H. and Lander, E. S. (1995). *Genomics*, **28**, 549.
30. Green, M. C. (1963). In *Methodology in Mammalian Genetics* (ed. Burdette, W. J.), p. 56. Holden-Day, San Francisco.
31. Rise, M. L., Frankel, W. N., Coffin, J. M. and Seyfried, T. N. (1991). *Science*, **253**, 669.
32. Warden, C. H., Fisler, J. S., Pace, M. J., Svenson, K. L. and Lusis, A. J. (1993). *J. Clin. Chem.*, **92**, 773.
33. Lander, E. S. and Botstein, D. (1989). *Genetics*, **121**, 185.
34. Haley, C. S. and Knott, S. A. (1992). *Heredity*, **69**, 315.
35. Darvasi, A., Winreb, A., Minke, V., Weller, J. I. and Soller, M. (1993). *Genetics*, **134**, 943.
36. Drudik, D. K., Pomp, D., Zeng, Z.-B. and Eisen, E. J. (1995). *J. Anim. Sci.*, **73**, S110.
37. Taylor, B. A. and Phillips, S. J. (1997). *Genomics*, **43**, 249.
38. Darvasi, A. (1998). *Nature Genet.*, **18**, 19.
39. Lander, E. and Kruglyak, L. (1995). *Nature Genet.*, **11**, 241.
40. Witte, J. S., Elston, R. C. and Schork, N. J. (1996). *Nature Genet.*, **12**, 357.
41. Doerge, R. W. and Churchill, G. A. (1996). *Genetics*, **142**, 285.
42. Dupuis, J. and Siegmund, D. (1998). *Genetics*, in press.
43. Frankel, W. N., Johnson, E. W. and Lutz, C. M. (1995). *Mammal. Genome*, **6**, 839.
44. Morel, L., Mohan, C., Yu, Y., Croker, B. P., Tian, N., Deng, A. and Wakeland, E. K. (1997). *J. Immunol.*, **158**, 6019.
45. Darvasi, A. and Soller, M. (1995). *Genetics*, **141**, 1199.
46. Bailey, D. W. (1971). *Transplantation*, **11**, 325.
47. Taylor, B. A. (1978). In *Origins of Inbred Mice* (ed. Morse, H. C., III), p. 423. Academic Press, New York.
48. Belknap, J. K., Mitchell, S. R., O'Toole, L. A., Helms, M. L. and Crabbe, J. C. (1996). *Behav. Genet.*, **26**, 149.
49. Bailey, D. W. (1985). *J. Hered.*, **76**, 107.
50. Baker, P. J., Bailey, D. W., Fauntleroy, M. B., Stashak, P. W., Caldes, G. and Prescott, B. (1985). *Immunogenetics*, **22**, 269.

51. Warden, C. H., Fisler, J. S., Shoemaker, S. M., Wen, P. Z., Svenson, K. L., Pace, M. J. and Lusis, A. J. (1995). *J. Clin. Invest.*, **95**, 1545.
52. Bailey, D. W. (1975). *Immunogenetics*, **2**, 249.
53. Demant, P. and Hart, A. A. M. (1986). *Immunogenetics*, **24**, 416.
54. Moen, C. J., Snoek, M., Hart, A. A. and Demant, P. (1992). *Oncogene*, **7**, 563.
55. Fijneman, R. J., de Vries, S. S., Jansen, R. C. and Demant, P. (1996). *Nature Genet.*, **14**, 465.
56. Moen, C. J., Groot, P. C., Hart, A. A., Snoek, M. and Demant, P. (1996). *Proc. Natl Acad. Sci. USA*, **93**, 1082.
57. Hudgins, C. C., Steinberg, R. T., Klinman, D. M., Reeves, M. J. and Steinberg, A. D. (1985). *J. Immunol.*, **134**, 3849.
58. Litt, M. and Luty, J. A. (1989). *Am. J. Hum. Genet.*, **44**, 397.
59. Routman, E. J. and Cheverud, J. M. (1995). *Mammal. Genome*, **6**, 401.
60. Reed, P. W. *et al.* (1994). *Nature Genet.*, **7**, 390.
61. Beier, D. R., Dushkin, H. and Sussman, D. J. (1992). *Proc. Natl Acad. Sci. USA*, **89**, 9102.
62. Taylor, B. A., Rowe, L. and Grieco, D. A. (1993). *Genomics*, **16**, 380.
63. Frankel, W. N., Stoye, J. P., Taylor, B. A. and Coffin, J. M. (1990). *Genetics*, **124**, 221.
64. Lueders, K. K., Frankel, W. N., Mietz, J. A. and Kuff, E. L. (1993). *Mammal. Genome*, **4**, 69.
65. Julier, C., de Gouyon, B., Georges, M., Guenet, J. L., Nakamura, Y., Avner, P. and Lathrop, G. M. (1990). *Proc. Natl Acad. Sci. USA*, **87**, 4585.
66. Hunter, K. W. *et al.* (1996). *Genome Res.*, **6**, 290.
67. Wang, D. G. *et al.* (1998). *Science*, **280**, 1077.
68. Taylor, B. A., Navin, A. and Phillips, S. J. (1994). *Genomics*, **21**, 626.
69. Asada, Y., Varnum, D. S., Frankel, W. N. and Nadeau, J. H. (1994). *Nature Genet.*, **6**, 363.
70. Darvasi, A. and Soller, M. (1994). *Genetics*, **138**, 1365.
71. Taylor, B. A. and Phillips, S. J. (1996). *Genomics*, **34**, 389.
72. Collin, G. B., Asada, Y., Varnum, D. S. and Nadeau, J. H. (1996). *Mammal. Genome*, **7**, 68.
73. Mohlké, K. L., Nichols, W. C., Westrick, R. J., Novak, E. K., Cooney, K. A., Swank, R. T. and Ginsburg, D. (1996). *Proc. Natl Acad. Sci. USA*, **93**, 15352.

5

Mapping genomes

PAUL DENNY and STEPHEN D. M. BROWN

1. Introduction

It is an exciting time in mammalian genetics, as the power of molecular genetic and genomic tools allows a genetic approach to many different biological problems in both mouse and man. The rapid pace of development means, however, that this chapter will quickly become out of date, so we advise the reader to consult the literature for novel and improved methods. The methods described here outline our approach to the generation of genetic and physical maps. There are alternative approaches but there is insufficient space to provide a completely comprehensive set of protocols. We describe the use of genetic and genomic resources to construct maps *de novo* as well as the use of published genetic maps for the construction of higher resolution physical maps such as for positional cloning of an unknown gene.

Mapping of mutant phenotypes is described in Chapter 4, together with an explanation of the types of breeding schemes used to produce animals suitable for such studies. However, it is pertinent to review briefly the types of cross typically used in genetic mapping of DNA markers, to place the technical approaches described in this chapter into a wider context. Genetic mapping depends on two basic biological phenomena:

- the process of recombination that occurs during meiotic cell division in the production of gametes;
- natural (or induced) genetic variation between mouse strains.

In a typical mapping cross, the parental strains would ideally be inbred (defined as sibling (sib) mated for at least 20 generations) which means that there is no heterozygosity in members of that inbred strain. After two inbred strains are crossed to produce offspring of the hybrid, F_1, generation, recombination in the germline of those hybrid animals gives rise to gametes carrying genetic material derived from both parents. Two main types of cross are useful in genetic mapping; backcross and intercross (see Chapter 4). Another useful mapping resource also described in Chapter 4 are recombinant inbred (RI) strains. The power to map any given locus in a genetic cross is dictated by the effective number of meioses in the cross. In a backcross this is simply n, the number of animals in the first backcross (N2) generation (as each N2 offspring has only one hybrid parent and so inherits the product of only one useful meiotic event) whilst for an intercross, it is $2n$, where n is the number of animals in the F2 generation (as each F2 offspring has two hybrid parents and so inherits the products of two useful meiotic events). For a set of RI strains, the number of effective meioses is approximately $4n$, where n is the number of RI strains.

It is usually unnecessary to construct a new mapping cross solely to position DNA markers, as publicly available crosses, together with genotype data and DNA, are readily available (1,2). Alternative approaches would be to collaborate with an academic lab maintaining a mapping cross (3–5) or to use a commercial service to carry out genotyping, such as Research Genetics (http://www.resgen.com/). The main publicly available crosses are described in *Table 1*. The Jackson Laboratory mapping panels and RI strains are extremely useful for moderate resolution mapping; one advantage is that DNA is generally available for typing markers by restriction fragment length variation (RFLV) or other similar blot hybridization based methods. The European Collaborative Interspecific Backcross (EUCIB) has a much higher potential resolution (0.3 cM at the 95% confidence level), but has limited DNA resources and so can only be used for polymerase chain reaction (PCR)-based genotyping. The other cross in *Table 1* is that used by the Massachusetts Institute of Technology/Whitehead Institute Centre for Genome Research (WI-CGR). This consists of only 92 meioses and is not available as DNA for genotyping assays. Nevertheless, this is an important map resource as it has been typed for more than 6000 simple-sequence length polymorphism (SSLP) markers. This has allowed construction of a marker-dense, intermediate-resolution genetic map of the whole mouse genome which has been used to anchor a yeast artificial chromosome (YAC) clone physical map based on sequence-tagged site (STS) content (6). Recently, 3368 of these SSLPs have been mapped at high resolution on the EUCIB resource (7). These maps are invaluable as guides in the first steps towards construction of high-resolution genetic and physical maps.

2. Applications of genetic and physical mapping

The main applications of genetic mapping are the testing of candidate genes for mutant phenotypes—the so-called 'positional candidate' approach (8) and the study of chromosome structure, both within and between species (9).

The applications of physical mapping are related to the resolution of the techniques used to produce the maps. These range from cytogenetic maps with resolution of a few megabases, through to physical maps consisting of contigs of overlapping clones, which form the substrate for determining the ultimate physical 'map', the DNA sequence itself. For a detailed description

Table 1. Mapping crosses

Source	Name(s) of crosses	Parental strains	Website URL (or e-mail address)
MRC UK Human Genome Mapping Project Resource Centre	European Interspecific Backcross (EUCIB)	*Mus spretus* and C57BL/6	http://www.hgmp.mrc.ac.uk/Research/eucib.html
Jackson Laboratory	BSS and BSB backcrosses	*Mus spretus* and C57BL/6J	http://www.jax.org/resources/documents/cmdata/bkmap/CMIntro.html
Jackson Laboratory	Recombinant inbred strains:[a]		http://jaxmice.jax.org/index.shtml
	AKXD	AKR/J x DBA/2J	
	AKXL	AKR/J x C57L/J	
	AXB	A/J x C57BL/6J	
	BXA	C57BL/6J x A/J	
	BXD	C57BL/6J x DBA/2J	
	BXH	C57BL/6J x C3H/HeJ	
	CXB	BALB/cBy x C57BL6By	
	NXSM	NZB/BLNJ x SM/J	
	SWXJ	SWR/Bm x SJL/Bm	
Institute for Experimental Animals, Hamamatsu, Japan (Dr Nishimura)	Recombinant inbred strain[a] SMXA	SM/J x A/J	mnisim@hama-med.ac.jp
MIT/Whitehead Centre for Genome Research	intercross	*Mus castaneus* and C57Bl/6-*ob*	http://www-genome.wi.mit.edu/genome_data/mouse/mouse_index.html

[a]This list of RI strains is not exhaustive—we have included only those sets of strains which have been typed with sufficient markers likely to allow the mapping of a locus.

of the utility of the various hierarchies of physical maps, the reader is advised to consult recent comprehensive reviews in the field (12).

3. Genetic mapping

3.1 Types of genetic marker

The primary requirement for a genetic marker is that it differs in some detectable manner between mouse strains. These differences can manifest themselves in many ways: simply as visible phenotypes (differences in coat colour were among the earliest recognized genetic markers) or as protein or nucleic acid variants. With the advent of molecular cloning and DNA amplification by PCR, the use of differences in nucleic acid sequence has revolutionized genetic mapping. The most commonly used molecular genetic markers are those based on:

- differences in sequence length, for example of a simple repeated dinucleotide such as $(CA)_n$, called simple sequence length polymorphism (SSLP) markers
- differences in sequence which alter restriction enzyme recognition sites and so allow detection of restriction fragment length polymorphism or variants (RFLP or RFLV)
- differences in sequence detected by other means, often now called single nucleotide polymorphism (SNP) markers

SSLP and SNP markers are usually amplified by PCR and allelic differences in the mobility of the PCR products detected using a gel-based system. In the near future, however, it is likely that gel analysis will be replaced by hybridization of SNP marker PCR products to very high density arrays of oligonucleotides.

In this section, we present methods for detecting allelic differences in SNP markers using single-strand conformation polymorphism (SSCP) and in SSLP markers using a fluorescent DNA sequencer.

3.1.1 Genotyping using silver-staining of SSCP gels

This method is based on the observation that denatured double-stranded DNA, when allowed to renature under appropriate conditions, can form a number of semi-stable structural conformations in addition to the classical Watson–Crick double helix (13). Sequence differences result in different alternative conformations that can be resolved in a non-denaturing acrylamide gel system and detected by silver-staining, for example. The ability to detect sequence differences as SSCP is dependent mainly on temperature, but as it is impractical to run gels under a range of conditions, we routinely only run gels in 1 × TBE at 4°C. It is also common to run SSCP gels at ambient temperature, with the addition of glycerol to 10% (v/v), to increase the

Table 2. Acrylamide concentrations used for different sizes of PCR product

Size of PCR product (bp)	Acrylamide concentration (%)
400–500	6
300–400	8
200–300	10

stability of certain conformations. The size of PCR product influences the choice of acrylamide concentration used in gels (see *Table 2*). We make and run gels as described in ref. 14. The silver-staining in *Protocol 1* is a modification of that used by the Regional Molecular Genetics Laboratory, St Mary's Hospital, Manchester, UK. It is advisable to check the yield and specificity of PCR products on a normal agarose gel prior to loading them on an SSCP gel, because spurious products may complicate analysis.

Protocol 1. Detection of PCR products in SSCP native acrylamide gels by silver-staining

Equipment and reagents

- Solution 1: 10% (v/v) ethanol, 0.5% (v/v) acetic acid
- Solution 2: 0.1% (w/v) $AgNO_3$
- Solution 3: 1.5% (w/v) NaOH, 0.1% formaldehyde; freshly prepared before use
- Solution 4: 0.75% (w/v) Na_2CO_3
- Large plastic tray (e.g. photographic developing tray)
- Shaking platform with speed control

Method

Note: It should be emphasised that gloves should be worn when handling the gels in this protocol, not only to protect the worker against toxic acrylamide and silver salts, but also to protect the gel. Fingerprints stain very effectively in this protocol!

1. Separate the gel plates and carefully transfer the gel into a clean plastic tray containing sufficient solution 1 to allow the gel to move freely. Agitate gently for 3 min to fix the DNA in the gel matrix.
2. Discard the fixing solution and replace with fresh solution 1; agitate the gel for a further 3 min.
3. Discard the fixing solution and replace with solution 2; agitate the gel for 15 min. Solution 2 can be re-used three times before disposal. N.B. As solution 2 contains silver, a toxic heavy metal, it should be discarded according to local safe practice.
4. Rinse the gel quickly with two changes of distilled water. Do not leave the gel in the water, but go immediately to the next step.
5. Immerse the gel in solution 3 (developer), agitate it gently for 5–20 min

Protocol 1. *Continued*

and allow the image to develop until bands are clearly visible. Discard the solution.

6. Immerse the gel in solution 4 and agitate it gently for 10 min. Discard the solution.
7. Seal the gel in plastic for viewing on a light box and for archiving.

3.1.2 Genotyping using fluorescently-labelled dCTP incorporation into PCR products from SSLP markers, for analysis on ABI sequencers

This protocol is based on methods developed during the EUCIB project (15) using the ABI fluorescent dCTP labelling kit (Perkin–Elmer). PCR products are labelled by incorporation of dCTP coupled to one of three fluorescent dyes. A fourth dye is used to fluorescently label a size standard that is included as an internal control in all samples electrophoresed on a gel. When excited at a single wavelength by the laser in a fluorescent sequencer, these dyes emit light at four different, resolvable wavelengths. This means that three different markers giving PCR products of the same size can be resolved in one lane, by making use of the different dyes. With judicious choice of markers that amplify alleles in non-overlapping size ranges, it is possible to pool multiple PCR products labelled with the same dye and resolve them in a single lane. Indeed, it is possible to run as many as 24 different markers in a single lane on a fluorescent sequencer, using the available fluorescent dyes (16). We find, however, that the main limiting factor is that the majority of mouse SSLP PCR products cluster in the size range 100–160 bp. In general, we pool up to nine markers in a lane, using the three different dye colours and markers that give non-overlapping allele sizes. Optimization of individual PCR assays, for example by titration of magnesium ion concentration, is particularly important as any spurious products can lead to inaccurate genotyping. Advice on PCR optimization can be found in ref. 17. There is insufficient space to include detailed protocols for the use of ABI sequencers and downstream analysis of genotype data, so we direct the reader to the appropriate user manuals. A combination of ethanol precipitation and size-exclusion chromatography is used to concentrate the PCR products and remove unincorporated fluorescently labelled dCTP.

Protocol 2. Labelling of PCR products by fluorescent dCTP incorporation[a]

Equipment and reagents

- ABI Prism 310, 373 or 377 fluorescent sequencer
- Thermal cycler (e.g. MJ Research PTC-200)
- *Taq* DNA polymerase (Perkin–Elmer)
- dNTP 2 mM solution diluted from 100 mM stock solutions (Pharmacia)

- Fluorescently labelled dCTP kit (Perkin–Elmer). Stock solutions supplied in the kit need dilution as follows: R6G (green), dilute 1 in 10; R110 (blue), dilute 1 in 20; TAMRA (yellow), dilute 1 in 5.
- 10× PCR buffer: 100 mM Tris–HCl pH 8.3, 500 mM KCl, 15 mM $MgCl_2$ (Perkin–Elmer)
- 20 mM $MgCl_2$ stock solution
- Genomic DNA (in 10 mM Tris–HCl, 0.1 mM EDTA, pH 8.0 at approximately 2.5–5.0 ng/μl)
- Oligonucleotide primers at 6.6 μM (Research Genetics)

Method

1. Prepare a single 10 μl reaction as follows:

10× buffer	1 μl
dNTP stock	1 μl
each primer	0.125 μl
fluorescently labelled dCTP	1 μl
Taq DNA polymerase	0.08 μl
genomic DNA	5 μl
sterile distilled water	1.67 μl

2. As a guide, we amplify using conditions as follows:
   (a) 95 °C for 3 min;
   (b) 25–35 cycles of: (i) 95 °C for 5 sec; (ii) 55 °C for 20 sec; (iii) 72 °C for 30 sec;
   (c) 72 °C for 7 min.

[a] Optimization is obviously important to obtain specific amplification, but is particularly important for fluorescently labelled products, as the sensitivity of the detection system is so high. We find that it is critical to check products by electrophoresis on the 377 sequencer, rather than simply on agarose gels. Consult Kidd and Ruano (17) for advice on optimization of PCR amplification. We have found that the fluorescently labelled dCTP kit marketed by Perkin–Elmer ABI gives consistently better results in terms of efficiency of labelling than the equivalent labelled dUTP kit (Elaine Hopes, unpublished observations).

Protocol 3. Concentration and purification of fluorescently labelled PCR products prior to electrophoresis[a]

Equipment and reagents

- Microcentrifuge (e.g. Eppendorf 5415C)
- Bench-top centrifuge with microtitre plate rotor (e.g. Beckman GS15 and S2096 rotor)
- 0.2 ml microcentrifuge tubes
- 10 M ammonium acetate
- Absolute ethanol
- 70% ethanol
- Formamide
- Sephadex G50, Fine DNA grade (Pharmacia), stored at 4 °C in 10 mM Tris–HCl, 1 mM EDTA, pH 8.0 (TE)
- 96-well filter bottom microtitre plates (Multiscreen-GV, Millipore)
- 2 mg/ml dextran blue in 50 mM EDTA
- ROX-labelled molecular size standard, e.g. GS350 (ABI)

Protocol 3. *Continued*

Method

1. Pool between 1.0 and 7.5 μl of each marker, depending on the fluorescent intensity of the products seen during optimization.
2. Add ammonium acetate to a final concentration of 2 M, mix by vortexing and precipitate the DNA by adding 2.5 vols of absolute ethanol, mixing thoroughly by vortexing. Chill on ice for 10 min.
3. Centrifuge for 30 min (e.g. in a Beckman GS15R centrifuge, 14000 r.p.m.) at 4°C to pellet the DNA.
4. Carefully remove the supernatant and discard, then wash the pellet with ice-cold 70% (v/v) ethanol and centrifuge for 10 min at 14000 r.p.m.
5. Carefully remove the supernatant and discard, leaving the pellet to air-dry at room temperature, then re-dissolve in 7.5 μl sterile distilled water. Store at –20°C.
6. Prepare microtitre format spin columns by filling the wells of filter bottom plates with Sephadex G50 slurry and centrifuging at 2500 r.p.m. for 2 min. Repeat centrifugation with additional Sephadex until there is a uniform packed bed volume of approximately 225 μl in each well.
7. Wash the spin columns with 100 μl of water, centrifuge at 2500 r.p.m. for 2 min. Repeat centrifugation twice to remove excess liquid.
8. Place a standard microtitre plate underneath the filter bottom plate to collect the purified products. Apply up to 15 μl of sample and then centrifuge at 2500 r.p.m. for 2 min.
9. Mix formamide and dextran blue solution in the ratio 5:1. Mix this loading solution with ROX-labelled size standard in the ratio 22:1 and then add 1.5 μl of the diluted size standard to 1 μl of sample.
10. Denature samples at 95°C for 2 min, and chill rapidly on ice prior to electrophoresis.

[a] The sensitivity of the ABI Prism 377 means that PCR assays optimized for this system will require re-optimization if using the ABI 373 or 310 sequencers.

3.2 Linking different maps together

Genetic maps produced from different crosses differ because of stochastic effects (which are reduced by increasing the number of animals in each cross) and because of true genetic differences between strains that affect rates of recombination. Areas of the genome containing recombinational 'hot-spots' will be genetically expanded relative to the rest of the genome. Furthermore, differences in chromosomal structure, such as inversions or deletions, suppress recombination, so that genetic maps will be compressed in such regions.

These phenomena conspire to make the construction of consensus maps fraught with difficulty. The basic approach to link genetic maps together is to use common markers in each map as anchors and to interpolate the likely positions of other markers that are typed only in one cross. A similar approach can be used to link physical and genetic maps.

3.3 Generating new genetic markers in specific genomic regions

The high density of SSLP markers mapped onto the mouse genome means that for many workers, there is no need to generate additional genetic markers to achieve the required quality of map in their region of interest. However, if a high resolution map is required and there are sufficient meioses in the cross to allow it, or if a cross is being made using closely related strains where there is a lack of polymorphism in existing markers, there may still be a need to produce new markers. We have found this to be necessary, for example, when using congenic strains derived from the parental strains NOD and C57BL/6 to map diabetes susceptibility loci (18). We developed new SSLP markers using a protocol based on a combination of methods described by others (19,20). It allows for selective cloning of sequences containing microsatellites from a vectorette 'library' constructed from any DNA source. The vectorette method is described in ref. 21. We have used this protocol to isolate novel microsatellites from mouse YAC clones to aid in characterization of congenic strains and in the construction of YAC contigs (18). The protocol is written to select microsatellite sequences from YACs, but can equally well be applied to any DNA source. Combining a primer specific for a mouse repetitive element (22) together with the vectorette primer improves the complexity of sequences and enriches for mouse-specific sequences when amplifying from a vectorette library prepared from total YAC DNA. An alternative approach also used in our lab is to sequence random subclones from YAC, bacterial artificial chromosome (BAC) and P1 artificial chromosome (PAC) clones, design PCR primers from these sequences and detect polymorphism between strains using SSCP, as described in *Protocol 1* above.

Protocol 4. Direct selection of microsatellite-containing sequences

Equipment and reagents

- Vectorette library: produced as in ref. 21
- 5′-end biotin-labelled $d(TG)_{20}$ and $d(AG)_{20}$ oligonucleotides (Genosys), 500 ng/ml working dilution
- Vectorette-specific primer (TCGCTAAGAGCATGCTTGCCAATGCTAAGC, from ref. 21)
- Mouse B1 repeat primer (CTGGAACTCACTCTGAAGAC, from ref. 22)
- Streptavidin-coated magnetic beads ('Dynabeads'; Dynal)
- STE buffer: 100 mM NaCl, 10 mM Tris–HCl pH 8.0, 1 mM EDTA

Protocol 4. *Continued*

- STET buffer: STE, as above, with addition of 0.1% (v/v) Tween-20 detergent
- 1 × SSC: 150 mM NaCl, 15 mM trisodium citrate
- Magnetic holder, e.g. Dynal MPC-E (Dynal)
- 2 × SSC
- 0.1 × SSC

A. *PCR amplification and hybridization*

1. Amplify the vectorette library with appropriate primers for your requirements: in this example, we assume the use of vectorette and B1 repeat primers. It may be necessary to determine appropriate amplification conditions empirically, but as a guide, we usually set up a 100 μl reaction and use these cycling parameters:
   (a) 94°C, 1 min;
   (b) 56°C, 1 min;
   (c) 72°C, 4 min;
   (d) repeat steps (a) to (c) for 20–25 cycles.
2. Add each biotinylated oligonucleotide to the finished PCR reactions to give a final concentration of 50 ng/ml, then denature the mixture at 94°C for 10 min.
3. Cool the hybridization to 37°C and incubate at this temperature for 15–30 min.

B. *Hybrid capture using streptavidin beads*

1. Before use, gently mix the Dynabeads by inversion, then transfer 12.5 μl of the bead suspension into a 1.5 ml microcentrifuge tube.
2. 'Capture' the beads using the magnetic holder, then remove the supernatant and wash the beads with 100 μl of STET by rotating the tube in the holder, forcing the beads to move through the liquid. Remove the supernatant and repeat the wash described above.
3. Add Tween-20 to 0.1% (v/v) to the hybridization mixture, then add the mixture to the magnetic beads and incubate at room temperature for 30 min, shaking gently.
4. Capture the beads using the magnet, and dispose of the supernatant.
5. Wash the beads with 1 ml of 2 × SSC at 65°C for 10 min; repeat.
6. Wash the beads with 1 ml of 1 × SSC at 65°C for 10 min; repeat.
7. Wash the beads with 1 ml of 0.1 × SSC at 65°C for 30 min.
8. Elute the hybrid-selected DNA by incubating the beads in 50 μl water or 10 mM Tris–HCl, 0.1 mM EDTA pH 8.0 at 70°C for 10 min. Store the eluted DNA at –20°C.
9. Use 2 μl of eluted DNA in a 100 μl PCR using the conditions described above. Check an aliquot of the reaction product on an agarose gel.

10. Repeat the hybridization, hybrid capture, elution and amplification steps described above with the remainder of the product from step 6. The final reaction product is then cloned in an appropriate vector for sequencing.

As an example: in one experiment, 11 randomly selected clones derived from two rounds of selection with biotinylated microsatellite oligonucleotides performed on a mouse YAC vectorette library were sequenced. Ten of them contained clear microsatellite repeats (P. Denny, unpublished data).

3.4 Criteria for making the decision to construct physical maps

One of the most important decisions in a positional cloning project is the stage at which to initiate physical mapping. There are no specific rules and we can only give general guidelines, based on our experience (23). The relationship between average genetic distance and physical distance in the mouse is about 1 cM = 2 Mb. We advise positional cloners to develop their genetic maps to at least 0.5 cM resolution and if possible, to better than 0.1 cM. This is because the effort involved in chromosome walking with YAC clones, even with an average size of 1 Mb (24), is considerable and local variations in recombination frequency can mean that physical distances are underestimated. The ideal is to have a pair of genetic markers that flank the location of the mutant locus, and other genetic (molecular) markers that do not recombine with the mutant locus. The non-recombinant markers can then be used to begin the process of physical mapping. It is likely that many workers will be using SSLP markers that were developed by the WI-CGR; as mentioned above, about 5100 of these markers have been typed on a large-insert YAC library (6). In some cases, therefore, the first step in physical mapping should be to check this database for YAC contigs containing the markers of interest. Those scientists working on the X chromosome should also consult the information on the UK Mouse Genome Centre web pages (http://www.mgc.har.mrc.ac.uk/xmap/xmap.html) as this provides a comprehensive YAC STS-content map of this chromosome.

4. Physical mapping

4.1 Introduction to physical mapping

In this section, we include protocols dealing with physical maps based on STS content. We do not discuss the myriad of other physical mapping methods, which are reviewed with particular emphasis on the mouse by Herman (26). The accuracy of physical maps based solely on STS content is dependent on depth of clone coverage and on marker density. Linkage of STS markers by single clones is sensitive to the level of chimerism in a clone library, whereas linkage by multiple clones (often referred to as double linkage) increases the likelihood that such linkage faithfully represents the genome. It is desirable, therefore, to use clone libraries with the highest possible depth of coverage (see *Table 3*). It is also desirable to have a high marker density which will increase the likelihood of obtaining a primary clone contig (see Section 4.2). Nevertheless, independent methods of detecting clone overlaps, for example restriction enzyme digest 'fingerprinting' of bacterial clones (25), will usually be needed to validate an STS-content map. The accuracy of STS-content maps can also be increased by the use of markers for which there is positional information, or that are derived from chromosome-specific sources.

The choice of clone type for map construction is necessarily dictated by the needs of the project; as mentioned above, YACs from the WI/MIT820 library (*Table 3*) have been typed for the majority of the MIT SSLP markers and substantial contigs exist across the genome. These contigs, especially where doubly linked, are invaluable in the first stage of physical mapping when determining the approximate size of a minimal genetic interval. We would recommend, however, that BAC and PAC clones be used in subsequent mapping, due to their greater stability, minimal chimerism and ease of handling by comparison with YACs.

4.2 Primary contig generation

A primary contig is the set of clones obtained from screening a clone library(s) with all the available markers in a region of interest. This may form a series of 'islands' rather than a single contiguous unit and so will require one or more rounds of generation of additional markers and re-screening to completely cover the critical region (Section 4.3). Clone screening is typically done either by PCR assay of a hierarchical set of DNA pools representing all the clones in a library or by probe hybridization to high-density arrays of clones on filters. The most likely source of these reagents for many laboratories is either a commercial supplier or academic genome centre (see *Table 3*). It is common for DNA pools to be arranged in a format that does not identify a unique set of clones, but a set of microtitre dishes that contain positive clones. This means that the worker has to perform a large number of assays to identify each positive clone. We would recommend using hybridization assays in preference to PCRs wherever possible to avoid this additional effort. This may necessitate designing novel hybridization probes for certain types of markers containing repetitive sequences, such as SSLP markers. We routinely prepare hybridization probes by designing a pair of partially complementary oligonucleotides from the available sequence (so-called 'overgos' (J. McPherson, personal communication) which, after annealing, can be labelled by polymerization (or 'fill-in') of the two complementary strands. We design pairs of oligonucleotides that are 24–35 nucleotides in length and anneal over 8–10 nucleotides at their 3′ ends, as shown schematically in *Figure 1*.

Table 3. Large insert genomic library resources

Clone type	Vector	Host strain	Mouse strain	Constructed by	Sources for screening and clones[a]	Average insert size (kb)	Genome coverage	Screening methods[b]	Reference
YAC	pRML1 + pRML2	J57D	♀ C57BL/6	WI-CGR	RG, RZPD, HGMP-RC, GS	820	10×	HF/PCR	24
	pYAC4	AB1380	♀ C57BL/6	WI-CGR	RG, RZPD	650	6×	HF/PCR	30
	pYAC4	rad52	♀ C57BL/10	St Mary's Hospital, London,	HGMP-RC, RZPD	240	3×	HF	31
	pYAC4	AB1380	C57BL/6 and C3H	ICRF, London	RZPD	750	4×	HF/PCR	32
	pYAC4	AB1380	♀ C57BL/6	Princeton, NJ,	-	260	2×	-	33
PAC	pPAC4	DH10B	♀ 129/SvevTACfBr	RPCI	RPCI, RZPD	137	12×	HF	unpublished
BAC	pBACe3.6	DH10B	♀ 129/SvEvTACfBr	RPCI	RPCI	154	11×	HF	unpublished
	pBACe3.6	DH10B	♀ C57BL/6J	RPCI	RPCI	197	11×	HF	unpublished
	pBeloBAC11	HS996[c]	CJ7 ES cells (129/Sv)	Caltech/RG	RG	125	9×	HF/PCR	unpublished
	pBeloBAC11	DH10B	RW4 cells (129/SvJ)	GS	GS	120	3×	HF/PCR	unpublished
	pBeloBAC11	DH10B	♂ C57BL/6	GS	GS	120	3×	HF	unpublished
P1	pAD10SacBII	NS3529	♀ C57BL/6	ICRF, London	RZPD	80	3×	HF	unpublished
	pAD10SacBII	NS3529	C127 cells (RIII)	GS	GS	85	2–3×	custom screen	34
	pAD10SacBII	NS3529	E14 cells (129/OLA)	GS	GS	85	3×	custom screen	unpublished

[a]See Table 7 for explanation of abbreviations.
[b] HF, high density filters.
[c] –, No longer widely used.
[d]Modified DH10B/r.

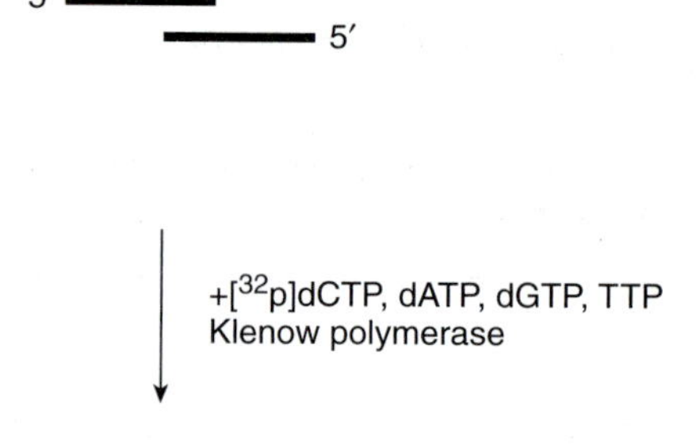

Figure 1. Annealed overlapping oligonucleotides being labelled with [^{32}P]dCTP to form a double-stranded 'overgo' probe.

4.2.1 Clone identification using hybridization assay

Protocol 5. Labelling oligonucleotide fill-in probes ('overgos')

Equipment and reagents

- Klenow fragment DNA polymerase (2 units/μl) (Promega)
- 10 × Klenow buffer (100 mM Tris–HCl pH 8.5, 100 mM $MgCl_2$)
- dNTP mix (1 mM each dATP, dGTP, dTTP) (Pharmacia)
- Sephadex G50 'Nick' columns (Pharmacia)
- [^{32}P]dCTP (3000 Ci/mmol, 10 mCi/ml) (Amersham)
- 'Left' oligonucleotide (10 pmol/μl in 50 mM NaCl)
- 'Right' oligonucleotide (10 pmol/μl in 50 mM NaCl)

Method

1. Mix equal amounts of 'left' and 'right' oligonucleotides and heat to 85°C for 5 min, then transfer to 37°C to anneal for 10 min. After annealing, keep the reaction on ice.
2. For a single fill-in labelling reaction mix together:

   2 μl 10 × Klenow buffer

   2 μl pre-annealed oligonucleotides, from step 1

   11 μl sterile distilled water

   1 μl dNTP mix

   3 μl [^{32}P]dCTP

   1 μl Klenow polymerase

   Incubate at room temperature (18–25°C) for 1 h.

3. Purify the labelled reaction products using Sephadex Nick columns according to the manufacturer's instructions.

We normally design oligonucleotide probes to contain 40–60% GC residues and use a hybridization temperature of 20°C below the T_m of the probe calculated with the formula:

$$T_m = 81.5 + 16.6(\log_{10}[Na^+]) + 0.41(\text{fraction G+C}) - (600/N).$$

Protocol 6. Hybridization of overgos to high-density filter grids

Equipment and reagents

- High-density gridded clone filters for library of interest (BAC/YAC; see *Table 3* for sources)
- Hybridization oven and hybridization bottles (e.g. Hybaid)
- Nylon gauze (Hybaid)
- 2 × SSC
- Hybridization solution: 0.5 M sodium phosphate pH 7.2, 7% SDS, 1 mM EDTA, 1% BSA); note: to achieve the correct Na^+ concentration in this solution, it is essential to make a stock 1 M solution of sodium phosphate as follows: dissolve 134 g $Na_2HPO_4 \cdot 7H_2O$ in 800 ml of water, add 4 ml 85% H_3PO_4 and adjust to 1 litre

Method

1. Pre-wet the filters and nylon gauze with 2 × SSC. Form a sandwich, interleaving the gauze and filters and carefully roll them up and insert them into the hybridization bottle. As many as eight filters can be put into a single bottle. Make sure that the direction of rotation of the bottle in the oven does not cause the filters to roll in on themselves.
2. Pour any excess 2 × SSC away and then add 50 ml of pre-warmed hybridization solution to the bottle.
3. Pre-hybridize the filters at the calculated hybridization temperature (*Protocol 5*) for at least 1 h. Background may be reduced on first use of filters by extending the pre-hybridization time to 4–16 h.
4. Denature the probe by incubation at 95°C for 5 min and chill on ice. Remove about 10 ml of the hybridization solution from the bottle, add the denatured probe, mix and return to the bottle(s).
5. Hybridize the filters overnight at 20°C below the T_m of the probe (as calculated in *Protocol 5*).
6. Pour off the radioactive hybridization solution and dispose safely and appropriately according to your local regulations.
7. Fill the bottles about two-thirds full with 2 × SSC, 0.1% SDS at room temperature, return to the oven and rotate them for 30 min.
8. Dispose of the 2 × SSC wash appropriately, then transfer the filters into a large tray on a shaking platform and wash for 30 min with approximately 1–2 L of 2 × SSC, 0.1% SDS pre-warmed to 37°C.

Protocol 6. *Continued*

9. Monitor the background radiation remaining on the filters: approximately 10–20 counts per second is a reasonable level when measured a few centimetres from a filter. If the level is significantly higher than this, repeat the wash, but raise the temperature of the wash solution to 50 °C. If the background remains high after washing at 50 °C, repeat the wash at 60 °C. Wrap the filters in cling film or seal into plastic bags and expose to autoradiographic film for 16–48 h.

4.3 Chromosome walking/gap filling

After identifying a set of positive clones, the next step in the construction of a map is to cross-check the clones with all available markers to identify overlaps. Given a sufficient density of markers and depth of clone coverage, the clones will form a single contig. If this is not the case, then it will be necessary to 'walk' along the chromosome to join contigs together. The most efficient way of doing this is to produce new markers from clone insert ends and use these to rescreen clone libraries. We do this using two PCR-based methods known as thermal asymmetric interlaced (TAIL)-PCR (27) and vectorette or 'bubble' PCR (21). An alternative approach is to sequence the ends of clone inserts directly. This is not yet routine in our laboratory, but protocols areavailable at the websites of the large sequencing centres, for example ACGT, University of Oklahoma (http://www.genome.ou.edu/proto.html) or GSC (http://genome.wustl.edu/gsc/protocols/protocols.shtml).

4.3.1 Isolating insert end sequences

TAIL-PCR is a modified PCR combining specific priming from a known sequence and non-specific priming from a flanking, unknown sequence (28). It differs from most related procedures in that it requires no manipulation of the template DNA prior to amplification. The basis of the method is the interspersion of cycles of high stringency and low stringency annealing. In high stringency cycles, the specific primer is favoured, whereas during low stringency cycles, the two primers have similar priming efficiency. Non-specific amplification products are reduced substantially by the use of a 'nested' set of two or three specific primers, as shown schematically in *Figure 2*. The most critical part of the method is the design of these specific primers, which should have a T_m at least 10°C higher than that of the non-specific primer. We have designed specific primers for the BAC vector pBeloBAC11 (28) (see *Table 5*) and we have used them to isolate insert end sequences from genomic BAC clones to aid in construction of a sequence-ready map of a region of the mouse X chromosome (P. Denny, S. D. M. Brown *et al.*, manuscript in preparation). We use the same set of non-specific primers described by Liu and Whittier (27), where they are called arbitrary, degenerate (AD) primers (see *Table 6*). The thermocycling conditions described in this protocol were optimized on an MJ Research PTC-225 Tetrad, but should work on similar machines, such as the Perkin–Elmer 9700.

Protocol 7. Thermal asymmetric interlaced (TAIL) PCR

Equipment and reagents

- Thermal cycler, preferably suitable for oil-free reactions (e.g. MJ Research PTC-200/225)
- dNTPs, 2 mM and 0.25 mM stock solutions (Pharmacia)
- 10 × PCR buffer: 100 mM Tris–HCl pH 8.3, 500 mM KCl, 15 mM $MgCl_2$ (Perkin–Elmer)
- *Taq* DNA polymerase (e.g. Perkin–Elmer)
- 1.5 μM and 2 μM working solutions of vector specific primers (for example, see *Table 5*)
- AD primer working solutions (see *Table 6* for concentration)

Method

1. Use the 'calculated' setting on MJ Research thermal cyclers; this uses an algorithm that calculates the temperature of the reaction based on the thermal characteristics of sample and plastic container, e.g. polypropylene or polycarbonate tubes of differing thickness. Similar approaches are used on other thermocyclers.
2. Set up the primary TAIL-PCRs as follows:
   3 μl 10 × buffer
   3 μl outermost vector specific primer (1.5 μM stock) (see *Table 5*)
   3 μl AD primer
   16.6 μl sterile distilled water
   3 μl 2 mM dNTP mix
   1 μl clone DNA (about 1 ng)
   0.4 μl *Taq* polymerase (2 units)
   Note: Set up four reactions for each template, each reaction using a different AD primer.
3. Amplify using conditions for primary TAIL-PCRs shown in *Table 4*.
4. After amplification, take 2 μl of the product and dilute with 150 μl of water. This diluted product is used in the secondary TAIL-PCR.
5. Set up the secondary TAIL-PCRs as follows:
   3 μl 10 × buffer
   3 μl internal specific primer (2 μM stock)
   3 μl AD primer (the same primer as used in primary reaction)
   16.6 μl sterile distilled water
   3 μl 0.25 mM dNTP mix
   1 μl diluted primary TAIL-PCR product
   0.4 μl *Taq* polymerase (2 units)

Protocol 7. *Continued*

6. Amplify using conditions for secondary TAIL-PCRs shown in *Table 4*.
7. After amplification, take 2 μl of the product and dilute it with 150 μl of water. This diluted product is used in the tertiary TAIL-PCR. Reserve 10 μl of undiluted secondary TAIL-PCR product for gel electrophoresis.
8. Set up the tertiary TAIL-PCRs as follows:

   10 μl 10 × buffer

   10 μl innermost specific primer (2 μM stock) (see *Table 5*)

   10 μl AD primer (same primer as used in primary reaction)

   49.2 μl sterile distilled water

   10 μl dNTP 0.25 mM mix

   10 μl diluted secondary TAIL-PCR product

   0.8 μl *Taq* polymerase (4 units)
9. Amplify using conditions for tertiary TAIL-PCRs shown in *Table 4*.
10. Run 10 μl aliquots of secondary and tertiary TAIL-PCR products on a 1.5% agarose gel to check product size and likely specificity.
11. Sequence undiluted tertiary products amplified from pBeloBAC11 using the ABI Prism BigDye M13 universal primer kit according to the manufacturer's instructions.

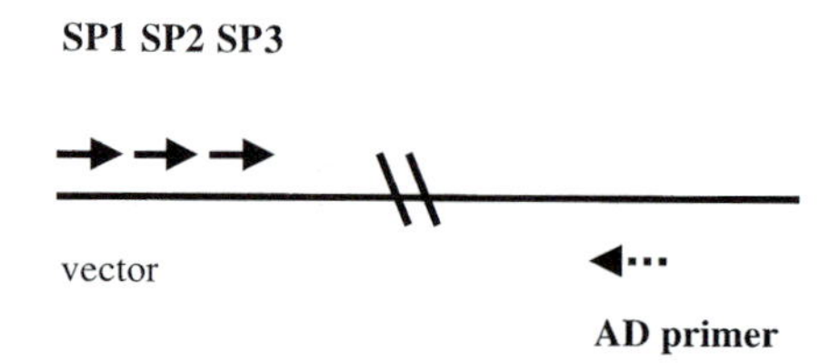

Figure 2. TAIL-PCR schematic, showing an imaginary template containing annealing sites for the nested, vector specific primers, SP1, SP2 and SP3 and for an arbitrary degenerate (AD) primer.

4.4 Validation of clone contigs

The last stage in construction of a physical map based on STS content should be its validation by a complementary method such as restriction digest 'fingerprinting' of the clones (25). This is necessary because of the possibility of artefactual clones being produced during either library construction or clone propagation. These fingerprinting methods also allow estimation of the degree

Table 4. Thermocycling conditions for TAIL-PCR

Reaction	No. of cycles	Thermal conditions
Primary (TAIL-1)	1	92°C (3 min)
	5	94°C (5 sec), 53°C (5 sec), 72°C (2 min)
	1	94°C (5 sec), 30°C (20 sec), ramping to 72°C at 0.3°C/sec, 72°C (2 min)
	10	94°C (5 sec), 44°C (20 sec), 72°C (2 min)
	16	94°C (5 sec), 53°C (5 sec), 72°C (2 min) 94°C (5 sec), 53°C (5 sec), 72°C (2 min) 94°C (5 sec), 44°C (20 sec), 72°C (2 min)
	1	72°C (5 min)
Secondary (TAIL-2)	1	92°C (3 min)
	17	94°C (5 sec), 53°C (5 sec), 72°C (2 min) 94°C (5 sec), 53°C (5 sec), 72°C (2 min) 94°C (5 sec), 44°C (20 sec), 72°C (2 min)
	1	72°C (5 min)
Tertiary (TAIL-3)	20–35	94°C (5 sec), 44°C (20 sec), 72°C (2 min)
	1	72°C (5 min)

Table 5. Vector-specific primers for pBeloBAC11

'Left' primers		'Right' primers	
primer	sequence	primer	sequence
L1	GCTACATCATTAGGCACCC	R1	GATGTGCTGCAAGGCGAT
L2	ATGCTTCCGGCTCGTATGTT	R2	AAGTTGGGTAACGCCAGGG
L3	CTTCCGGCTCGTATGTTGTG	R3	TTTTCCCAGTCACGACGT

Note that primers L1, L2 and L3 and R1, R2, and R3 anneal to sequences progressively closer to the BAC vector cloning site. This means that L1 or R1 should be used in primary TAIL-PCR reactions, L2 or R2 in secondary TAIL-PCR reactions and L3 or R3 in tertiary TAIL-PCR reactions.

Table 6. Arbitrary degenerate primers

Primer	Sequence[a]	Concentration (μM) of stock solution		
		1° PCR	2° PCR	3° PCR
AD1	TG(A/T)GNAG(A/T)ANCA(G/C)AGA	50	30	30
AD2	AG(A/T)GNAG(A/T)ANCA(A/T)AGG	50	30	30
AD3	CA(A/T)CGICNGAIA(G/C)GAA	25	15	15
AD4	TC(G/C)TICGNACIT(A/T)GGA	25	15	15

[a] I = inosine.

Table 7. Website URLs for suppliers of physical mapping resources

Supplier	Abbreviation	Website URL
Research Genetics	RG	http://www.resgen.com
German Genome Project Resource Centre	RZPD	http://www.rzpd.de
MRC UK Human Genome Mapping Project-Resource Centre	HGMP-RC	http://www.hgmp.mrc.ac.uk
Genome Systems	GS	http://www.genomesystems.com/
Roswell Park Cancer Institute	RPCI	http://bacpac.med.buffalo.edu/
Imperial Cancer Research Fund	ICRF	http://www.rzpd.de

overlap of clone inserts. A fluorescent labelling version of this protocol was recently published (29).

Acknowledgements

We would like to thank Gonzalo Blanco for his work in optimizing the TAIL-PCR method and designing primers for the pBeloBAC11 vector, Elaine Hopes for advice on fluorescent genotyping, Paul Lyons for improvements to the microsatellite selection method and Rachael Bate for advice on the oligonucleotide fill-in method. We would also like to thank workers at the MRC UK Mouse Genome Centre for their comments and support.

References

1. Rowe, L. B., Nadeau, J. H., Turner, R., Frankel, W. N., Letts, V. A., Eppig, J. T., Ko, M. S., Thurston, S. J. and Birkenmeier, E. H. (1994). *Mamm. Genome*, **5**, 253–274.
2. Breen, M., Deakin, L., Macdonald, B., Miller, S., Sibson, R., Tarttelin, E., Avner, P., Bourgade, F., Guenet, J. L., Montagutelli, X. *et al.* (1994). *Hum. Mol. Genet.*, **3**, 621–627.
3. Copeland, N. G. (http://nmrweb.ncifcrf.gov/abl/mgl/index.html). .
4. Kozak, C. A. (http://www.niaid.nih.gov/research/labs/lmm5.htm). .
5. Seldin, M. F. (http://trc.ucdavis.edu/Departments/BiolChem/Seldin.htm). .
6. WI-CGR. http://www-genome.wi.mit.edu/genome_data/mouse/mouse_index.html. (MIT/Whitehead Institute Centre for Genome Research).
7. Rhodes, M., Straw, R., Fernando, S., Evans, A., Lacey, T., Dearlove, A., Greystrong, J., Walker, J., Watson, P., Weston, P. *et al.* (1998). *Genome Res.*, **8**, 531–542.
8. Collins, F. S. (1995). *Nature Genet.*, **9**, 347–350.
9. Carver, E. A. and Stubbs, L. (1997). *Genome Res.*, **7**, 1123–1137.
10. McGinnis, W. and Krumlauf, R. (1992). *Cell*, **68**, 283.
11. Pieler, T. and Bellefroid, E. (1994). *Mol. Biol. Rep.*, **20**, 1.
12. Ballabio, A., Brown, S. and Fisher, E. (eds) (1998). In *Genome Analysis: A Laboratory Manual*, Vol 2 (eds Birren, B., Green, E., Klapholz, S., Myers, R. & Roskams, J.). Cold Spring Harbor Laboratory Press, NY.

13. Orita, M., Iwahana, H., Kanazawa, H., Hayashi, K & Sekiya, T. (1989). *Proc. Natl Acad. Sci. USA*, **86**, 2766–2770.
14. Michaelides, K., Schwaab, R., Lalloz, M., Schmidt, W. & Tuddenham, E. (1995). In *PCR2: A Practical Approach* (eds. McPherson, M. J., Hames, B. D. and Taylor, G. R.), pp. 255–288. IRL Press.
15. Rhodes, M., Dearlove, A., Straw, R., Fernando, S., Evans, A., Greener, M., Lacey, T., Kelly, M., Gibson, K., Brown, S. D. *et al.* (1997). *Genome Res.*, **7**, 81–86.
16. Reed, P. W., Davies, J. L., Copeman, J. B., Bennett, S. T., Palmer, S. M., Pritchard, L. E., Gough, S. C., Kawaguchi, Y., Cordell, H. J., Balfour, K. M. (1994). *Nature Genet.*, **7**, 390–395.
17. Kidd, K. and Ruano, G. (1995). In *PCR2: A Practical Approach* (eds. McPherson, M. J., Hames, B. D. and Taylor, G. R.), pp. 1–38. Oxford University Press.
18. Denny, P., Lord, C. J., Hill, N. J., Goy, J. V., Levy, E. R., Podolin, P. L., Peterson, L. B., Wicker, L. S., Todd, J. A. & Lyons, P. A. (1997). *Diabetes*, **46**, 695–700.
19. Korn, B., Sedlacek, Z., Manca, A., Kioschis, P., Konecki, D., Lehrach, H. & Poustka, A. (1992). *Hum. Mol. Genet.*, **1**, 235–242.
20. Karagyozov, L., Kalcheva, I. D. and Chapman, V. M. (1993). *Nucleic Acids Res.*, **21**, 3911–3912.
21. Morrison, J. and Markham, A. (1995). In *PCR2: A Practical Approach* (eds. McPherson, M. J., Hames, B. D. and Taylor, G. R.), pp. 165–196. Oxford University Press.
22. Simmler, M.C., Cox, R. D. and Avner, P. (1991). *Genomics*, **10**, 770–778.
23. Gibson, F., Walsh, J., Mburu, P., Varela, A., Brown, K. A., Antonio, M., Beisel, K. W., Steel, K. P. & Brown, S. D. (1995). *Nature*, **374**, 62–64.
24. Haldi, M. L., Strickland, C., Lim, P., VanBerkel, V., Chen, X., Noya, D., Korenberg, J. R., Husain, Z., Miller, J. & Lander, E. S. (1996). *Mammal. Genome*, **7**, 767–769.
25. Gregory, S. G., Soderlund, C. and Coulson, A. (1997). In *Genome Mapping: A Practical Approach* (ed. Dear, P.), p. 396. Oxford University Press.
26. Herman, G. E. (1998). *Methods*, **14**, 135–151.
27. Liu, Y.G. and Whittier, R. F. (1995). *Genomics*, **25**, 674–681.
28. Kim, U. J., Birren, B. W., Slepak, T., Mancino, V., Boysen, C., Kang, H. L., Simon, M. I. & Shizuya, H. (1996). *Genomics*, **34**, 213–218.
29. Gregory, S. G., Howell, G. R. and Bentley, D. R. (1997). *Genome Res.*, **7**, 1162–1168.
30. Kusumi, K., Smith, J., Segre, J., Koos, D. & Lander, E. (1993). *Mammal. Genome*, **4**, 391–392.
31. Chartier, F. L., Keer, J. T., Sutcliffe, M. J., Henriques, D. A., Mileham, P. & Brown, S. D. (1992). *Nature Genet.*, **1**, 132–136.
32. Larin, Z., Monaco, A. P. and Lehrach, H. (1991). *Proc. Natl Acad. Sci. USA*, **88**, 4123–4127.
33. Burke, D. (1991). *Mammal. Genome*, **1**, 65.
34. Pierce, J. C., Sternberg, N. and Sauer, B. (1992). Mammal. Genome, 3, 550–558.

6

Mouse cytogenetics and FISH

1. Introduction

There are several situations in which it may be helpful to look directly at mouse chromosomes. One may wish to screen a specific living mouse for possible chromosome rearrangements, or to determine precise cytogenetic breakpoints without compromising the ability of the mouse to breed. One may need to know the approximate chromosomal location of a newly cloned mouse gene, or the site of insertion of a foreign transgene. Occasionally it may be helpful to have a quick screen for whether an animal is homozygous or heterozygous for a foreign transgene. This first question is most likely to be asked in a laboratory skilled in cytogenetics and answered by blood culture and G-banding, bearing in mind that methods used routinely in human cytogenetics frequently cannot be used directly for the mouse. Some of the other questions are now often asked by molecular biologists whose expertise is in DNA hybridization rather than visual pattern recognition of banded chromosomes. For some of these studies fluorescent *in situ* hybridization (FISH) can provide an increasing range of useful tools.

This chapter is divided into two sections corresponding broadly to the needs and expertise of these two communities, although of course in practice there is much overlap. Where small details of procedure differ between the two sections, this is a reflection of actual working practice in the two different laboratories.

6A

Analysing mouse chromosomal rearrangements with G-banded chromosomes

ELLEN C. AKESON and MURIEL T. DAVISSON

1. Introduction

This section describes the preparation of metaphase chromosomes from mouse peripheral blood lymphocytes for chromosomal typing of living mice. Included is a list of mouse strains and their relative response to phytohaemagglutinin (PHA), a list of the requirements necessary for the production of well-spread metaphase chromosomes with good morphology for banding, a Giemsa banding (G-band) method for these chromosomes, and their classification into a karyotype. The identification of mouse chromosomes is presented with a detailed process for determining aberrant chromosomes and describing the common pitfalls of misclassification. A precise description of the characteristic banding patterns (landmarks) at different stages of chromosome condensation is presented elsewhere (1). The cell-harvesting part of the procedure also can be used to prepare mitotic chromosomes directly from fetal liver or bone marrow.

2. Metaphase chromosomes

The preparation of mitotic chromosomes from mouse peripheral blood involves the following steps (2,3):

(a) Cells are cultured with a mitogen to stimulate lymphocyte proliferation. These mitogens include the T-cell stimulators PHA and concanavalin A (Con-A) and the B-and T-cell stimulators lipopolysaccharide (LPS) and pokeweed mitogen (4,5). Con-A, PHA and pokeweed mitogen are lectins, agglutinating proteins that have been extracted from plant seeds.

(b) Cells are arrested in metaphase by disrupting the formation of spindle fibres with colchicine (50 μg/ml) or colcemid (0.03 μg/ml).

(c) Cells and nuclei are swelled by treatment with hypotonic KCl (0.075 M); this also improves the spreading of chromosomes and allows G-banding.

(d) Cells are fixed in Carnoy's fixative (3:1 methanol:glacial acetic acid) to preserve the chromosome structure and enhance Giemsa staining.

(e) The cell suspension is dropped onto a clean slide and rapidly dried to obtain flattened and well-spread metaphase chromosomes.

Protocol 1. Mitotic chromosomes from mouse peripheral blood

Equipment and reagents

- RPMI 1640 medium containing glutamine and Hepes (Gibco–BRL); supplement each 100 ml of medium with 1 ml of gentamicin solution (10 mg/ml, Sigma)
- PHA (Murex); for optimum results construct a dose–response curve with each new lot of PHA to assay the mitotic index and strain response. We have found that 7.5 μg/ml of purified PHA with mitogenic units in the same range gives the best results regardless of the mitogenic units marked on the vial or the strain used.
- 750 μg/ml LPS (Sigma, from *Escherichia coli* serotype 0111:B4)
- 500 USP units/ml of sterile sodium heparin (Sigma), diluted in RPMI 1640 medium)
- fetal calf serum (FCS; Sigma)
- 50 μg/ml colchicine (Sigma) in saline
- 0.56% (0.075 M) KCl
- 70% ethanol
- Carnoy's fixative (3:1 methanol:glacial acetic acid)
- Pressurized air (e.g. Dust Off Plus, Polysciences, or from most photographic suppliers)
- Cleaned microscope slides
- Razor blade
- Haematocrit tubes
- Snap-cap tubes
- Conical glass centrifuge tubes
- Bench-top centrifuge
- 16 × 125 mm Corning culture tubes

A. *Culture procedure*

1. Use aseptic technique throughout until the harvest procedure (part C).
2. For each mouse, set up one 16 × 125 mm Corning culture tube containing:
   - 0.95 ml of RPMI 1640 medium with gentamicin
   - 0.1 ml of PHA (concentration as determined by dose–response curve, see Section 3)
   - 0.1 ml of 750 μg/ml LPS
   - 0.15 ml of FCS
3. Refrigerate the tubes until they are needed.
4. Obtain two tubes of blood per mouse from the retro-orbital sinus using heparinized 70 μl micro-haematocrit capillary tubes. Those unskilled in this procedure may obtain blood from the tail vein (see *Protocol 1B*). In the UK, it is easier to obtain a Home Office licence for the tail-vein method.
5. Inoculate the two 70 μl haematocrit tubes of blood immediately into a 12 × 75 mm Fisher snap-cap tube containing 0.1 ml of heparin.

Protocol 1. *Continued*

6. Cap the tubes tightly and swirl gently to mix the blood and heparin. Blood should be held at room temperature (for not more than 2 h) until inoculation into the culture medium.
7. Add 0.2 ml of blood and heparin mixture to each prepared culture tube. Cap the tubes tightly and swirl gently to mix the cells and medium.
8. Incubate the culture tubes at an angle of ~45° angle in a shaking water bath at 37 °C. Set the shaking speed so that the basket moves back and forth about 32–35 times/minute. If an incubator is used to incubate the cultures, then tilt the rack of tubes at an angle by resting it on two 1/2 to 5/8 inch high Petri dishes to allow maximum surface area exposure. Hand-shake the tubes three times daily to resuspend the cells.

B. *Collecting blood by the tail-vein method*

1. Pre-warm the mouse in a large jar placed beneath a desk lamp for 1–2 min. The mouse is warm enough when it rubs its nose or shows excessive activity. **Caution:** overheated mice can go into shock and die.
2. Place the mouse in a restraint so that the tail is free. Wash the tail with 70% ethanol and wipe it twice with a clean tissue.
3. Gently cut across the vein on the side of the tail about one inch from the base of the tail with a razor blade wiped with 70% ethanol.
4. Collect at least 0.15 ml of blood (five drops), letting it run down the side of the tube containing 0.1 ml of heparin. Be careful not to touch the tail to the mouth of the tube.
5. Follow *Protocol 1A*, steps 6–8.

C. *Harvest procedure*

1. At 41–43 h add 0.15 ml of colchicine solution (50 μg/ml) to each culture tube and incubate the tubes at 37 °C for a further 15–20 min.
2. Transfer each culture to a 5 ml conical glass centrifuge tube and centrifuge at 400*g* for 10 min at room temperature in a clinical bench-top centrifuge.
3. Remove the supernatant and gently add 2–3 ml of warm (37 °C) 0.075 M (0.56%) KCl. Gently suspend the cells by pipetting with a Pasteur pipette.
4. Incubate the tubes at 37 °C for 15 min.
5. Centrifuge at 500*g* for 10 min.
6. Remove the supernatant without disturbing the pellet, being careful not to remove the buffy coat (white layer) on top of the pellet. Gently

add 3–4 ml of Carnoy's fixative down the side of the tube. Pipette gently but rapidly to prevent the cells from clumping.

7. Stopper the tubes and allow them to sit for at least 30 min at room temperature. (Refrigerate if tubes are to be held for >30 min.) The procedure can be interrupted at this point for 1–2 h.
8. After 30 min, centrifuge the cells at 400*g*, remove the fixative and resuspend the cells in fresh fixative.
9. Repeat step 8 twice more.
10. Centrifuge the cells at 400*g* and resuspend the cells in about 0.3–0.5 ml of fresh fixative. Slides are prepared from this cell suspension by placing the entire sample on one slide. If slides are not made immediately, save this last wash until just before making slides. Slides can be made the next day but wash the cells at least twice before leaving overnight.

D. *Slide making*

1. Immerse pre-cleaned slides in fixative at least 15 min prior to use. Wipe slides dry with a lint-free tissue.
2. Drop small drops of cell suspension onto the slide surface and allow it to spread. If too much sample has been used, it will bubble at the edges.
3. As soon as the drop begins to contract and Newton's rings (rainbow colours around the edge of the drop) are visible, blow on the slide surface to accelerate drying. Slides may also be dried by using tubing connected to an air supply or by lightly spraying with pressurized air. Rapid drying is critical to obtaining well-spread metaphases.
4. The 'bomb' method of dropping one drop of fixative onto the cell suspension once it has started to dry may be used to increase the spreading of the chromosomes. Use a very small drop of fixative and keep the slide in a flat position while adding the fixative.
5. Repeat steps 2 and 3 until all of the sample is on the slide. Monitor cell concentration by phase microscopy.

The length of culture depends on the strain of mice used and is usually 41–43 h. A large number of blast cells seen on the slide indicates that the cultures were harvested too soon. If mostly late metaphases are seen, it indicates that the cultures were harvested too late. In general, females tend to provide better cultures than males because they have a better response to PHA. The optimum age is 7–10 weeks. Cells from mice as young as 5 weeks to as old as 2 years have been cultured but the success rate and the number of metaphases per slide are usually greatly reduced in such cases. LPS is used in conjunction with PHA to boost metaphase numbers in older mice.

3. Stimulation of peripheral mouse lymphyocytes by phytohaemagglutinin and lipopolysaccharide

3.1 Phytohaemagglutinin

According to Heiniger *et al.* (6), the response of lymphocytes to PHA is governed by up to five major genes and there is considerable variation in PHA response among mouse strains. Because of this inter-strain variability, a dose–response curve should be determined for each strain to be cultured. Pool blood from each strain and prepare two tubes for each concentration. The recommended dose–response curve range is between 3 and 9 μg/ml. Concentrations of <3 μg/ml appear not to stimulate proliferation and show a low mitotic index, whereas concentrations of >9 μg/ml appear toxic with a reduced mitotic index. Mice of the C57BL/6 strain are recommended as a control because of their good response to PHA.

The DBA family of mice (DBA/1J, DBA/2J) and DBA hybrid mice are the poorest responders. *Table 1* (6) lists some common inbred strains according to their response to PHA stimulation.

Table 1. Strain response to PHA stimulation of lymphocyte proliferation, adapted from ref. 6

Poor	Intermediate	Good
DBA/1J	A/J	C57BL/10Sn
DBA/2J	LP/J	WC/Re
LG/J	A/HeJ	WH/Re
SEA/GnJ	CBA/J	C57BL/6J
SM/J		BALB/cJ
C3H/HeJ		129/J
P/J		PRO/Re
SWR/J		C57BR/cdJ
CBA/CaJ		BUB/BnJ
I/LnJ		RIII/J
SEC/ReJ		WB/Re
NZB/BINJ		CE/J
SJL/J		C57L/J
LT/ChRe		PL/J
CBA/H-T6J		AKR/J
RF/J		C58/J
MA/J		C57BL/10J
ST/bJ		C57BLKS/J
Au/SsJ		HRS/J
BDP/J		
CE/J		

3.2 Lipopolysaccharide

LPS is extracted from the cell wall of certain Gram-negative bacteria. It is used in the preparation of mitotic chromosomes to enhance the metaphase yield of poor responders to PHA. Whereas PHA stimulates the proliferation of T-lymphocytes only, LPS stimulates both B- and T-lymphocytes. Davisson and Akeson (2) have found that the addition of LPS to the culture medium containing a T-cell stimulator (PHA) improves the metaphase yield from older mice or low responders to PHA. We routinely add both PHA and LPS to our culture medium.

4. G-bands

The acetic-saline-Giemsa (ASG) and ASG/trypsin techniques are the most commonly used G-band methods. These techniques involve relatively mild conditions that remove only about 9% of the chromosomal DNA and produce characteristic banding patterns in mouse chromosomes that are distinctive for each telocentric chromosome (7). A band is a part of the chromosome that is clearly distinguishable from its adjacent segments. G-positive bands can be described as pale, medium or heavy depending on staining intensity. These contain late replicating DNA, which is rich in A and T and appears to contain relatively few genes (7). These positively stained bands are separated by unstained regions of the chromosome (G-negative bands), which are early replicating, contain G/C-rich DNA, and appear to contain most of the house-keeping genes. Important in chromosome identification is the recognition of invariant banding patterns (or primary and secondary landmarks) that are characteristic features of each chromosome. Except for band A1 (the centromeric heterochromatin), the G-banded pattern for each chromosome is consistent among all laboratory mouse strains (1). The mouse karyotype contains 19 autosomal pairs (with chromosome 1 being the largest and chromosome 19 the smallest) and two sex chromosomes. The centromeric heterochromatin (C-band) in all the autosomes is darkly stained and ranges in size from very small (or absent) to extra-large. There is considerable polymorphism in C-band size among the chromosomes and among different subspecies; the Y chromosome lacks a C-band and usually appears stained uniformly grey. Often the inactive X chromosome in female mice stains uniformly and more darkly than the active X and can be distinguished by this property.

The Nesbitt and Francke (8) nomenclature system discriminates between major and minor bands using the following system:

(a) Identify the chromosome by number.
(b) Label the major bands (or regions) starting at the centromeric end with a capital letter, starting with A. (The centromeric heterochromatin band is designated A1).

(c) Subdivide the major bands into minor bands and give these minor bands a number within the lettered group.

Protocol 2. G-banding

Equipment and reagents

- 2 × SSC (8.8 g NaCl, 4.4 g trisodium citrate, in 500 ml distilled water)
- 0.9% NaCl
- Gurr buffer (Gurr pH 6.8 phosphate buffer plus an equal volume of distilled water)
- Coplin staining jars
- Trypsin–Giemsa solution (1.0 ml Gurr's Improved Giemsa R66 (Bio/medical Specialities), 45 ml Gurr pH 6.8 phosphate buffer (Bio/medical Specialities), four drops of 0.0125% trypsin (Sigma)

Method

1. Make air-dried mitotic preparations by dropping small droplets of cell suspension onto slides and blowing dry (Protocol 1). Bands are sharper if slides are aged 7–10 days at room temperature, but this is not essential.
2. Incubate the slides in Coplin jars (five or six per jar) in 2 × SSC at 60–65°C for 1.5 h.
3. Transfer all slides to 0.9% NaCl at room temperature, then rinse each slide in fresh NaCl and drain. Thorough rinsing is critical.
4. Stain slides for 5–7 min in the trypsin–Giemsa solution. Remove the metallic film which forms on the stain surface with a cotton ball before placing the slides in the Coplin jar or float the film off with running water before removing the slides.
5. Transfer all the slides in the jar to fresh phosphate buffer.
6. Rinse slides individually in two changes of buffer. Thorough rinsing is critical. Shake off excess liquid and blow dry with an air jet (see *Protocol 1*, step D3).

There are some extrinsic factors involved in the preparation and staining of the metaphase chromosomes that greatly affect the quality of the results. Freshly prepared slides tend to band poorly so it is best to age the slides to dehydrate and 'harden' the chromosomes. The length of time for which slides are incubated in 2 × SSC is critical: over-treatment can lead to C-banding only and loss of G-bands. The exposure time to trypsin is also critical; under-exposure gives poor band resolution, whereas over-exposure destroys chromosome morphology giving a fuzzy appearance to the chromosome. Some factors that affect chromosome sensitivity to trypsin include:

- chromosome length: contracted chromosomes are more sensitive than elongated ones.

- chromosome dryness: recently made slides are more sensitive than aged slides.
- chromosome fixation time: softly fixed chromosomes are more sensitive than 'hard' chromosomes (those that appear very dark under phase microscopy).

5. Identifying and karyotyping mouse chromosomes

The G-banded karyotype is useful for the identification of absent, additional or aberrant chromosomes (involving deletions, inversions, reciprocal translocations, Robertsonian chromosomes, duplications and insertions), and for the linear placement of loci on the chromosome following *in situ* hybridization. The karyotyping process consists of four distinct steps (9):

(a) The homologous chromosomes are paired together based on their similar banding pattern.

(b) Then the sex chromosomes are identified. A female karyotype will contain two large X chromosomes, one of which may be very darkly stained (the inactive X chromosome). A male karyotype will contain one large X chromosome and one very small uniformly stained Y chromosome.

(c) The paired chromosomes are arranged in descending order by size, dividing the chromosomes into loosely organized groups.

(d) Individual chromosomes are classified on the basis of their unique banding patterns. This process can be made easier with the help of a known standard karyogram with which to compare the unknown chromosomes.

The classification of mouse chromosomes is more difficult than that of human chromosomes. As mouse chromosomes are all telocentric and show less variation in length than human chromosomes, they cannot be initially classified by size and centromere position (as human chromosomes can). Thus, identification of individual mouse chromosomes is based primarily on recognition of the characteristic G-band pattern of that particular chromosome. The following chromosomes have similar banding patterns that can lead to their misidentification: chromosomes 1 and X; chromosomes 1 and 6; chromosomes 5 and 7; chromosomes 8 and 12; chromosomes 9 and 13; and chromosomes 15 and 18. The variability of chromosome condensation and staining between metaphase spreads and the similar G-band pattern of some mouse chromosomes also contributes to their frequent misidentification. Cowell's article (1), which shows a representation of each chromosome at several stages of contraction, is a good photographic guide for identifying mouse chromosomes. *Figure 1* shows a standard mouse G-band karyogram.

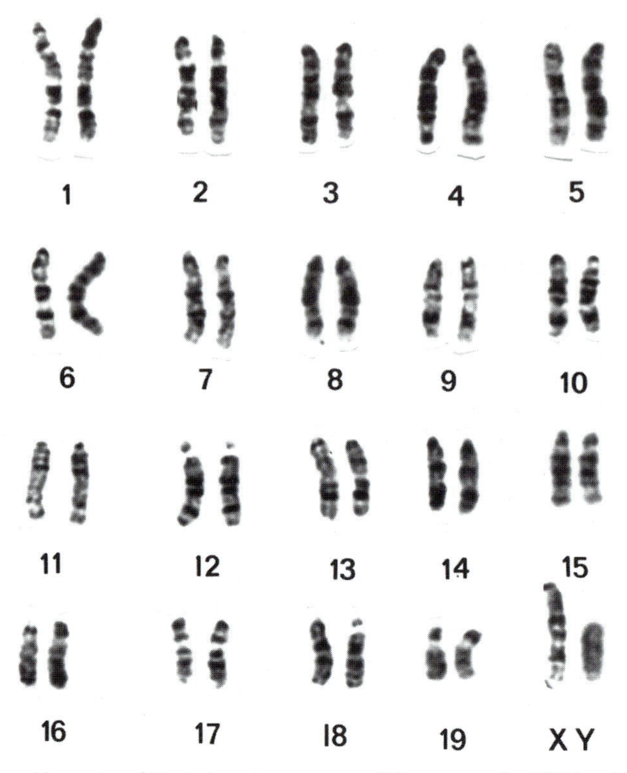

Figure 1. Normal karyote of the laboratory mouse (*Mus musculus*) ($2N = 40$).

6. Recognizing aberrant chromosomes

To identify aberrant chromosomes in a mouse karyotype, the first step is to count the number of chromosomes in several metaphase spreads. Metaphase spreads from diploid cells should contain 40 chromosomes (19 pairs of autosomes and the sex chromosomes). In some instances a normal spread will contain fewer than 40 chromosomes due to chromosome scatter following nuclear disruption. Long-term cell cultures often show great fluctuation in chromosome number. Robertsonian translocations, which are produced by the centromeric fusion of two chromosomes, are easily recognized as X-shaped metacentrics among the background of V-shaped telocentrics. A true metacentric will be present in every metaphase spread from a cell harbouring a Robertsonian translocation. Sometimes overlapping chromosomes may resemble a metacentric but will not be seen in every spread. At least ten metaphase spreads with well-banded chromosomes should be examined. Changes in banding patterns are more difficult to discern. The human eye detects altered patterns rather than changes in individual bands. Thus, translocations that do not significantly alter the banding pattern of the chromosomes involved are less discernible. Small duplications or deletions are not detectable cytologically unless a G-positive band is missing or the banding pattern is altered. Confirmation of translocations is now made easier by the use of chromosome paints and FISH (see Chapter 6B)

Acknowledgement

We thank Cecilia Schmidt for providing technical support.

References

1. Cowell, J. C. (1984). *Chromosoma*, **89**, 294.
2. Davisson, M. T. and Akeson, E. C. (1987). *Cytogenet. Cell Genet.*, **45**, 70.
3. Evans, E. P. (1987). In *Mammalian Development: a Practical Approach* (ed. Monk, M.), p. 93. Oxford University Press.
4. Stobo, J. D., Rosenthal, A. S. and Paul, W. E. (1972). *J. Immunol.*, **108**, 117.
5. Andersson, L. and Hayry, P. (1972). *Experientia*, **28**, 81
6. Heiniger, H., Taylor, B. A., Hards, E. J. and Meier, H. (1975). *Cancer Res.*, **35**, 825.
7. Comings, D. E., Avelino, E., Okada, T. A. and Wyandt, H. E. (1973). *Exp. Cell Res.*, **77**, 469.
8. Nesbitt, M. N. and Francke, U. (1973). *Chromosoma*, **41**, 145.
9. Lee, J. J., Warburton, D. and Robertson, E. J. (1990). *Anal. Biochem.*, **189**, 1.

6B

Fluorescent *in situ* hybridization (FISH) to mouse chromosomes

MARGARET FOX and SUE POVEY

1. Introduction

The technique of fluorescent *in situ* hybridization (FISH) has added a new dimension to human and mouse karyotyping; its applications are numerous and varied. This section covers its use at the chromosome level in the mouse. Techniques are described for the purpose of placing genes or cloned DNA of interest onto chromosomes. Similar techniques are useful for detecting the integration site in transgenics using the transgene insert as a FISH probe. The insert is detectable in interphase nuclei as well as on metaphase chromosomes.

The first part of this section contains a general overview of the tools available for identification of individual mouse chromosomes by FISH. The first detailed protocol is for obtaining chromosomes from mouse spleen by short-term culture, which in our hands is easier than obtaining chromosomes from peripheral blood, and is convenient if the particular mouse can be killed. This is followed by some detailed protocols for labelling probes and *in situ* experiments. We also have some comments about the quality and quantity of DNA supplied as a probe as we have found this to be critical.

2. FISH and chromosome identification

2.1 Repeat DNA

Species-specific repeat sequences have been harnessed to identify mouse chromosomes and to co-localize genes by FISH. Interspersed repeats in the mouse are of two types, namely long and short interspersed sequences (LINES and SINES). The location of LINES (e.g. L1) and SINES in the genome has been shown to correspond to AT- and GC-rich areas respectively. These areas in turn correspond with late replicating, Giemsa-positive and early replicating R-band areas. It is the L1 repeat which, when used as probe, gives a banding pattern and, when co-hybridized with a gene probe, enables localization of the unknown probe sequence on a chromosome (1,2).

The fluorochrome 4′,6-diamidino-2-phenylindole (DAPI) stains chromatin and gives banding similar to G banding on chromosome arms. When used as a counterstain in FISH, it can aid the identification of chromosomes.

2.2 Mouse chromosome paints

Chromosome paints are complex mixtures of unique DNA sequences that derive from the entire length of the chromosome. In spite of the difficulties predicted for the generation of mouse paints by fluorescence-activated cell sorting (FACS) of mouse chromosomes because of their similar sizes, this has been achieved with relative ease (3). When isolated chromosomes were subjected to high-resolution flow sorting, most resolved into separate peaks. The DNA from these peaks was amplified by PCR, and the chromosomes identified by PCR and FISH. Some peaks were unresolved and comprised two pairs of homologues.These were further resolved by using Robertsonian translocations in mice homozygous for the translocation. By flow sorting the Robertsonian chromosomes it was possible to separate them. Strain differences in chromosome size were also used. In this fashion it has been possible to obtain DNA paints from all mouse chromosomes, thus opening up the way to the identification of individual chromosomes using FISH. The paints are commercially available (see Section 4.3).

2.3 P1 probes

Most recently a resource of P1 probes for each chromosome, both centromeric and telomeric, has been developed (4). For each chromosome clones were selected which mapped to a centromeric and a telomeric position and could therefore aid in mapping in a similar way to chromosome paints. In addition each P1 probe contains a PCR polymorphism which makes it possible to link the physical and genetic maps. The probes are obtainable from Genome Systems.

3. Obtaining mouse chromosomes

Cultures of mouse spleen have proved to be a convenient method for obtaining normal mouse chromosomes. The key ingredient is lipopolysaccharide (LPS), which gives a high rate of mitoses. This is the method we routinely use for examining the karyotype of a mouse strain or, in the case of a transgene, its particular insert location in the genome. We have used direct preparations from 10–11 day whole embryos to obtain interphases as well as metaphases for genotyping in transgenics. Tissue culture lines from embryonic tissue have also been used for obtaining mouse chromosomes for FISH (4). The mouse spleen culture method is given in *Protocol 1* and subsequent harvesting of the cultures is described in *Protocol 2*.

Protocol 1. Mouse spleen culture

Equipment and reagents

- RPMI 1640 medium buffered with 25 mM HEPES (obtained from Gibco-BRL)
- RPMI 1640 medium buffered with $NaHCO_3$ (Gibco-BRL)
- Glutamine, penicillin and streptomycin (GPS), 100x solution (Gibco-BRL)
- Fetal calf serum (FCS), inactivated by heating to 56°C for 30 min (it is also obtainable ready-inactivated)
- Lipopolysaccharide (LPS), 1 mg (Sigma)
- Scalpel
- Mouse spleen freshly dissected from adult mouse
- Cell strainer, pore size 70 μm (Falcon)
- Petri dishes (Falcon)
- 25 cm^2 tissue culture flasks (Falcon)
- 50 ml centrifuge tubes
- Bench-top centrifuge
- Haemocytometer
- CO_2 incubator

A. *Preparation of media*

1. Prepare the collection medium: to a 100 ml bottle of Hepes-buffered RPMI 1640, add 1 ml of GPS and 2 ml of heat-inactivated FCS.
2. Prepare the growth medium:
   (a) To 100 ml of $NaHCO_3$-buffered RPMI 1640 add 1 ml of GPS. Take 36 ml of this solution.
   (b) Add 4 ml heat-inactivated FCS (final concentration 10%).
   (c) Add the contents of a 1 mg LPS vial (final concentration 20 μg/ml). This is sufficient for at least one spleen. It is usually prepared on the day of use.

B. *Obtaining a suspension of spleen cells*

1. Place the intact spleen directly after dissection into cold collection medium. Place this on ice and immediately transport on ice.
2. Using sterile conditions, place the spleen in a sterile Petri dish and rinse with about 5 ml of collection medium. Draw this off and place another 1 ml of medium in the dish.
3. Using a scalpel, cut open the spleen lengthways to release the spleen cells into the medium, scraping the cells free with the tip of a pipette.
4. Pass the contents through a cell strainer by placing the cell strainer above a 50 ml centrifuge tube which will hold it. Rinse out the contents of the Petri dish with additional collection medium and pass through the cell strainer.
5. Spin at 250 g for 5 min and remove the supernatant.
6. Re-suspend in about 10 ml cold collection medium. Leave the tube on ice; it can be transported to the laboratory at this stage.
7. Count the cells using a haemocytometer and calculate the total number of cells present.[a]

8. Spin at 250 *g* for 5 min. Remove the supernatant.
9. Add growth medium, adjusting the cell density to 5×10^6 cells/ml.
10. Place cells into T25 tissue culture flasks, 5 ml per flask.
11. Incubate the cells in a 5% CO_2 incubator for 48 h.[b]
12. Harvest the cultures (see *Protocol 2*).

[a] It is not necessary to do cell counts once the technique becomes routine. The density can be judged by examination of the culture flask under an inverted microscope. The yield of spleen cells depends on the age of the mouse (the more mature, the larger the spleen) and whether the spleen remains intact during handling and before cutting open.
[b] Satisfactory results can also be obtained by shortening the culturing time to 24 h.

Protocol 2 Harvesting mouse spleen cultures

Equipment and reagents

- 10 μg/ml colcemid (Gibco-BRL)
- Hypotonic KCl, 0.075 M
- Freshly prepared Carnoy's fixative (3:1 methanol:acetic acid); use dried methanol and glacial acetic acid, Analar grade (see Chapter 6A, *Protocol 1*).
- 70% aqueous solution of acetic acid
- 10 ml centrifuge tubes
- Bench-top centrifuge
- Pre-cleaned microscope slides

Method

1. Add 5 μl colcemid solution per ml of medium to the flasks (to give a final concentration of 0.05 μg/ml) and incubate for a further 30 min.[a]
2. Transfer to 10 ml conical centrifuge tubes and centrifuge at 200*g* for 5 min.
3. Add 5 ml of pre-warmed hypotonic KCl and incubate for 8 min at 37 °C.
4. Centrifuge at 200*g* for 5 min.
5. Draw off the supernatant using a 1 ml pipette leaving 0.1–0.2 ml. Resuspend the pellet very gently with the pipette.
6. Add a drop of freshly prepared fixative. Mix by shaking the tube gently. Add additional fixative drop by drop, slowly, progressively increasing the amount and mixing, over a period of 2–4 min. Top up the tube with fixative to 5 ml.
7. Leave at 4°C for at least half an hour. Store overnight at 4°C if necessary.
8. Spin down at 300*g* for 5 min. Draw off old fixative and replace with 5 ml freshly prepared fixative.
9. Centrifuge again and draw off the supernatant, taking care not to disturb the pellet. Add fresh fixative to the pellet and mix, adding sufficient fixative to make a milky solution, roughly 0.5–1 ml.

Protocol 2 *Continued*

10. Drop one or two drops of the suspension onto a precleaned, cold, and damp microscope slide held at a 30° angle. Let the drop run down the slide and allow to dry.
11. When dry, add 0.5–1 ml of an aqueous solution of 70% acetic acid to the surface of the slide, held horizontally using a pipette. Let it drain off and place the slide vertically on its narrow side to dry.
12. Examine the slide using phase contrast microscopy under low power to assess its quality and that of the culture. There should be a sufficient number of mitoses present, which gives an indication of the success of the culture; the mitoses should be in metaphase and sufficiently spread to allow the chromosomes not to overlap one another, but still to be intact.
13. Make the required number of slides, check them for quality, then pass them through an alcohol series of 70%, 90% and absolute ethanol, 3 min each. Air dry and store them at 4°C (for short-term storage, of a few days) or –20°C (up to about 2 months).
14. Top up the remaining suspension with fixative up to 5 ml and store this at –20°C.

[a] Overcontracted chromosomes are an indication of too long an exposure to colcemid. This relatively short exposure time will ensure long chromosomes which will give enough detail within the arms. Too short an exposure will lead to insufficient accumulation of mitoses.

4. Probes and labelling

4.1 DNA

Probe DNA is obtained from phage, cosmid, PAC or YAC clones using methods described elsewhere (5,6). Our experience is confined to FISH using DNA from these clones. DNA is dissolved in TE buffer (10 mM Tris–HCl, 1 mM EDTA, pH 7.5) and its concentration measured using a fluorometer (Pharmacia). The concentration needs to be sufficient, at least 25 ng/μl, to ensure the correct reaction volume during DNA labelling. For FISH the whole probe is used and it is not necessary to cut out the insert from the vector DNA.

4.2 Labelling by nick translation

Mouse chromosome paints are supplied already labelled, but it is necessary to label the probe used for mapping. This is done by using one of two haptens, namely digoxygenin (DIG) and biotin. These substances are conjugated to nucleotides which are in turn incorporated into the probe DNA by nick translation. Since both can be detected separately, it is often advantageous to use them simultaneously. Detection takes place indirectly, as will be seen in Section 5. The DIG labelling method is given in *Protocol 3A*. Since in single FISH experiments biotinylated unique sequence probes have been used extensively, the biotin labelling protocol is also given in *Protocol 3B*. In both methods incorporation of either DIG or biotin takes place by nick translation. Randon priming is also suitable for labelling probes. For nick translation, commercial kits are used and conditions are standard. Under these conditions, the size of the DNA fragments of the probe will be 200–500 bp in length, which is suitable for FISH. If the probe fragments are too large, more background will result and there will be less accessibility to the target chromosomes. If they are too small, there will be poor hybridization efficiency and sensitivity.

We usually label 400 ng of probe in one reaction (less than the 1 μg recommended by the manufacturers). One hybridization requires 200 ng; the 400 ng quantity labelled will be sufficient for two hybridizations. In other respects the protocols are virtually as given by the manufacturers of the kits.

4.3 Chromosome paints and direct labelling

Chromosome paints are available from Cambio Ltd and are ready labelled with a choice of three systems, namely biotin, and the fluorochromes fluorescein isothiocyanate (FITC) and Cyanin3 (Cy3). Because of this choice, the application is wide and opens up many possibilities of combinations for two or more colours. Paints directly labelled with fluorochromes are fluorescent and can therefore be used in direct FISH, as distinct from indirect FISH. Detection methods using antibodies and binding are not necessary so that the procedure is shorter. Because the target is large, namely a whole chromosome, direct FISH is feasible; in the case of small probes such as phages, direct FISH is not justifiable and would depend not only on the size of the target but also on the sensitivity of the imaging equipment used for analysis. The DIG–biotin dual FISH method will be described in Section 5.

Protocol 3A. Labelling DNA with digoxygenin

Equipment and reagents

- DIG nick translation mix, 5× concentrate (Roche). Store this at −20°C and always keep on ice when in use. It contains: DNA polymerase I, DNase I, 0.25 mM each of dATP, dCTP and dGTP, 0.17 mM dTTP and 0.08 mM DIG-11-dUTP
- Stop buffer: 0.3 M EDTA, pH 8 (autoclaved, stored at room temperature)
- Microcentrifuge tubes (e.g. Eppendorf)
- Microcentrifuge

Method

1. Pipette the following components into a microcentrifuge tube on ice to give a final volume of 20μl:
   - 400 ng probe DNA

- Milli-Q or equivalent water (autoclaved) to bring volume to 16 μl
- 4 μl of DIG nick translation mix

2. Mix and spin briefly to collect the reaction to the bottom of the tube
3. Incubate at 15°C for 90 min.
4. Add 2 μl of stop buffer and mix.
5. Store the probe at –20°C before use.

Protocol 3B. Labelling DNA with biotin

Equipment and reagents

- Bionick labelling system (Gibco-BRL), comprising (a) 10x dNTP mix (0.2 mM each dCTP, dGTP and dTTP; 0.1 mM dATP and 0.1 mM biotin-14-dATP; (b) 10x enzyme mix which contains the same enzymes as the DIG nick translation (0.5 units/μl DNA polymerase I and 0.0075 units/μl DNase I); (c) stop buffer (0.3 M EDTA, pH 8)
- Microcentrifuge tubes (e.g. Eppendorf)
- Microcentrifuge

Method

1. Pipette into a microcentrifuge tube on ice to give a final volume of 50μl:
   - 400 ng DNA
   - volume of supplied water to 40 μl
   - 5 μl 10x dNTP mix
   - 5 μl 10x enzyme mix.
2. Mix and spin briefly to collect the reaction to the bottom of the tube.
3. Incubate at 15°C for 90 min.
4. Add 5 μl stop buffer.
5. Store at –20°C before use.

5. FISH

5.1 Probe preparation

The labelled probe needs to be purified from the unincorporated nucleotides and additional DNA added. Cot-1 DNA is the highly repetitive fraction of genomic DNA. It is used as competitor DNA to suppress hybridization of repeat signals which are present in the probe, both mapping (genomic) and paint. It is necessary to bear in mind that repeats are species specific and mouse Cot-1 DNA should be used for mouse FISH. Herring sperm DNA also reduces background and acts as carrier DNA; it needs to be fragmented because of its high molecular weight. A product suitable for FISH is available for purchase. Probe precipitation is described in *Protocol 4*.

After precipitation of the probe, it is combined with further components which promote hybridization. An essential ingredient is formamide, which gives best results when deionized. Formamide becomes ionized when exposed to the atmosphere; using a freshly opened bottle will prevent this.

Protocol 5 describes the preparation of hybridization mix which is added to the dried probe. The quantity of hybridization mix depends on what combination is hybridized. When used alone on one slide, 10 μl is sufficient. If it is used in combination, dissolve in 6 μl hybridization mix and use 4 μl of paint. The probes can be denatured separately as described below. Alternatively they can be combined before denaturation. If denatured separately, they can be combined on the slide just before hybridization, as given in *Protocol 6*. This allows for combinations of mapping probe and paint to be made.

The paints are supplied at up to five-fold concentration from Cambio Ltd. This allows five probes to be hybridized together. Here we describe combination not by using concentrated probe but by adjusting the volume of hybridization mix. Because the paint has a large target and the resulting signal is strong, less is used.

Protocol 4. Probe precipitation

Equipment and reagents

- Labelled probe DNA
- Mouse Cot-1-DNA (Gibco-BRL), 1 mg/ml
- Herring sperm DNA for hybridization (Sigma), 10 mg/ml
- 3 M sodium acetate pH 5.2
- 100% ethanol, stored at –20 °C
- Microcentrifuge

Method

1. Mix together:
   - 200 ng probe, i.e. half volume: 11μl
   - 10 μl mouse Cot-1-DNA
   - 2 μl herring sperm DNA
2. Add 0.1 vol. sodium acetate and mix.
3. Add at least 2 vols of ice-cold absolute ethanol; mix.
4. Place in a –70 °C freezer for at least 30 min, or at –20 °C overnight.
5. Spin for 5 min in a microcentrifuge at 15 000*g*.
6. Aspirate or pour off the supernatant.
7. Freeze-dry to dry off remaining liquid, or leave on bench undisturbed at room temperature until dry.

Protocol 4 *Continued*

8. Add 6–10μl hybridization mix (see *Protocol 5* and Sections 5.1 and 5.2) and dissolve. Normally, the probe will dissolve readily. It is preferable to use the probe for FISH on the same day. If necessary it can be stored at −20 °C until the slides are ready.

Protocol 5. Hybridization mix preparation

Equipment and reagents

- Deionized formamide (Analar) (To deionize: take 100 ml and add 5 g Amberlite mixed bed resin (Bio-Rad). Agitate gently for 1 h then filter through two thicknesses of Whatman no. 1 filter paper).
- 20 × SSPE (3.0M NaCl, 0.2M NaH_2PO_4, 0.02M EDTA, pH 7.4) (Sigma)
- Dextran sulfate (BDH)
- Microcentrifuge tubes (e.g. Eppendorf)

Method

1. Combine the following in a Universal tube:
   - 5 ml deionized formamide
   - 1 ml 20 × SSPE.
   - about 3 ml water (purified Milli-Q or equivalent, autoclaved)
   - 1.0g dextran sulfate
2. Dissolve the dextran sulfate by incubating at 70 °C, shaking occasionally for 3 h or until dissolved.
3. Adjust pH to 7.2 using concentrated HCl or 10 N NaOH.
4. Adjust the volume to 10 ml with purified water.
5. Aliquot into 1.5 ml microcentrifuge tubes and store at −20 °C. Each aliquot can be thawed and re-frozen between use.

5.2 Hybridization

Probe denaturation is performed as follows:

(a) Prepare the mapping probe as in *Protocol 4*. Take the paint, mix briefly using a vortex and centrifuge briefly to collect. Pipette 4μl (see Section 5.1) into a 0.5 ml microcentrifuge tube.

(b) Denature both tubes at 75 °C, e.g. in a fan oven for 5 min.

(c) Transfer to a 37 °C waterbath to pre-anneal for 30 min or until the slides are ready.

Probe denaturation and slide pretreatment (*Protocol 6*) should be carried out so that they coincide. Following the full FISH protocols will result in satisfactory chromosome banding with minimum background fluorescence, in addition to providing successful probe hybridization. The RNase and proteinase K steps enhance the chromosome banding if it is not necessary to identify the chromosomes then these steps can be omitted. Paint probes and larger probes like YACs will hybridize in any case, but smaller probes will not.

Protocol 7 describes dual detection. For single detection using biotin, use only FITC–avidin and FITC anti-avidin, in two stages. FITC and avidin are supplied conjugated to form FITC–avidin. Avidin has a strong affinity for biotin and binds to the biotin present on the chromosome slide after hybridization. The signal is amplified using FITC anti-avidin which in turn binds to FITC–avidin on the slide. For single detection using digoxygenin, use one step only. *Figure 2* shows an example of dual FISH using these methods. Further examples from transgenic mice are given in *Figures 3* and *4*.

Protocol 6. Slide pre-treatment and hybridization

Equipment and reagents

- RNase A (Sigma). Make up a 100× stock solution (10 mg/ml) in 2 × SSC,[a] boil for 10 min and store at –20°C. This solution can be thawed repeatedly. Just before use make up 1× working solution (100 μg/ml), sufficient for 100 μl/slide.
- Proteinase K (Roche): make up a stock solution 50 μg/ml in proteinase K buffer and store in single-use aliquots at –20°C. Just before use make up working solution: add 50 μl to 50 ml proteinase K buffer to give a final concentration of 50 ng/ml. Pre-warm at 37°C until slides are ready.
- Proteinase K buffer 10x concentration: 0.2 M Tris–HCl, 0.02 M $CaCl_2$, pH 7.4
- Equilibrated 37% formaldehyde saturated with $NaHCO_3$ which is allowed to settle.
- 1% formaldehyde: make up 50 ml PBS, 0.5g $MgCl_2 \cdot 6H_2O$ and add 1.3 ml of equilibrated 37% formaldehyde just before use.
- 70% formamide in 2 × SSPE (see *Protocol 5*). Make 1 ml just before use: mix together 200 μl purified water, 100 μl 20× SSPE, 700 μl formamide drawn from freshly opened bottle.
- 70% ethanol pre-cooled in –20°C freezer.
- Non-hardening rubber solution such as Cow Gum (Cambio Ltd)

Method

1. Treat the slides with RNase A. Lay the slides horizontally in a humidified box, pipette RNase A solution onto the surface, cover with a large (22 × 50 mm) coverslip, close the box and incubate at 37°C for 1 h.
2. Using a Coplin jar, rinse the slides in two changes of 50 ml of 2 × SSC at room temperature.
3. Dehydrate through an ascending ethanol series (70%, 90% and 100%) for 3 min each.
4. Air dry.
5. Place the slides in 50 ml proteinase K buffer, pre-warmed to 37°C, for 10 min.
6. Transfer the slides to 50 ml proteinase K solution at 37°C for 7 min.
7. Rinse in 50 ml PBS at room temperature for 5 min.
8. Transfer to 1% formaldehyde solution at room temperature for 10 min.

Protocol 6 *Continued*

9. Rinse in PBS as before at room temperature for 5 min.
10. Dehydrate as before.
11. Denature in 70% formamide: place the slides horizontally, pipette 100 μl solution onto the surface, cover with large (22 × 50 mm) coverslip and place in a fan oven at 75°C for exactly 3.5 min.[b]
12. Slip off the coverslip and quickly plunge the slide into ice cold 70% ethanol for 2 min.
13. Dehydrate further through cooled 90% and 100% ethanol, 2 min each.
14. Air dry.
15. Pipette the contents of the tubes of prepared probe (both mapping probe and paint) onto a 22 × 22 mm coverslip, avoiding air bubbles.
16. Lift the coverslip onto the slide by inverting the slide and touching the droplet of probe which will spread out under the coverslip.
17. Apply rubber solution over coverslip edges and place the slide in a humidified box. Allow to hybridize at 37°C overnight or longer.

[a] 1 × SSC is 0.15 M NaCl, 0.015 M trisodium citrate.
[b] Denaturation temperature and timing are crucial in maintaining good chromosome morphology. Longer denaturation will result in good hybridization but chromosomes will give dull fluorescence.

Protocol 7. Washes and detection

Equipment and reagents

Method

- 50% formamide in 2 × SSC
- 5% non-fat milk powder in 4 × SSC (from Difco or any supermarket)
- Blocking agent (Roche)
- TNB: Use 1 × TN made up from 10 × TN solution (10 × TN: 1 M Tris–HCl, 1.5 M NaCl, pH7.5) and add 0.5% blocking agent. Heat slowly to 60°C and incubate, shaking occasionally for 2–3 h or until dissolved. Centrifuge in microcentrifuge at 15 000*g* for 5 min; use supernatant. Store at –20°C.
- SSCT: 0.05% Tween-20 detergent (Sigma) in 4 × SSC
- FITC–avidin (Vector Labs), 1 mg/ml, divided into 100 μl aliquots and stored at –20°C. Keep one aliquot at 4°C for use. Protect from light. On day of use dilute 1 in 200 in TNB to give a final concentration of 5 μg/ml. Make up 100 μl per slide; protect from light.
- FITC anti-avidin (Vector Labs), 1 mg/ml. Aliquot into 100 μl amounts. Store at –20°C and keep an aliquot at 4°C for use. Protect from light.
- Anti-DIG–rhodamine (Roche) 200 μg. Add 1 ml TNB and aliquot into 100 μl amounts. Store at –20°C. On the day of use make up a combined dilution of FITC anti–avidin diluted 1 in 200 together with anti-DIG rhodamine diluted 1 in 20 to give a final concentration of 10 μg/ml. Make up 100 μl per slide, e.g. 0.5 μl FITC anti–avidin, 5μl anti-DIG–rhodamine, 95μl TNB. Protect from light.
- Vectashield antifade mounting solution containing DAPI (Vector labs)
- Coplin jar

1. Peel the sealant off the coverslips using forceps.
2. Place the slides in a Coplin jar containing 50 ml of pre-warmed 50% formamide in 2 × SSC at 42 °C;[a] let the coverslip loosen in the liquid then remove it gently without it floating into the jar. Wash in a total of three changes for 5 min each.
3. Wash in two changes of 2 × SSC for 2.5 min each at 42 °C.[a]
4. Wash stringently in two changes of 0.1 × SSC for 2.5 min each at 42 °C.[a]
5. Transfer the slides to 4 × SSCT and place on horizontal shaker for 5 min at room temperature.
6. Transfer the slides to a Coplin jar containing 50 ml of 4 × SSC containing non-fat milk powder for 20 min at room temperature.
7. Drain the slides briefly and place them horizontally. Pipette 100 μl of FITC–avidin solution onto each and drop a large (22 × 50 mm) coverslip on top. Incubate for 20 min in the dark at 37 °C. Perform this step and all subsequent ones protected from light because they are light sensitive.
8. Place the slides in a Coplin jar and wash them on a horizontal shaker in three changes of 4 × SSCT for 5 min each at room temperature.
9. Drain the slides briefly and place them horizontally; pipette 100 μl of FITC anti-avidin anti-DIG-rhodamine onto each. Cover them as before and incubate for 20 min at 37 °C.
10. Wash in 4 × SSCT twice for 5 min each
11. Rinse in PBS once for 5 min.
12. Dehydrate through ethanol series, 70%, 90% and 100%, 3 min each.
13. Air dry.
14. Mount in Vectashield antifade containing DAPI. Use 10–20 μl of antifade and mount under a large (22 × 50 mm) coverslip. Keep at 4 °C protected from light.

[a] Carry out steps 2–4 at 42 °C in a waterbath containing the pre-warmed solutions.

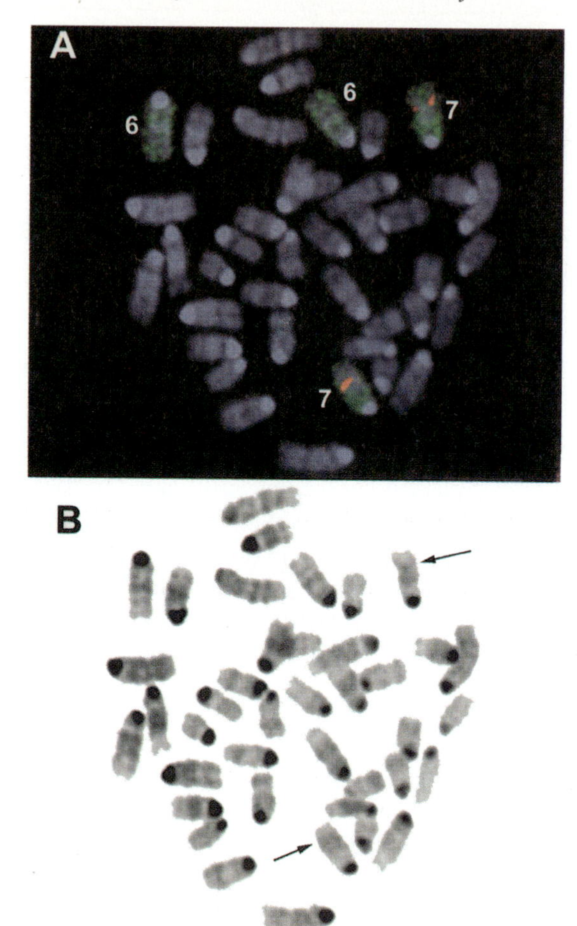

Figure 2. (A) FISH on metaphase chromosomes showing localization of a mouse lambda genomic clone (red) on mouse chromosome 7 (painted green). The lambda DNA was labelled with DIG and co-hybridized with paints 6 and 7. Chromosomes 6 and 7 were distinguished by their banding patterns and are indicated. Signals were detected using rhodamine–anti-DIG and FITC–avidin (see Section 5). Slides were examined on a Zeiss Axioskop fluorescence microscope equipped with a 50 W mercury ultraviolet light source, using a 20× Achroplan objective to scan for suitable metaphases. A 100x Plan-Neofluar oil objective was in place when capturing images, which was done using the attached cooled CCD (Photometrics) and Smartcapture software supplied by Digital Scientific. The three colours were captured separately, with the filter wheel containing filters, developed by Photometrics Inc., designed for visualizing FITC, Texas Red and DAPI counterstain (green, red and blue respectively); the images were combined automatically. The colours were adjusted and enhanced, whilst DAPI counterstain was lightened. (B) The inverted image with signals removed showing G-like banding. Arrows point to the position of the hybridization signal on chromosome 7.

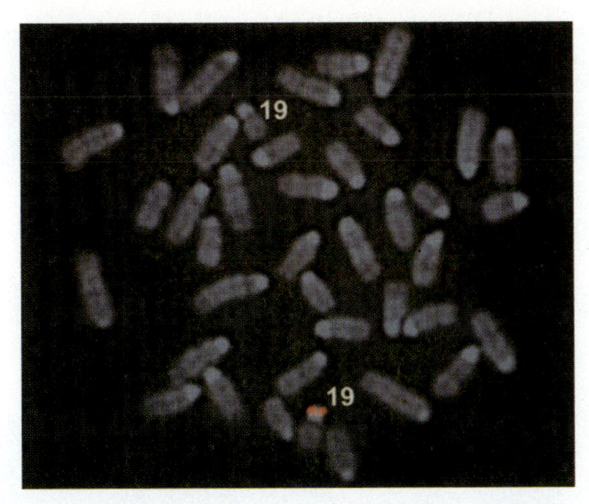

Figure 3. Metaphase chromosomes from a female transgenic mouse containing a human CD2 gene construct and probed with the same construct. FISH was carried out using biotin-labelled probe DNA. The signal was detected by FITC–avidin using equipment as described in *Figure 2*, except that only green and blue filters were used. The FITC signal was converted from green to red pseudocolour. The centromeric location is evident; this is the smallest chromosome but identification would require confirmation with painting or a second probe of known location.

6. Analysis and microscopy

View slides using a fluorescent microscope. Requirements for a fluorescent microscope are a UV light source, suitable objectives and a minimum of three filters for fluorescent microscopy. High magnification objectives should have a high numberical aperture and should be compatible for visualizing chromosome spreads. Filters with three different wavelengths are required because three colours (red, green and blue) must be made visible, i.e. the DAPI-counterstained chromosomes and nuclei as well as the signals (rhodamine for the mapping probe and FITC for the paint). The progress seen in the improvement of fluorescent dyes has been matched by developments in filter technology.

Recording the information is best done using digital imaging . Conventional microscopy is also possible. It is satisfying to see the signal under the microscope but imaging can also be used to reveal signals which could be missed by optical microscopic examination alone.

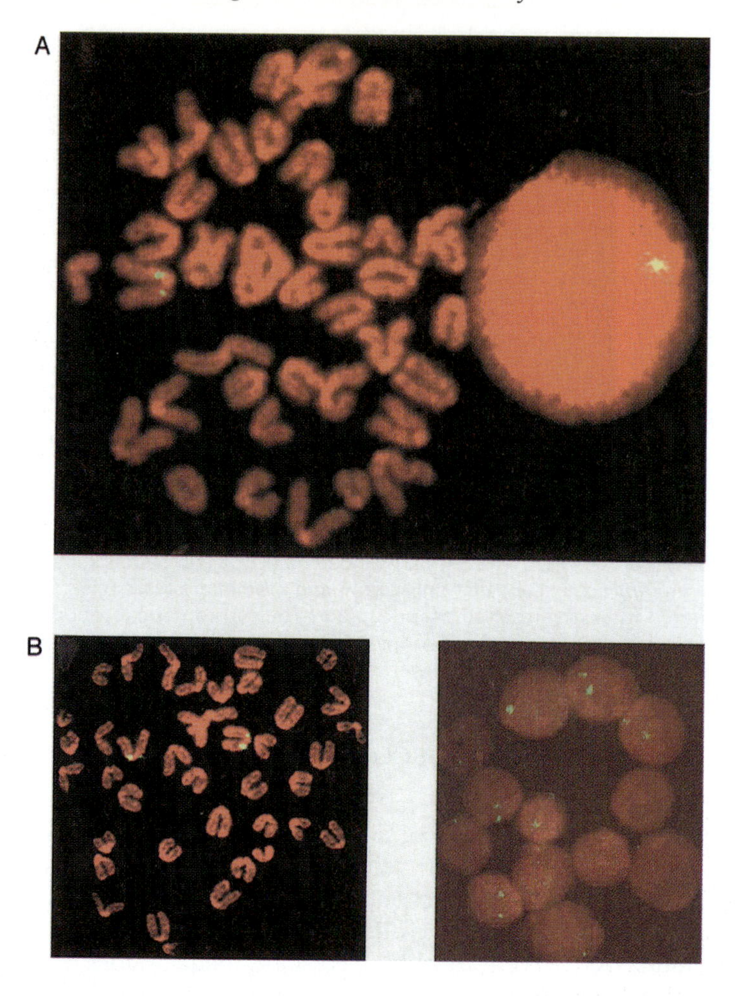

Figure 4. FISH on chromosomes from transgenic mice. (A) Metaphase and neighbouring interphase nucleus from a transgenic mouse containing the PI Z allele of the human alpha-1-antitrypsin gene. The yellow signal from the human cosmid DNA used as probe locates the transgene integration site. (B) Interphase nuclei and metaphase from a homozygous transgenic mouse hybridized with the same probe, showing two signals in the nucleus, and both homologous chromosomes containing the transgene. Probe DNA was labelled with biotin and detected with FITC–avidin, and chromosomes and nuclei were counterstained with propidium iodide (red). Images were captured using a MRC 600 confocal microscope. Samples were obtained from whole embryos and directly prepared; 30 interphase nuclei were examined to assess the homozygous or heterozygous status after heterozygous crosses. Metaphases confirmed the result. (C) Dual FISH on transgenic mouse chromosomes probed with cosmid DNA as in (A) and (B), detected with FITC (green) and hybridized simultaneously with a chromosome 5 cosmid labelled with DIG and detected with rhodamine (red). The proximity of red and green signals confirms that the transgene has integrated into chromosome 5, region A3. The homologue contains only the endogenous cosmid (red) signal.

c

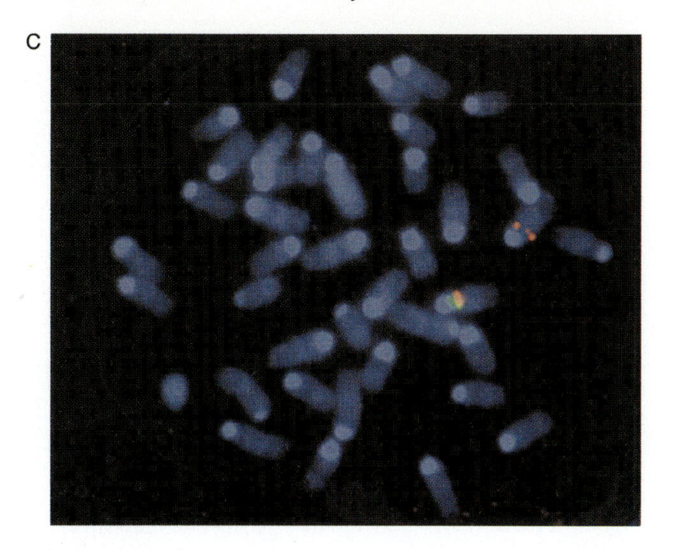

References

1. Boyle, A. L., Ballard, S. G. and Ward, D. C. (1990). *Proc. Natl Acad. Sci. USA*, **87**, 7757.
2. Boyle, A. L., Feltquite, D. M., Dracopoli, N. C., Housman, D. E. and Ward, D. C. (1992). *Genomics*, **12**, 106.
3. Rabbitts, P., Impey, H., Heppell-Parton, A., Langford, C., Tease, C., Lowe, N., Bailey, D., Ferguson-Smith, M. and Carter, N. (1995). *Nature Genet.*, **9**, 369.
4. Shi, Y.-P., Mohapatra, G., Miller, J., Hanahan, D., Lander, E., Gold, P., Pinkel, D., and Gray, J. (1997). *Genomics*, **45**, 42.
5. Sambrook, J., Fritsch, G. F. and Maniatis, T. (1989). *Molecular Cloning: a Laboratory Manual*, 2nd edn. Cold Spring Harbor Laboratory Press, Cold Spring Harbor, NY.
6. Silverman, G. A. (1996). In *Methods in Molecular Biology*, Vol. 54: *YAC Protocols* (ed. Markie, D.), p. 65. Humana Press Inc., Totowa, NJ.

7

Electronic tools for accessing the mouse genome

JANAN T. EPPIG

1. Introduction

The mouse is the premier mammalian model for human genetics and disease. It is unique among mammals in that it already has an extensive high-resolution genetic map and sophisticated techniques to manipulate its genome are well-developed. Here we explore the mouse-specific on-line tools that give the scientific community access to the wealth of mouse genomic and biological data and specialized resources. A list of major websites is given in *Table 1*, with additional websites included in the text.

2. Databases of genomic information for the mouse

2.1 The Mouse Genome Database (MGD): http://www.informatics.jax.org

The unifying database for mouse genetic, genomic, and biological information is the mouse genome database (MGD), accessible via the World Wide Web (WWW) from The Jackson Laboratory and at several sites worldwide (*Table 2*) (1,2). The MGD contains a wide variety of data and views from gene identification, to genomic structure and function, and to the manifestation of genotype as individual (variant) and strain phenotypes. It provides basic information about genes such as symbol, name, chromosomal location and subchromosomal placement (i.e. centimorgan and/or cytogenetic positions), and an array of expanded data, including descriptions of molecular clones, PCR primers, polymorphisms, homologies between mouse genes and genes of other mammalian species, gene product and function(s) associations, and descriptions of phenotypes for known mutant alleles. Maps can also constitute primary data. Genetic maps are available as consensus views, or for specific data sets, such as the MIT genetic map or the Copeland–Jenkins map. Cytogenetic maps, physical maps and comparative maps for mouse with human or other mammalian species can be displayed and printed. Users also can selectively create

Table 1. Major websites for mouse-specific information

Type of information	Name of site	URL	Synopsis
Genomic data	Mouse Genome Database (MGD)	http://www.informatics.jax.org/ (see *Table 2* for mirror sites)	Genes, markers, chromosomal aberrations; maps; homologies among mammals; experimental mapping data; polymorphisms and mutant alleles; gene and strain phenotypes; clones and ESTs; Chromosome Committee reports; nomenclature guidelines; data submission tools.
	MRC Harwell MouseDb	http://www.mgu.har.mrc.ac.uk	Lists of genes, mutants, chromosomal aberrations; Human–mouse homologies; imprinting maps, atlas of mouse comparing genetic and cytogenetic maps.
Expression data	Gene Expression Database (GXD)	http://www.informatics.jax.org/menus/expression_menu.shtml	Expression assays annotated for embryonic development; literature index for expression; cDNA/EST expression data.
	MRC HGU Mouse 3D Atlas	http://genex.hgu.mrc.ac.uk/	3D reconstructions of mouse embryos; anatomical dictionary and fate tables.
Transgenic/knockout information	TBASE	http://tbase.jax.org/	Transgenic and knockout information, including construct. phenotype, experimental details.
	Mouse Knockout Database (MKM)	http://www.biomednet.com	Synopsis of knockout phenotypes for genes reported in the literature.
	Induced Mutant Resource (IMR)	http://www.jax.org/resources/documents/imr/	Information on transgenics/knockouts/ other induced mutations, phenotypes, strains.
Nomenclature	International Committee for Standardized Nomenclature for Mice	*Nomenclature guidelines*: http://www.informatics.jax.org/support/nomen/table.shtml *Submit a new gene symbol*: http://www.informatics.jax.org/support/nomen/nomen_submit_form.shtml *Laboratory code registration:* http://www2.nas.edu/labcode/	Find nomenclature rules for naming genes and strains; submit a new symbol for approval; look up lab codes.

Strains/stocks	International Mouse Strain Resource (IMSR)	http://www.jax.org/pub-cgi/imsrlist; http://imsr.har.mrc.ac.uk	Catalog of strains/stocks/mutants from multiple institutes with links; acts as an aid to finding mice of particular genotypes anywhere in the world.
	European Mouse Mutant Archive (EMMA)	Montorotondo: http://www.emma.rm.cnr.it/ Orleans: http://cdta.cnrs-orleans.fr /Anglais/Acceuilcdtagb.html Harwell: http://www.mgu.har.mrc.ac.uk/stocklist/stocklist.html Karolinska: http://www.ki.se/kfc/meg/Murine/Murine.html Oeiras: http://www.pen.gulbenkian.pt/	EMMA is composed of a consortium of five European instutitions for maintenance, distribution, and cryopreservation of mouse strains/stocks and mutants, with the main node in Montorotondo.
	The Jackson Laboratory Resources	*Live and cryopreserved stocks*: http://jaxmice.jax.org/	The Jackson Laboratory has the largest collection of inbred strains, hybrids, congenics, recombinatnt inbreds, mutants, chromosomal aberrations, transgenics, and tartgeted mice. Approximately 2300 genetically distinct stocks are distributed.
		DNA resource: http://www.jax.org/resources/documents/dnares/index.html	The DNA Resource provides aliquots of DNA from most of the stocks available from The Jackson Laboratory and from many that are extinct.
Websites: guides to rodent information	Whole Mouse Catalog	http://www.rodentia.com/wmc/	These sites provide lists of links to information on mouse, ranging from genomic data to animal care to anecdotes.
	Electronic Zoo: Rodent Page	http://netvet.wustl.edu/rodents.htm	
Mouse electronic bulletin board	MGI-list	http://www.informatics.jax.org/support/lists.shtml	URL for registering on the mouse electronic community bulletin board. Messages are forwarded to all subscribers to the list.

maps using particular data sets and particular selected genetic markers (e.g. showing only genes associated with eye defects or only genes with homologues in rat), or for particular chromosomal regions. Additional important data sets include M. F. Festing's descriptions of inbred strain characteristics, International Mammalian Genome Society Chromosome Committee Reports, and the guidelines of the International Committee for Standardized Nomenclature for Mice. A primary function of the MGD is the integration of these disparate data types from different sources into a comprehensive information resource that places information in contextual relationships and provides a rich set of links to external supplementary data. For example, phenotypes are linked with human phenotype data in 'Online Mendelian Inheritance in Man' (OMIM), bibliographic data to PubMed, clone and gene data to the DNA sequence databases (GenBank, EMBL, DDBJ) and to SwissProt, and homology information to the databases of other species. Integration and linking allow one to query the MGD for answers to questions such as:

- What homeobox genes on mouse chromosome 2 have known homologues in rat?
- Are there ESTs for these genes and what are their sequences?
- What genes in mouse and human are known to affect melanocytes and what are the respective phenotypes of mutations in these genes?
- Are there primers for acetylcholine receptor gamma that differentiate C57BL/6J and DBA/2J strains?
- What genes are within 2 cM of *pi* (*pirouette*) and have known human homologues, e.g. as potential candidate genes?

As a resource for the scientific community, the MGD provides a variety of entry points for data searches and an intuitive interface for viewing search results. Electronic data submissions and pre-publication nomenclature registry are encouraged to assist with making the MGD a true community resource containing the most comprehensive information possible. As genome research continues to change, the MGD is evolving to place greater emphasis on

Table 2. Primary and mirror sites for the Mouse Genome Database (MGD)

Site	URL
Primary site	http://www.informatics.jax.org
Mirror sites	
Australia	http://mgd.wehi.edu.au:8080
France	http://www.pasteur.fr/Bio/MGI
Israel	http://bioinfo.weizmann.ac.il/mgd/
Japan	http://mgd.niai.affrc.go.jp
UK	http://mgd.hgmp.mrc.ac.uk

sequence-based data, molecular phenotypes (including sequence variation and mutant alleles) and disease models.

2.2 The MRC Mammalian Genetics Unit: http://www.mgu.har.mrc.ac.uk/

A distinct spectrum of mouse information is actively collected and assembled for the scientific community in MouseDb at the Mammalian Genetics Unit at MRC (Harwell, UK). This site includes data and maps on imprinted genes of the mouse (3), chromosome anomaly maps (4), the atlas of the mouse genome, presenting side-by-side comparisons of cytogenetic and genetic maps (5), and mouse/human homology data linked to potential candidate human diseases (6,7).

3. Specific data sets for genetic, radiation hybrid and physical mapping

3.1 Genetic data

Several genetic mapping resources have been developed that use DNA from large numbers of progeny of genetic crosses where quantities of DNA have been saved, such that data from typing new loci are continually being added. Thus, maps of high resolution with thousands of markers can be developed. The data sets described below are integrated into the MGD, as well as being available from the orginator's website (see also Chapter 5).

3.1.1 The Whitehead Institute/MIT map: http://www-genome.wi.mit.edu/

The Genome Center of the Whitehead Institute for Biomedical Research at the Massachusetts Institute of Technology developed the first large-scale mouse mapping resource, using a (C57BL/6J-*Lep*^{ob} × CAST)F2 cross. Although they did not distribute DNA or incorporate data from other laboratories into their data set, they provided a key community resource by developing and mapping 6331 simple sequence length polymorphism (SSLP) markers. This map and the SSLP markers are widely used by the community as a reference dataset (8,9).

3.1.2 EUCIB: http://www.hgmp.mrc.ac.uk/MBx/MBxHomepage.html

The European Collaborative Interspecific Backcross Group (EUCIB) (10) distributes DNA and maintains a database for results of typing subsets of the 982 DNA samples derived from animals of an interspecific backcross between C57BL/6 and *Mus spretus* (all backcrosses derived from F1 females; approximately half of the progeny represent backcrosses to C57BL/6 and half to *M. spretus*). An interval mapping strategy has been used in developing the EUCIB map. All mice have been typed for a few anchor markers and these markers are precisely placed; other markers are placed by interval, with only a small subset of the animals typed. Interval mapped markers are, in general, binned and localized to a map 'range', and marker order among them may be ambiguous.

3.1.3 Jackson Laboratory backcross: http://www.jax.org/resources/documents/cmdata/

The Jackson Laboratory backcross DNA panel mapping resource (11), like the EUCIB panel, supplies DNA from reciprocal interspecific backcrosses between C57BL/6 and *M. spretus*. Each panel (i.e. each of the reciprocal backcrosses) is available as a set of 94 samples, that with two controls are convenient for microtitre plate format. The 188 mice of the Jackson Laboratory backcrosses are completely typed (i.e. every animal is typed for every locus), thus providing precise localization of all crossovers and reliable distance determinations. Complete data also allow analysis of other phenomenon such as segregation distortion and map compression or expansion.

3.1.4 Copeland–Jenkins interspecific backcross: http://www.informatics.jax.org/searches/crossdata_form.shtml

A unidirectional interspecific backcross, (C57BL/6J × *M. spretus*) × C57BL/6J, was made at the Frederick Cancer Research and Development Center (12). DNA from this cross has been developed primarily through collaborative efforts, with an emphasis on mapping known genes and uses a largely complete typing strategy. The Copeland–Jenkins map has been integrated with the Whitehead/MIT map by mapping selected SSLPs onto the Copeland–Jenkins backcross panel.

3.1.5 M. Seldin interspecific backcross: http://www.informatics.jax.org/searches/crossdata_form.shtml

An interspecific backcross including a segregating immunological mutation $Fasl^{gld}$, (C3H/HeJ-$Fasl^{gld}$ × *M. spretus*) × C3H/HeJ-$Fasl^{gld}$ was developed by Seldin and co-workers (13). It utilizes the interval mapping strategy for mapping markers, with near-complete typing of animals for anchor loci and typing of selected animals for other loci. Markers that have been typed on this cross include SSLPs, genes and ESTs.

3.2 Radiation hybrid data

While recombination mapping provides accurate genetic placement of loci, radiation hybrid (RH) mapping can be used for high throughput mapping and high resolution of locus order. The T31 mouse radiation hybrid panel consists of 100 cell lines from the fusion of a 3000 rad irradiated mouse embryo primary cell line and a hamster cell line (14). Two complementary databases have been established to serve as community resource points for gathering, distributing and analysing the T31 RH data. These two resources also are prepared to handle data from future RH panels when they are developed.

3.2.1 The Jackson Laboratory mouse radiation hybrid database: http://www.jax.org/resources/documents/cmdata/rhmap/

The Jackson Laboratory mouse radiation hybrid database is designed to provide web-based access to a comprehensive interactive database for the mouse T31 RH panel data. It includes an electronic submission interface for depositing RH typing data from users, data error checking and quality control, technical support, data analysis and the development of RH maps. All data, with references and experimental notes, can be viewed or downloaded. Data are shared with the MGD and the European Bioinformatics Institute data repository (RHdb, see below).

3.2.2 The European Bioinformatics Institute radiation hybrid database: http://www.ebi.ac.uk/RHdb/index.html

The European Bioinformatics Institute (EBI) radiation hybrid database (RHdb) is a repository for the raw data for constructing radiation hybrid maps, STS data, scores and experimental conditions (15). It is designed to be a species-neutral database, and currently contains human, mouse and rat RH data. Maps are not assembled from the accumulating data, but may be submitted by data developers.

3.3 Physical mapping data

Mouse physical mapping data on the Internet is currently limited and comes from the genome-wide effort at the Whitehead/MIT and the X-chromosome effort at the Medical Research Council Genome Centre group in Harwell, UK (see Chapter 5). Most of these data are integrated into the MGD, as well as being available from the orginators' sites.

3.3.1 Genome-wide physical map: http://www-genome.wi.mit.edu/

The first genome-wide physical mapping data for mouse have been developed by the Whitehead Institute/MIT. These data include contigs and STS-content mapping across the entire mouse genome and utilize the previously developed SSLP markers that characterized the MIT genetic map of the mouse. Thus the physical map is intimately tied to the genetic map. The Whitehead Institute website makes available data on the markers, PCR primers, probe/YAC hit/miss data and contig localization as well as map figures.

3.3.2 X-Chromosome physical map: http://www.mgc.har.mrc.ac.uk/

The first detailed physical mapping project of a large chromosomal region in mouse is that of the X-chromosome region between markers *Ids* (iduronate 2-sulfatase) and *Dmd* (dystrophin, which is mutated in muscular dystrophy). A website at the UK Mouse Genome Centre, Harwell, UK (http://www.mgc.har.mrc.ac.uk/xmap/xmap.html) describes the project and its status. The physical maps resulting from these data are displayed through the Human Genome

Mapping Project (HGMP) Resource Centre at Hinxton, Cambridge, UK (http://www.hgmp.mrc.ac.uk/MBx/MBxPhysicalChromXStatus.html). Sequencing of a 3 Mb region of mouse chromosome X and its equivalent human chromosomal region is being carried out as a collaborative effort between the HGMP, MRC Harwell, the Genomic Sequencing Centre (Jena), and Children's Hospital (Columbus, Ohio) (http://www.mgc.har.mrc.ac.uk/comp_seq/).

4. Gene expression data

4.1 The Gene Expression Database (GXD): http://www.informatics.jax.org/menus/expression_menu.shtml

The Gene Expression Database (GXD) stores and integrates expression information, concentrating on mouse embryonic development (16,17). The many types of expression assays and data, in their sum, describe expression in normal and mutant genotypes. Expression descriptions are standardized through use of an anatomical dictionary and hierarchy developed jointly with the mouse 3D atlas project and supported by digitized two-dimensional images where feasible. The GXD is integrated with the MGD to provide connections with genotype and phenotype information. The GXD index provides entry into the literature through notation of age, gene and assay performed.

4.2 The mouse 3D atlas: http://genex.hgu.mrc.ac.uk/

The 3D Atlas of Mouse Development is a collaborative work from the MRC Human Genetics Unit at Western General Hospital, Edinburgh, and the University of Edinburgh (18). The atlas is being created using digitized data from serial sections of C57BL/6 embryos for each of the 26 Theiler stages of embryonic development (19). Sections are computationally aligned and corrected using warping software to adjust images for distortion due to fixation or experimental handling. Images are reconstructed into three-dimensional objects that can be rotated and 'virtually' sliced in any plane. Anatomists are 'painting' each embryo to associate all embryonic structures in the computerized three-dimensional reconstructions with the appropriate anatomical terms from the collaboratively developed anatomical dictionary (20). Thus the mouse three-dimensional atlas provides an ideal educational tool for studying mouse development. It is also a vehicle for collection and display of expression information on the three-dimensional images. In collaboration with the GXD, the mouse three-dimensional atlas is part of a larger resource being built as a gene expression information resource that will combine the detailed information from *in situ* experimental data from the GXD with the three-dimensional display of appropriate embryonic developmental stages to produce an interactive view of expression layered on anatomy (21).

5. Databases of transgenics, knock-outs and other induced mutations

Molecular manipulation of the mouse genome can be accomplished using many techniques. Transgenic mice are created when exogenous DNA is inserted into the genome, often in multiple copies and at multiple random chromosomal sites (see Chapter 9). Targeted mutations created by homologous recombination are very specific, replacing the host gene sequence with the target construct (see Chapter 10). Other means of inducing mutations include chemical and radiation mutagenesis (see Chapter 8). Radiation is known to break chromosomes, and can result in major chromosomal rearrangements (22). Chemical mutagens have varying effects depending on the specific mutagen. Chlorambucil is used to produce chromosomal deletions and rearrangements (23) and *N*-ethyl-*N*-nitrosourea (ENU) is being used to uncover new point mutations to develop new phenotypes and allelic series throughout the genome (24,25). Websites for these ENU mutagenesis sites are available at the MRC, Harwell, UK (http://www.mgu.har.mrc.ac.uk/mutabase/) and at the Insititue of Mammalian Genetics, Neuherberg (http://www.gsf.de/isg/groups/enu-mouse.html).

5.1 The transgenic/targeted mutation database (TBASE): http://tbase.jax.org

TBASE presents organized information on transgenic animals and targeted mutations generated and analysed worldwide. It has become a major resource for data describing the constructs, mutant strains and phenotypic results of such genetic manipulation. TBASE originated at Oak Ridge National Laboratory (26), was transferred to the Johns Hopkins University (27) and currently resides at The Jackson Laboratory.

5.2 Mouse Knockout and Mutation Database (MKMD): http://www.biomednet.com/

The mouse knockout and mutation database (MKMD) is available by subscription through BioMedNet. Data provided include for each mutation a brief textual phenotypic description and statement of gene mutation type, function, lethality and background. This resource was built on the original contributions of Brandon *et al.* (28), but has been extended to include all classical mutations which have been defined at a molecular level. Follow-up studies in which mutant mice have been used are also included.

5.3 Induced Mutant Resource (IMR): http://www.jax.org/resources/documents/imr/

The Induced Mutant Resource (IMR) at The Jackson Laboratory imports, maintains and distributes mice created by mutagenesis, transgenesis and targeted technologies. In conjunction with the physical mouse stocks, the IMR maintains a database to provide information about these strains, including how the mutant was constructed, a description of mutant phenotype, husbandry, typing of alleles and references.

6. Animal resources lists

6.1 International mouse strain resource (IMSR): http://www.jax.org/pub-cgi/imsrlist and http://imsr.har.mrc.ac.uk/

A continual issue for the research community is the availability and location of important mouse strains and stocks needed for experimental studies. Major inbred strains can often be obtained from commercial breeders, but the vast majority of specialized strains and mutants are held by non-profit-making institutions and laboratories. The international mouse strain resource (IMSR) provides a WWW interface to the scientific community for 'finding a mouse' (29). Its goal is to act as a universal catalogue, pointing the researcher to the holder of various mouse stocks and providing links to additional information about particular mutations. The first implementation of this catalogue consists of a joint listing of all mouse stocks, strains and mutants available from The Jackson Laboratory and the MRC Harwell site (UK), one of the European Mouse Mutant Archive (EMMA) nodes (see below). The WWW interface, available identically at both sites, allows searching for gene, mutation and strain background, and provides the user with strain and genotype information and whether the stock is available live or cryopreserved. Links then take the user further, to data on the stock of choice and information on how to obtain the animals. Further expansion of the IMSR will include adding strains from other EMMA nodes (Italy, France, Sweden, Portugal), Oak Ridge National Laboratory (USA) and major Japanese mouse distribution centres.

6.2 The Jackson Laboratory: JAX mice IMR, MMR, DNA resource: http://jaxmice.jax.org/index.shtml

The stocks and strains available from The Jackson Laboratory consist of a continually increasing collection of inbreds, special stocks, mutants (spontaneous, induced and targeted) and chromosomal anomalies, available as live mice, cryopreserved embryos and gametes, or DNA samples. The Laboratory holds more than 2300 different stocks and strains. It is rapidly expanding beyond this number as new targeted mutations and ENU-induced alleles are developed. Distinct colonies at The Jackson Laboratory hold JAX mice (the main commercial mouse resource and colonies for distribution), the Induced Mutant Resource (IMR; dedicated to handling targeted knockouts, transgenics and other targeted mutations as well as mutations induced by chemical or radiation treatment), and the Mouse Mutant Resource (MMR; dedicated to the maintenance and study of spontaneous mutations). The DNA resource maintains DNA from nearly all stocks and strains available in these resources, as well as DNA from some extinct stocks and mutations.

6.3 European Mouse Mutant Archive (EMMA): http://www.emma.rm.cnr.it/

EMMA is an international European effort to coordinate mouse production, distribution, cryopreservation and recovery. Its main node in Montorotondo, Italy is responsible for both managing a substantial mouse colony resource and for coordinating resource development among the nodes, including a co-ordinating database and website. Each node of EMMA represents particular local scientific expertise and will thus handle those mouse stocks appropriate to its resident research expertise. While EMMA is in its infancy, it has great potential for providing needed animal resource management as the number of new targeted, transgenic, and induced (ENU) mutations rapidly expands.

7. mgi-list, an electronic bulletin board for the mouse community: http://www.informatics.jax.org/support/lists.shtml

The MGD project has established a mouse genomics community bulletin board, mgi-list, which uses a ListServer system to reflect e-mail messages received and to forward these to all subscribers. The MGI-list bulletin board has been used, for example, to publicize the mouse YAC screening service from Baylor College of Medicine, to announce The Jackson Laboratory gene mapping resource and to publish a call for papers from *Mammalian Genome*. Frequently it is used to query for special mouse mutants and to seek sources for various biological reagents. It is also used to discuss research topics, such as the best reference for mouse brain anatomy, the evidence for recombination reflecting chiasmata and the experience of other researchers with particular vectors or ES cell lines. The main bulletin board for announcements to the community and general information exchange has over 1100 subscribers (mgi-list@informatics.jax.org). Traffic on the mgi-list bulletin board exceeds 1000 messages/year.

8. Summary

The mouse is a focus of genomics research as the best understood mammal for comparative analysis with human. This chapter has outlined major WWW resources that are specific for mouse genomics and biology. Other general resources for molecular reagents, sequence, protein, metabolism and so forth have not been reviewed here, but most of these are found as supporting hypertext links within the mouse-specific WWW resources. As the body of knowledge about the laboratory mouse continues its rapid growth, it is critical that the central core of mouse-specific informatics resources to coordinate and integrate these data be maintained and expanded to support the biomedical research community.

References

1. Blake, J. A., Richardson, J. E., Davisson, M. T., Eppig, J.T. and the MGD Database Group (1999). *Nucleic Acids Res.*, **27**, 95.
2. Eppig, J. T., Blake, J. A., Davisson, M. T. and Richardson, J. E. (1998). *Methods*, **14**, 179.
3. Beechey, C. V. (1998). In *Genomic Imprinting: Results and Problems in Cell Differentiation* (ed. R. Ohlsson). Springer Verlag (in press).
4. Beechey, C. V. (1997). *Mouse Genome*, **95**, 789.
5. Lyon, M. F., Cocking, Y. and Gao, X. (1997). *Mouse Genome*, **95**, 731.
6. Searle, A. G., Edwards, J. H. and Hall, J. G. (1994). *J. Med. Genet.*, **31**, 1.
7. Evans, C. D., Searle, A. G., Schinzel, A. A. and Winter, R. M. (1996). *J. Med. Genet.*, **33**, 289.
8. Dietrich, W. F., Katz, H., Lincoln, S. E., Shin, H. S., Friedman, J., Dracopoli, N. C. and Lander, E S. (1992). *Genetics*, **131**, 423.
9. Dietrich, W. F. *et al.* (1996). *Nature*, **380**, 149.
10. European Backcross Collaborative Group (1994). *Human Mol. Genet.*, **3**, 621.
11. Rowe, L. B., Nadeau, J. H., Turner, R., Frankel, W. N., Letts, V. A., Eppig, J. T., Do, M. S., Thurston, S. J. and Birkenmeier, E. H. (1994). *Mammal. Genome*, **5**, 253.
12. Copeland, N. G. *et al.* (1993). *Science*, **262**, 57.
13. Watson, M. L., D'Eustachio, P., Mock, B. A., Steinberg, A. D., Morse, H. C., III, Oakey, R. J., Howard, T. A., Rochelle, J. M. and Seldin, M. F. (1992). *Mammal. Genome*, **2**, 158.
14. McCarthy, L. C., Terrett, J., Davis, M. E., Knights, C. J., Smith, A. L., Critcher, R., Schmitt, K., Hudson, J., Spurr, N. K. and Goodfellow, P. N. (1997). *Genome Res.*, **7**, 1153.
15. Rodriguez-Tome, P. and Lijnzaad, P. (1999). *Nucleic Acids Res.*, **27**, 115.
16. Ringwald, M., Mangan, M. E., Eppig, J. T., Kadin, J. A., Richardson, J. E. and the GXD Database Group (1998). *Nucleic Acids Res.*, **27**, 106.
17. Ringwald, M., Baldock, R., Bard, J., Kaufman, M., Eppig, J. T., Richardson, J. E., Nadeau, J. H. and Davidson, D. (1994). *Science*, **265**, 2033.
18. Kaufman, M. H., Brune, R. M., Davidson, D. R. and Baldock, R. A. (1998). *J. Anat.*, **193**, 323.
19. Theiler, K. (1989). *The House Mouse: Atlas of Embryonic Development.* Springer-Verlag, New York.
20. Bard, J. B. L., Kaufman, M. H., Dubreuil, C., Brune, R. M., Burger, A., Baldock, R. A. and Davidson, D. R. (1998). *Mech. Dev.*, **74**, 111.
21. Davidson, D., Baldock, R., Bard, J., Kaufman, M., Richardson, J. E., Eppig, J. T. and Ringwald, M. (1998). In *In Situ Hybridization. A Practical Approach* (ed. Wilkinson, D.), pp. 190–214. Oxford University Press.

22. Russell, W. L., Bangham, J. W. and Russell, L. B. (1998). *Genetics*, **148**, 1567.
23. Rinchik, E. M., Flaherty, L. and Russell, L. B. (1993). *BioEssays*, **15**, 831.
24. Brown, S. D. and Nolan, P. M. (1998). *Human Mol. Genet.*, **7**, 1627.
25. Hrabe de Angelis, M. and Balling, R. (1998). *Mutat. Res.*, **400**, 25.
26. Woychik, R. P., Wassom, J. S., Kingsbury, D. and Jacobson, D. A. (1993). *Nature*, **363**, 375.
27. Jacobson, D. and Anagnostopoulos, A. V. (1996). *Trends Genet.*, **12**, 117.
28. Brandon, E. P., Idzerda, R. L. and McKnight, G. S. (1995). *Curr. Biol.*, **5**, 873.
29. Eppig, J. T. and Strivens, M. (1999). *Trends Genet.*, **15**, 81.

8

Mutagenesis of the mouse germline

MONICA J. JUSTICE

1. Introduction

Concern about genetic risk from radiation and chemicals has, in the past, directed the research of germline mutagens in the mouse. Now, however, the Human Genome Project compels functional studies of the mammalian genome (1). This requires new mutations, changing the focus for mouse mutagenesis from studies of genetic risk to isolating new variants at a high frequency (2,3). The value of mutagenesis in the mouse is compounded by the highly developed comparative genetic linkage map between mouse and human, which allows mammalian gene functions to be defined in conserved regions and provides mouse models of human diseases.

Mouse germline mutagenesis can be achieved by a number of methods (4–8):

- treatment of male or female mice with chemicals or radiation;
- insertional mutagenesis; or
- gene targeting.

The goal of this chapter is to highlight the most efficient method for obtaining mutations in mice with phenotype-driven screens. This is accomplished by chemical mutagenesis of male mice using *N*-ethyl-*N*-nitrosourea (ENU). ENU treatment causes the highest mutation rate of any mouse mutagen, allowing efficient screening for mutations in particular processes by identifying altered phenotypes. Other mutagens, including radiation and chlorambucil, are briefly mentioned. Breeding schemes, mutation recovery and other factors to consider in a mutagenesis experiment are discussed to help an investigator conduct a successful screen.

2. Mouse mutagenesis

2.1 Spontaneous mutations

Variants in mouse populations have been recognized since ancient times. Albino mice were used by Chinese priests in auguries as early as 1100 BC, and 'fancy' waltzing mice were recorded in Japan as early as 80 BC. Since then, a variety of mutations that produce visible phenotypes have been collected by mouse fanciers and geneticists. Because of the ease of visualizing alterations in the coat, many spontaneous mouse mutations exist for pigmentation loci (9). Numerous alleles at the *white spotting* (*W*, now called *Kit*) and *steel* (*Sl*, now *Mgf*) loci, detected by their dominant effect on the coat, have been used elegantly to dissect the properties of the Kit and MGF proteins (10,11). Through a standard weaning protocol during routine stock maintenance, scientists at the Jackson Laboratory have isolated numerous spontaneous mutations that alter coat colour, skin, hair, behaviour, neurological function, blood cells or skeletal morphology (12).

The nature and incidence of spontaneous mutations varies. In historical control groups at Oak Ridge National Laboratory, the Medical Research Council Mammalian Genetics Unit, Harwell, and the Jackson Laboratory, the incidence of spontaneous mutations has been catalogued (13). They occurred on average at one in every 100 000 offspring screened per locus (mutation rate $= 1 \times 10^{-5}$). The molecular nature of some spontaneous mouse mutations has been determined, and the lesions include point mutations affecting coding properties or RNA splicing, deletions, and insertions of transposable elements (14–18).

Although spontaneous mutations are a valuable resource, their acquisition is rare and by chance. Further, because of the cost involved in mouse maintenance and breeding, phenotypes are limited to viable, visible mutations. A cost-effective genetic strategy in the mouse requires powerful mutagens that allow for the selection of desired mutants at specific loci.

2.2 Systems for assessing mutation rate

Many systems have been employed to assess germline mutation rate in the mouse. Some detect a variance in the phenotype of the offspring, such as the specific locus test, while others screen for dominant visible phenotypes or assay for health, fertility and viability. In addition, transgenic mice containing bacterial reporter genes are used to determine the nature and rate of mutations in both somatic and germ cells (19, reviewed in 20). However, the nature of germline mutations induced in artificial systems can vary from those selected for in the whole animal (21) (*Table 1*, for example).

2.2.1 Reporter genes

The Big Blue (22) and MutaMouse (23) transgenic assays have been widely used in genetic toxicology. Their advantage is that the target locus can be shuttled into bacteria where the selection for the mutated phenotype takes place. The Big Blue mouse uses a *lacI* reporter gene, whereas the MutaMouse uses a *lacZ* reporter gene for blue/white colour selection after shuttling into *Escherichia coli*. Because these are whole-animal transgenic assays, mutations can be detected in any organ or cell type, including somatic and germ cells.

Table 1. Summary of sequenced mutations caused by ENU

Mutation	*E. coli*	*Drosophila*	Mouse/rat soma	Big Blue mouse soma/germline	Mouse germline
A·T to T·A	1/145 (1)	5/38 (13)	45/87 (52)	24/73 (33)	18/44[a] (41)
A·T to G·C	26/145 (18)	5/38 (13)	18/87 (21)	5/73 (7)	19/44 (43)
G·C to A·T	108/145 (74)	25/38 (67)	7/87 (8)	28/73 (38)	3/44 (7)
G·C to C·G	1/145 (1)	1/38 (3)	3/87 (3)	2/73 (3)	2/44 (5)
A·T to C·G	4/145 (3)	1/38 (3)	13/87 (15)	2/73 (3)	1/44 (2)
G·C to T·A	1/145 (1)	1/38 (3)	1/87 (1)	8/73 (11)	1/44 (2)
Other	4/145[b] (3)	0/38 (0)	0/87 (0)	4/73[c] (5)	0/44 (0)

Numbers in parentheses are the percentage of total mutations in that column. From Marker *et al.*, 1997, with additions (15,21,61–64).
[a]One of these mutations was found after ENU and radiation treatment.
[b]These mutations were three insertions and one deletion.
[c]These mutations were small deletions.

Traditionally, transgenic reporter assays have been used to test the potential carcinogenicity of compounds. However, their efficacy in detecting germline mutation rates has been established and compared with the more traditional mutation detection assays listed below (24).

2.2.2 Specific locus test

The specific locus test (SLT) uses T (tester) stock mice, which carry seven recessive, viable mutations affecting visible traits. Six influence coat colour: non-agouti (*a*, chromosome 2), brown ($Tyrp1^b$, chromosome 4), chinchilla (Tyr^{ch}, chromosome 7), dilute ($Myo5a^d$, chromosome 9), pink-eyed dilution (*p*, chromosome 7) and piebald-spotting ($Ednrb^s$, chromosome 14), while one controls ear morphology: short-ear ($Bmp5^{se}$, chromosome 9). In an SLT using the T stock, wild-type males are divided into two groups: one set treated with radiation or chemicals, and one set an untreated control. The treated males are crossed to T-stock females. Because the T-stock mutations are recessive, the progeny appear wild-type ($a/+$; $Tyrp1^b/+$; $p\ Tyr^{ch}/+\ +$; $Myo5a^d\ Bmp5^{se}/+\ +$; $Ednrb^s/+$), unless a mutation has occurred at any one of the specific loci. For any treatment, the number of specific mutations is recorded to calculate a germline mutation rate. The value of the SLT is to provide a rapid and defined assay to address the parameters affecting mutagenesis in the whole animal. One parameter measured in the SLT is the timing of mutation recovery after treatment, providing data on the germ-cell stage that is affected by the mutagen. After treatment, males are mated to two new females each week for 7 weeks. If mutations are derived from females mated 1 week after treatment, the mutations occurred in spermatozoa; if 1–3 weeks after treatment, in spermatids; and if more than 7 weeks after treatment, in spermatogonia.

The SLT can be used to assay the effects of treatment on either the male or female germline. Wild-type female mice can be treated with a mutagen, while the germ cells are at various stages of development, and mated to T-stock males. However, the recovery of mutations from the female germline is not efficient.

Although each system has its own benefits, the SLT provides the most power for assaying factors that affect the timing and frequency of germline mutations (see ref. 25, for review). The SLT has revealed that the chemical mutagens ENU and chlorambucil, as well as X-rays, can induce mutations at a rate significantly above the spontaneous mutation rate in male mice (5).

2.2.3 Dominant phenotype assays

Dominant phenotypes are a useful indicator of a successfully mutagenized germline in a mutagenesis experiment (26). Specific classes of dominant phenotypes such as eye and skeletal defects can be assayed (27–30). As early as the 1930s, investigators noted a reduction in litter sizes from irradiated males, which was attributed to embryo death. Today, scoring resorption moles comprises the dominant-lethal test (DLT) as an indicator of the mutagenicity of an environmental agent. Unlike the SLT, this assay does not provide quantitative data for the parameters affecting mutagenesis.

2.2.4 Other assays

Reciprocal translocations can be induced by the irradiation of male germ cells, and some of the offspring from the treated males appear 'semi-sterile' due to the death of about half of the offspring (31). G. D. Snell postulated that death was caused by unbalanced segregation products resulting from a translocation. This observation, with some refinements, is the basis for the heritable translocation test (HTT) (see ref. 32, for review).

Another assay for genetic damage is based on a principle proposed by J. B. S. Haldane, that every mutation causes the same decrease in population fitness regardless of the severity of the mutant effect (33). This system counts mutations in a population in which each mutant leads to one extinction. Thus, the impact of mutation in the Haldane–Muller principle is measured by reduced Darwinian fitness. The supporters of this method argue that any approach for measuring mutation rate that is based on observable phenotypes would ignore a significant proportion of induced mutations.

2.3 Developing a mouse mutant resource

Mutagenesis has been a powerful tool for understanding gene function and disease in organisms ranging from bacteria to fish. The mouse is also a powerful genetic model organism for the human with advantages that include genetic malleability, inbred genetic backgrounds, extensive comparative genetic and physical maps, chromosome engineering and functional similarity to the human. Of the 80 000–120 000 mammalian genes being identified through the Human Genome Project, we know the function of relatively few. The number of novel mutations isolated by chemical mutagenesis (26,34–40), and germline gene disruptions (41–43), suggest that the potential for mammalian functional information is extensive. The comparative genetic linkage map between the mouse and human genomes is a valuable tool for extending functional information derived by mutagenesis in mice into the human (44). Therefore, mouse mutagenesis is an important tool for defining mammalian gene function, generating human disease models and allele series, and dissecting mammalian biological systems.

2.3.1 Determining functional complexity of genomic regions

In a few years, a significant proportion of the nucleotide sequence of the human genome will be available in databases (45). The sequence of *Saccharomyces cerevisiae*, *Caenorhabditis elegans* and several microorganismal genomes is complete, and sequencing efforts are tackling the genomes of other microorganisms, *Drosophila* and the mouse (46,47). Sequence analysis, however, often falls short of revealing function. The post-sequencing challenge is to define the function of the genes, a task that is already being addressed in yeast and worms (48). Gene function has been ascribed using a number of strategies:

- by homology with genes that have been analysed by mutation in other organisms;
- by identification of functional motifs;
- by assigning a gene to a specific human disease or mouse mutation;
- by examining expression patterns; and
- from biochemical interactions *in vivo*.

However, in complex genomes, the evolution of gene families requires more information than can be obtained from sequence comparisons. Thus, functional genetic studies of the mammalian genome are required, and large-scale, high-throughput mutagenesis is the best avenue for determining function.

2.3.2 Human disease models

As new mouse mutations are generated, the comparative sequence information will allow predictions of gene function in humans. Many of these new mutations will be powerful models of human diseases because mouse and human biological systems are similar. For example, health-related genes such as those involved in cardiovascular and neurological diseases, haematopoiesis, asthma, inflammation, obesity, osteoporosis, sex determination and fetal–maternal circulation can be studied only in mammals. Further, the phenomenon of imprinting only occurs in mammals, making the mouse an appropriate system for investigating disruptions in imprinted loci. The power of the mouse for modelling human disease has already been demonstrated by a number of mutations, including loss-of-function mutations in *sonic hedgehog*, which cause cyclopia or holoprosencephaly in mouse and human (49–50). Many human diseases, however, may be caused by a more subtle mutation, such as a missense mutation, rather than a gene disruption. For example, human familial hypertropic cardiomyopathy can be recapitulated in the mouse by introducing a missense point mutation into the α-cardiac myosin heavy-chain gene, whereas complete gene disruption does not model the disease (51). Increasing the mouse mutant resource will increase the number of disease models for use as gene therapy or pharmaceutical treatment targets.

2.3.3 Allelic series

Although a knock-out database of the mouse genome is a laudable goal, it should only be considered as a starting point for functional analyses, since variant alleles of a locus are more likely to reveal the full range of functional information. Subtle variants of phenotypes can produce a fine genetic dissection of a gene when coupled with molecular and protein information (9). ENU-induced alleles of a mutation can also confirm the functional identity of candidate genes, as was done by creating new alleles of the *kreisler*, *quaking* and *eed* mutations (15,52,53). Even further, generating a series of alleles may be the simplest and least expensive genetic screen to carry out (see Section 3.3).

2.3.4 Unravelling biochemical or developmental pathways

A genetic dissection of a biological pathway is a powerful screen that can use two approaches:

(a) sensitizing the pathway, such that recessive loci within the pathway can be recovered in a single generation; or
(b) recovering dominant modifiers (suppressors or enhancers) of a trait.

In the mouse, only one sensitized pathway screen has been successfully carried out. In this example, mutations in the phenylalanine metabolic pathway were targeted (37). To recover mutations in this biochemical pathway, mice were administered a phenylalanine load to 'sensitize' the biochemical pathway. Mutations were recovered in the first-generation offspring that were heterozygous for a new mutation and had a reduced ability to clear the phenylalanine load, resulting in high phenylalanine levels in the blood. Although this pathway was not saturated by mutagenesis, new mutations in the phenylalanine hydroxylase structural gene, as well as mutations leading to aminoacidosis, were recovered in a small number of offspring (37,54,55). Additional mutagenesis experiments are likely to uncover developmental pathways unique to mammals.

3. High-efficiency mutagenesis with *N*-ethyl-*N*-nitrosourea

3.1 Mode of action

3.1.1 ENU: chemical properties, stability and half-life

The ethylating agent *N*-ethyl-*N*-nitrosourea (ENU) is a yellowish-pink crystal with a molecular weight of 117.10. It is highly sensitive to humidity, pH and light, and should be stored below –10 °C at an acid pH. At pH 5.0, ENU has a half-life of about 100 h, whereas at pH 9.0, its half-life is <3 min. Because ENU is highly labile at an alkaline pH, it can be safely handled with short-term chemical hazard containment.

ENU is a potent mutagen in many systems ranging from bacteria to mammals, and is a potent carcinogen in mammals. Its ethyl group can be transferred to oxygen or nitrogen radicals within nucleophilic bases, and a number of reactive sites have been identified *in vivo* (56). These include the N^1, N^3 and N^7 groups of adenine, the O^6 and N^3 of guanine, the O^2, O^4 and N^3 of thymine, and the O^2 and N^3 of cytosine. In addition, the phosphate groups of the DNA backbone can be ethylated (56). In bacteria and flies, the most commonly ethylated sites that escape DNA repair are the O^6 of guanine and the N^3 of adenine (57).

Isocyanic acid, which contains a carbamoyl group, is an intermediate degradation product of ENU. It is more stable than the ethylating moiety, and participates in the carbamoylation of protein amino acids. Carbamoylation of histones may have an effect on histone–histone and DNA–histone interactions (58), potentially resulting in non-heritable alterations. Eventually, ENU is degraded into diazoethane and urea under alkaline conditions.

3.1.2 Types of DNA lesion caused by ENU

Many ENU-induced lesions have been sequenced in numerous organisms isolated from several mutation-detection systems. Interestingly, the primary type of lesion induced by ENU appears to differ depending upon the organism and the assay system (*Table 1*). ENU is primarily a point mutagen, causing single base-pair changes; however, small deletions have occasionally been observed. In *E. coli*, *Drosophila* and the Big Blue mouse system, G·C to A·T transitions are common. However, in non-reporter mouse and rat

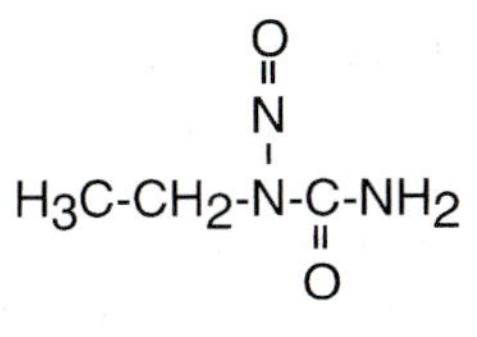

O
‖
N
|
H_3C-CH_2-N-C-NH_2
‖
O

Figure 1. Chemical structure of the alkylating agent ENU.

somatic cells and in the mouse germline, A·T to T·A transversions and A·T to G·C transitions occur most frequently. This difference may be due, in part, to the high rate of spontaneous G·C to A·T mutation in *E. coli* (59). Thus, the primary method of alteration in *E. coli* and flies appears to be alkylation of O^6 guanine. However, in the mouse germline and somatic cells, the A·T to G·C transitions and A·T to T·A transversions may be due to mispairing of O^4-ethyl T and O^2-ethyl T, respectively. Since mammalian O^6-alkylguanine–DNA alkyltransferase (AGT) can remove the alkyl group from O^6-ethylguanine in DNA, it is possible that this lesion is repaired easily in the mouse germline. It will be interesting to compare ENU-induced lesions found in mice with a disrupted AGT gene with those in AGT wild-type mouse strains. For a more thorough review of the reactivity of ENU to DNA, and its mutational capacity in organisms other than mice, refer to ref. 60.

3.1.3 Effects on protein products

One of the potential uses of ENU is to generate allele series at a single locus, especially where expression studies indicate multiple gene functions throughout the development and adult life of the mouse. The lesions induced by ENU may have a variety of effects on the protein product. ENU may eliminate the function of the protein product, either by changing a crucial amino acid, or by introducing a stop codon early in the protein. Alternatively, ENU may cause a hypomorphic change in the protein product, possibly by introducing a missense mutation. Many of the specific locus alleles isolated after ENU treatment were intermediate in phenotype and few were homozygous embryolethal (65,66). A hypomorphic allele may allow later-acting functions to be revealed. For example, *eed* is a mouse mutation that causes early embryonic lethality in the null allele. A hypomorphic allele of *eed* induced by ENU, however, allows the mouse to survive embryogenesis, showing skeletal transformations along the vertebral column, and providing insight into *eed* as a regulator of homeotic genes (53). Even further, ENU can cause neomorphic and antimorphic changes in other organisms, and will probably cause them in mice. An ENU-induced allele of *quaking* (*qk*) has a semi-dominant phenotype that can be explained by a gain of function for the protein product (67,68). Further studies of allelic series will probably identify many types of changes in the mouse.

3.2 Induction of mutations in the mouse germline

3.2.1 Doses and treatment protocols

The action of ENU is most potent in spermatogonial stem cells. Spermatozoa are not effectively mutagenized, but in a large sample size they do occasionally have new mutations (69). Treated post-spermatogonial stem-cell stages yield mutations at a rate one order of magnitude lower than stem-cell spermatogonia at the same dose of ENU (65). Mutation frequencies in the offspring of females treated with ENU are also low (70). Therefore, the most efficient method for obtaining mutations using ENU is from treated male mouse spermatogonia. Because ENU is a spermatogonial stem-cell mutagen, the lesions can be sampled in sperm derived from the treated stem cells. After treatment with ENU, males undergo a temporary sterile period due to the depletion of spermatogonia in the testes. Mutagenized spermatogonia that can survive the treatment repopulate the testis and undergo mitosis and meiosis in the seminiferous tubules, eventually giving rise to clones of mutagenized sperm.

Several doses of ENU have proven to be effective for mutation induction. The dose chosen may depend upon the genetic screen being used to isolate mutations. A very high mutation rate is desired for specific locus experiments and deletion saturation experiments, whereas a lower but effective mutation rate might be desired for a genome-wide, three-generation experiment. The most effective dose of ENU is a fractionated dose of 400 mg/kg body weight administered in four, weekly injections of 100 mg/kg each. However, most inbred mouse strains and some F1 hybrid animals cannot tolerate this dose, and may not regain fertility, or may die (see Section 4.2.1). Lowering the total dose to 360 mg/kg, administered in four, weekly injections of 90 mg/kg each, allows many inbred strains and most F1 hybrids to recover from the treatment, and gives an extremely high mutation rate. A total dose of 300 mg/kg, administered as a fractionated dose of 100 mg/kg every 3 weeks also results in a very high mutation rate. Moreover, a single dose of 100 mg/kg ENU is effective at inducing mutations, even though an optimum mutation rate is not achieved (*Table 2*). Different inbred strains and F1 hybrids respond differently to ENU (71) (M. J. Justice, unpublished results). Therefore, for comparison purposes, data on (101 × C3H)F1 hybrids are presented. It is recommended that an optimum dose protocol be determined for each inbred strain to be treated with ENU (71).

Table 2. Mutation rates

Dose regimen	Amount of ENU per injection (mg/kg)	Total dose (mg/kg)	Mutation rate[a]
Single	100	100	1 in 2323
	150	150	1 in 1687
	200	200	1 in 1456
	250	250	1 in 1484
Fractionated	100	300	1 in 800
	100	400	1 in 655

For comparison, all data are taken from W. L. Russell specific locus tests on (101 x C3H)F1 males (65,66,72,73). Other mutagenesis experiments yield similar rates, although some are higher.
[a]The mutation rate given is the reciprocal of that commonly shown. It is represented as the number of times a single mutation at a single locus would be encountered in that number of gametes, on the average.

Protocol 1. Dissolving ENU and determining concentration by spectrophotometry

Equipment and reagents

- *N*-Ethyl-*N*-nitrosourea, 1 g ISOPAC containers (Sigma)
- 95% ethanol, solvent
- Phosphate/citrate buffer, diluent: 0.1 M dibasic sodium phosphate, 0.05 M sodium citrate, pH 5.0
- Inactivating solution: see *Protocol 3*
- Disposable plastic cuvettes
- 10 ml, 30 ml plastic syringes
- 1 ml tuberculin syringes with 26 gauge, 3/8 inch needles
- 18 gauge needles
- 0.45 μm filter-sterilizing unit

A. *Dissolving ENU*

1. Prepare the phosphate/citrate buffer, and adjust to a pH of 5.0 with phosphoric acid. Filter sterilize using a 0.45 μm unit.
2. In an efficient chemical hood, inject 10 ml of 95% ethanol into the ISOPAC container. Gently agitate the suspension until the ENU goes into solution. Use the warmth from your hands to warm the containment vessel.[a] **Caution:** wear plastic gloves, lab coat and mask.
3. Inject 90 ml of phosphate/citrate buffer into the ISOPAC container vented with an 18 gauge needle.[b] Mix thoroughly.

B. *Spectrophotometric determination of ENU concentration*[c]

1. Dilute 400 μl of the suspended ENU to 2000 μl with phosphate/citrate buffer (1 in 5 dilution) in a disposable plastic cuvette.
2. Determine the OD_{398nm} relative to a blank containing a 1 in 50 dilution of 95% ethanol in the phosphate/citrate buffer (also in plastic disposable cuvette).
3. Calculate the concentration of the ENU in the solution based upon the observation that a 1 mg/ml solution gives an OD_{398nm} of 0.72. A wavelength scan from 350 to 450 nm can be performed if desired.

[a] ENU is very sensitive to light, humidity and pH. Therefore, dilute a new ISOPAC of ENU prior to each weekly injection, protect the ENU from light using a foil wrap, and inject mice within 3 h of diluting the ENU.

[b] If the amount of ENU injected per mouse is going to be small, you can halve the concentration of ENU by removing 5 ml from the ISOPAC prior to dilution, then adding 95 ml of buffer. Place the removed 5 ml of ENU directly into 50 ml of inactivating solution, and discard after a minimum of 2 h.

[c] The Sigma ISOPAC vials contain approximately 1 g of ENU, but this amount is not accurate, and ranges from about 0.7 g to 1.2 g. This step also controls for instability of the ENU and dilution errors.

Protocol 2. Treatment of male mice with ENU

Equipment and reagents

- Freshly diluted ENU of known concentration (see *Protocol 1*)
- Male mice of appropriate strain, 8–12 weeks old
- Inactivating solution (see *Protocol 3*)
- 1 ml tuberculin syringes with 26 gauge 3/8 inch needles

Injections may be a single dose of ENU or a fractionated dose. Determine which is appropriate for your experiment and strain of mice. Fractionated doses are administered weekly at approximately the same time each week.

A. *Weighing mice*

1. Mice should be weighed prior to each injection. Mark each animal to distinguish it on subsequent days. Determine and record live weight.
2. Calculate the appropriate amount of ENU to be injected into each mouse depending on the desired dose (see *Table 2* and *Protocol 5*). Dose can be calculated in mg/kg body weight, or μg/g body weight.

B. *Injecting ENU*

1. In an efficient chemical hood, administer the appropriate amount of ENU intraperitoneally to each mouse. The concentration of ENU should be such that <1 ml is injected into each animal. Animals may appear wobbly and lethargic for about 30 min after injections because of the alcohol diluent.
2. Clean all spills and soak all equipment and gloves coming into contact with ENU with the inactivating solution. Inactivate any remaining ENU and discard appropriately.
3. Keep the mice in an efficient chemical hood for at least 24 h after injections, or change bedding at least 24 h after injections into a plastic bag containing paper saturated with inactivating solution. This usually requires marking the cages containing treated animals with appropriate warnings.

Protocol 3. Inactivating and disposing of ENU

Equipment and reagents

- Inactivating solutions:[a]
  (i) 0.1 M KOH, prepared by dissolving 5.6 g KOH pellets in 1000 ml double-distilled water.
  (ii) alkaline sodium thiosulphate: 0.1 M NaOH, 20% (w/v) $Na_2S_2O_3$. Dissolve 200 g of $Na_2S_2O_3$ and 4 g NaOH to a final volume of 1000 ml with double-distilled water.
- Chemical hazardous waste container
- Plastic bags

Protocol 3. *Continued*

A. *Inactivating ENU*

1. Inject at least 50 ml of inactivating solution into the remaining ENU in ISOPAC. Leave in chemical hood exposed to light for at least 24 h.
2. Remove the seal from the ISOPAC and discard inactivated ENU into a chemical waste container; record appropriately.
3. Rinse the ISOPAC with water, and discard contents into the chemical waste container.
4. Discard the emptied ISOPAC in glass waste.

B. *Treating items that have been in contact with ENU*

1. Draw inactivating solution into all needles that have been in contact with ENU.
2. Soak any beakers, vials or any other equipment that have been in contact with ENU with inactivating solution.
3. Clean all spills and soak all equipment coming into contact with ENU with the inactivating solution.
4. Put the treated items in a plastic bag and discard in waste.

[a] ENU has a very short half-life under alkaline conditions; either inactivating solution can be used.

3.3 Genetic screens to isolate mutations

ENU induces lesions throughout the germline of the treated male. To isolate mutations that give the desired phenotypes, a variety of genetic screens can be employed. Phenotype assays depend upon the experiment, and will not be discussed here.

3.3.1 Dominant mutations

The simplest genetic screen is a search for visible dominant mutations, which can be collected as a by-product in the first-generation offspring of any ENU-mutagenized male (74). To increase the likelihood of isolating novel dominant mutants, new phenotype assays are being used. *Clock*, a mutation affecting circadian rhythm, was isolated by assaying for abnormal wheel-running activity in a light-free environment (75). Other groups are assaying for novel neurological phenotypes using the SHIRPA protocol (76), or for abnormal blood parameters (M. Hrabe de Angelis and R. Balling, unpublished). Since genetic backgrounds can change the probability of isolating certain phenotypes, care should be given to perform dominant screens in a variety of mouse strains, and to assay the ancestral strain for the screening characteristics.

3.3.2 Single-locus screens for recessive mutations

Additional alleles of an existing mutation can be recovered using the mutation of interest in an assay similar to the SLT. To isolate alleles of a viable phenotype, a mutagenized male is crossed to a female homozygous mutant for the test locus (*m*/*m*; *Figure 2*). ENU mutations specific to the locus (*) can be recognized in the first (G1) generation (*/*m*). If the phenotype is homozygous sterile or lethal, the analysis requires a linked marker and a two-generation breeding scheme to isolate additional lethal or sterile alleles. A single-generation breeding protocol can be used to isolate viable alleles of lethal mutations, although this scheme may be inefficient on a per-locus basis. Phenotypic screens may bias the classes of mutations isolated. For example, in a multitude of mutations at the *Myo5a* (previously called *dilute*) locus that affect melanocyte pigment formation and neurological function, some mutations affect pigment only while others influence both pigment and neurological function. However, no variant isolated affected neurological function without a concomitant coat-colour phenotype (63,64). This is probably because the phenotypic screen was designed to look for colour variants rather than neurological variants.

3.3.3 Three-generation breeding scheme for recessive mutations

A three-generation breeding protocol may be used to isolate recessive mutations in the mouse genome. Isolating viable traits is the simplest approach for this breeding scheme. Mutagenized male mice are mated with wild-type females (*Figure 3*). Each G1 offspring represents a gamete sampled from the mutagenized male, which may carry a new mutation of interest. A pedigree is established from individual G1 animals by mating to a wild-type animal. In this scheme, it is most efficient to mate the G1 males to wild-type females. The reciprocal cross (female G1 animals) can be used, but male G1s are preferred since females are reproductively limited. Two or three G2 daughters are then mated back to their fathers to produce G3 offspring, which are screened for

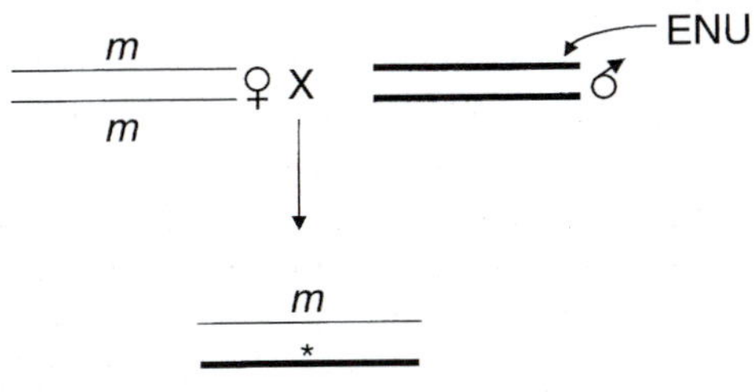

Figure 2. Generating allelic series at single loci. Males are mutagenized with ENU, and mated to females carrying a viable mutation (*m*) of interest. New alleles (*) are identified if they fail to complement the original allele.

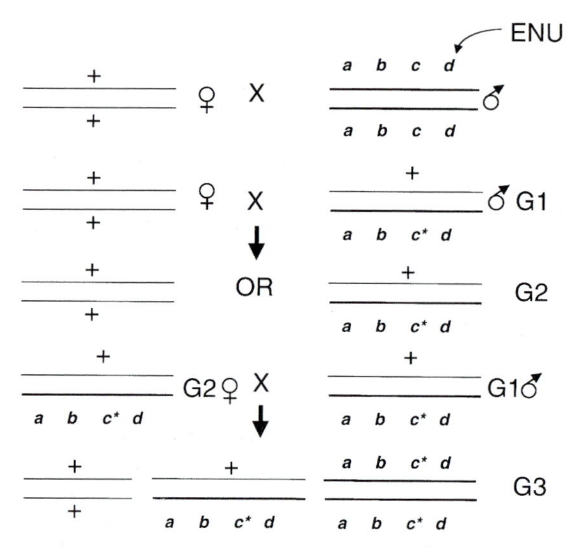

Figure 3. A three-generation breeding scheme to isolate recessive mutations. Potentially mutable loci in a region on a chromosome are designated *a, b, c* and *d*. New recessive mutations can be identified by in a three-generation pedigree without genetic markers. Following the initial cross between mutagenized males and untreated females (top row), G1 male offspring, potentially carrying new mutations (shown here as *c**), are crossed again with untreated females (second row) to produce G2 offspring which are wild type or carriers for the mutation (third row). G2 females are taken at random and backcrossed to their fathers (the fourth row shows the case where a carrier daughter has been used). If a new recessive mutation has been produced, then half of these crosses will have 25% of the offspring as homozygotes for the mutation (*c**/*c**) and can be identified in the G3 offspring (fifth row).

variation. Alternatively, brothers and sisters from the G2 generation can be intercrossed. However, as G2 animals may be wild type or heterozygous for a new mutation, then the likelihood that pairs of heterozygous mutants are randomly picked to uncover a particular mutation is lower than when fathers are mated to their daughters (*Figure 3*).

Many mutations cause the animal to die as an embryo. This lethal class of mutations is difficult to isolate in a three-generation breeding scheme, unless linked markers are used. One study in the *t*-region of chromosome 17 used linked dominant visible, recessive visible and serological markers to isolate lethal mutations in an 11 cM region (36), and then used these same markers in mutant stock maintenance. Alternatively, an embryo-lethal phenotype can be isolated by phenotypes after dissection of litters from females (40; A. Peterson, unpublished results). In these studies, the mutation must be maintained using flanking molecular markers, limiting the number of mutations than can be isolated and maintained in a single laboratory.

As an alternative, marked balancer chromosomes can be engineered in the

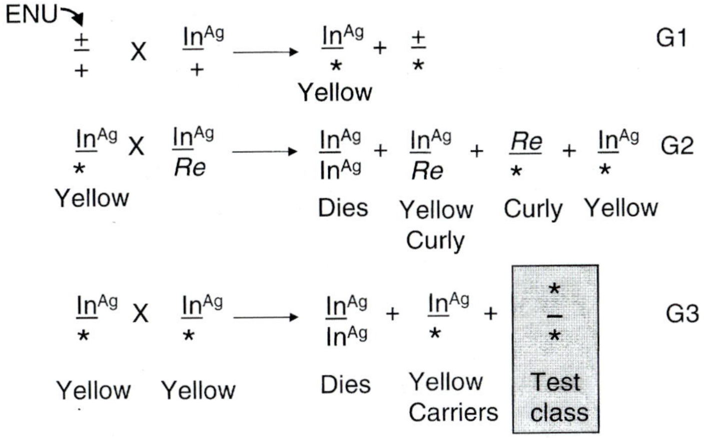

ENU
+/+ X In^{Ag}/+ → In^{Ag}/* + +/* G1
Yellow

In^{Ag}/* X In^{Ag}/*Re* → In^{Ag}/In^{Ag} + In^{Ag}/*Re* + *Re*/* + In^{Ag}/* G2
Yellow
Dies Yellow Curly Curly Yellow

In^{Ag}/* X In^{Ag}/* → In^{Ag}/In^{Ag} + In^{Ag}/* + */* G3
Yellow Yellow Dies Yellow Carriers Test class

Figure 4. Isolating recessive mutations with a marked balancer chromosome. The example shown is being used for saturation mutagenesis of mouse chromosome 11 (A. Bradley and M. Justice, unpublished). An inversion (In) tagged with a K14-*agouti* transgene (*Ag*) produces mice with a dominant yellow coat. One end of the inversion disrupts the *Wnt3* locus on mouse chromosome 11, causing mice homozygous for the insertion to die. By introducing the dominant curly coat marker *Rex* (*Re*) on chromosome 11 into the scheme, each class of G2 progeny can be distinguished. G2 animals that potentially carry a new recessive mutation on chromosome 11 have yellow coats (In^{Ag}/*). By mating yellow G2 brothers and sisters, interfering and background mutations can be avoided. A new recessive mutation (*/*) can be detected in the G3 offspring. If the mutation is lethal (i.e. */* are not seen, and all G3 offspring are yellow) the mutation can be rescued by the yellow-tagged carriers at G3 (In^{Ag}/*).

mouse using *Cre–loxP* recombination, to facilitate screening for recessive lethal mutations and stock maintenance (77) (*Figure 4*). Balancer chromosomes have been used extensively in *Drosophila* for mutation screens and stock maintenance. An ideal balancer chromosome suppresses recombination over a reasonable interval, carries a dominant visible marker, and is homozygous lethal. One scheme using an inversion marked with a dominant K-14-*agouti* transgene allows a screen of a large genetic interval on mouse chromosome 11 for recessive lethal mutations. In this scheme, daughters potentially carrying the mutation can be readily identified so it is not necessary to mate multiple daughters back to the fathers. Alternatively, G2 brothers carrying a potential new mutation can also be identified to mate to sisters. Mating marked G2 animals ameliorates the problem of co-segregating or background mutations. The use of genetic markers also reduces the number of animals that must be screened for variation.

3.3.4 Two-generation breeding scheme using deletions

Mutation screening using deletions exploits the concept of regional haploidy (*Figure 5*). ENU-treated males are mated to wild-type females to generate G1

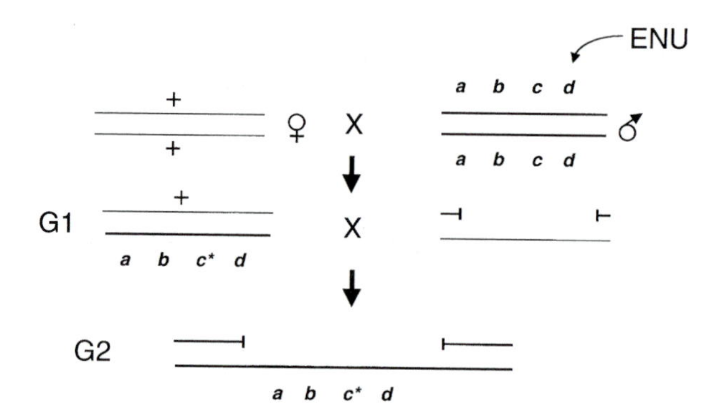

Figure 5. A two-generation breeding scheme using a deletion. Recessive loci in a region on a chromosome are designated *a, b, c* and *d*. A deletion creates regional haploidy, so a new recessive mutation (*c*/c**) can be isolated in the second generation. Either male or female G1 animals can be mated to male or female deletion carriers.

animals, which are then mated to females or males carrying deletions. The ENU-induced lesion may be identified in the G2 animals if it lies in the deletion interval. Incorporating genetic markers into the breeding scheme can further classify the potential mutant, carrier and non-informative offspring. The problem with such a screen is that only a small region of the genome can be screened for mutations in a single experiment. If G1 males are mated to multiple females carrying contiguous deletions, a larger chromosomal region can be 'scanned' for mutations (77). Because the mutation is selected by a failure of the lesion to complement the deletion, the mutation is localized to the deleted region as it is isolated. The two generation breeding scheme reduces the problem of co-segregating and background mutations. Pedigrees carrying new mutations can be maintained as stocks and analysed for their homozygous phenotype.

3.3.5 Modifiers and sensitized pathways

One of the most powerful genetic approaches in *Drosophila* is to dissect a developmental pathway with a modifier or sensitizer screen (78). In a modifier screen, a dominant mutation (*) can be isolated by its ability to modify (either enhancing or suppressing) a known recessive (*Figure 6a*) or dominant phenotype (not shown). Alternatively, in a sensitized pathway screen, recessive mutations in genes that interact in a developmental or biochemical pathway may fail to complement each other completely, a phenomenon called non-allelic non-complementation. A locus of interest (*m*) is included in the genetic scheme. A new mutation at a different locus (*) may fail to genetically complement it (*/+ *m*/+), yet yield a phenotype similar to the original homozygous mutation (*Figure 6b*). Such a screen may yield additional alleles of the original mutation, as well as interacting factors. This approach is ideal

(a)

m ♀ X m ♂ ← ENU

m m

m OR m *

m * m +

Select for suppressed
or enhanced phenotype

(b)

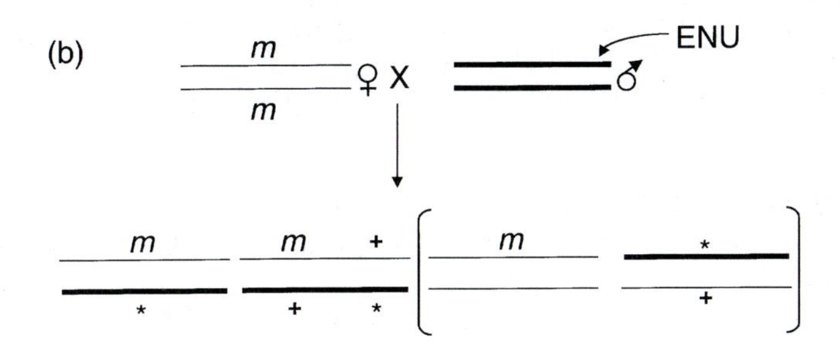

Select for similar phenotype

Figure 6. Modifier and sensitized pathway screens. (a) To isolate a dominant modifier of a recessive mutation (*m*), males homozygous for the mutation must be mutagenized, and mated to females homozygous for the mutation. Suppressors or enhancers may be isolated on the same chromosome or a different chromosome. (b) Sensitized pathways exploit the phenomenon of non-allelic non-complementation. A new mutation (*) fails to complement an existing mutation (*m*) while both are heterozygous.

to dissect a single biological system, and may be one of the most exciting approaches to take in the mouse.

4. Practical considerations for ENU mutagenesis

4.1 Breeding considerations

4.1.1 Male rotations

ENU-treated males should be mated frequently after they recover fertility to increase the likelihood of mutation recovery, because the lifespan of males may be shortened after ENU treatment. This can be accomplished by rotating the males weekly to two new females for 7 weeks (*Figure 7*). Because the mouse gestation period is approximately 3 weeks, and mice are weaned from the mothers at 3 weeks, the females bred during week 1 will be available for mating after 7 weeks. This protocol ensures the most productive matings from a mutagenized male.

Protocol 4. Male rotations

Equipment and reagents

- ENU-mutagenized male mice
- Female mice of appropriate genotype

A. *Cage set-up*

1. Males should be rotated in a seven-cage experiment in a format that is compatible with the animal colony. Males are placed with two new females each week on the same day. Space must be allocated for separation of the pregnant female to a new cage. This may be accomplished by leaving space for a cage next to the male cage. The females can be separated when pregnant, or when the male leaves the cage.

B. *Male rotations*

1. Place the male in the cage with two females on week 1. Rotate on the same day each week to a new cage with two females until the desired number of gametes have been sampled.
2. After 7 weeks, the males can be returned to the first cage of females. Alternatively, new males can be introduced into the cages.

C. *Female generation*

1. For adequate male rotation, enough females of appropriate breeding age must be generated. If eight ENU males are to be used in an experiment, two females are needed per male. This would require two females per male per week, or 112 females. If females are not available, the number of cages or number of females per cage can be modified.
2. Females can be used until their reproductive capacity is reduced. New, younger females should replace older females when necessary.

D. *Record keeping*

1. Good record keeping is essential. Record the date on which the male goes into the female cage, the date he leaves the cage, the date the litter is born and the number of pups in the litter. When the animals are weaned, record the date of weaning, the number and sexes of the pups weaned, and any deviants that are detected.

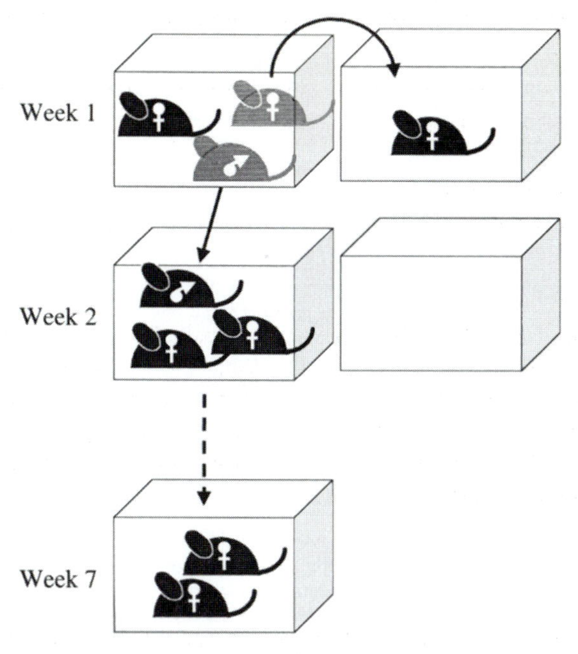

Figure 7. Rotating males in a seven-shelf experiment. Males are placed with two females each week. After mating, or when the female looks pregnant, she is removed to another cage for giving birth. Males are moved to a new cage on the same day each week.

4.1.2 Gamete sampling for spermatogonial stem-cell mutagenesis

After injecting ENU into male mice, the spermatogonial stem cells are depleted in the testis, causing sterility after 3 weeks. If some of the spermatogonia can repopulate the testis, the males will regain fertility in 11–22 weeks. The number of spermatogonia required to repopulate the testes to achieve fertility is not clear, and the number that do repopulate varies according to the dose protocol. Some males are presumably semi-sterile after treatment with ENU, and give small litter sizes as well as missed litters. Males that do not regain fertility show empty seminiferous tubules in histopathology sections of the testes (V. G. Bode, A. P. Davis and M. J. Justice, unpublished results).

Mutagenized spermatogonia undergo mitosis prior to meiosis during spermatogenesis. Therefore, the testes will be repopulated with clones of mutagenized sperm, making the occurrence of 'clusters' or repeat isolates of a single mutation common. In one ENU experiment, the gamete number and order of isolation of presumed clusters was recorded (*Figure 8*) (34). Two isolates from one of these presumed clusters (male 4) have been sequenced, and have the same base-pair change, confirming their derivation as a cluster (68). It is prudent, therefore, in a mutagenesis experiment to save only one mutation isolate per locus from a single male. This phenomenon of clustering

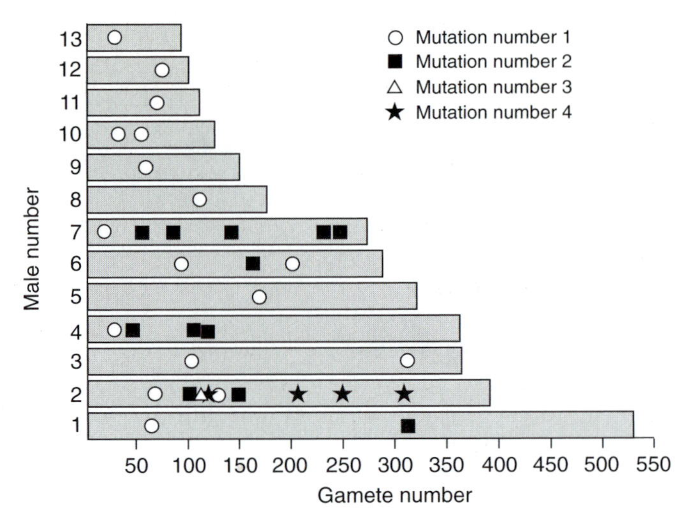

Figure 8. Gamete sampling for mutation isolation. In an ENU experiment where males were given a single dose of 250 mg/kg ENU, offspring were examined for dominant and recessive mutations (34). Eighteen new mutations were isolated in this experiment that represent alleles of nine different loci. Thirteen of the 17 males sampled produced new mutations. From the four males that did not produce new mutations, 227, 131, 113 and 111 gametes were sampled. The length of the horizontal bar represents the number of gametes sampled from each male on the vertical axis. The circle represents the first mutation isolated from each male, and could represent any one of the nine loci. A repeat isolate of that mutation would also be designated as a circle, since it represents the same mutation. The second mutation isolated represents a mutation from a different locus, and is designated as a box. A third mutation (triangle) and fourth mutation (star) were isolated from one male. The symbol is placed at the number where that gamete was sampled. For example, the first mutation was isolated from male 7 at gamete number 12. Male number 4 had one mutation isolated at gamete 21, but no repeat isolate of this mutation was observed in 357 gametes sampled. A second mutation was isolated from male 4, and a cluster of three was observed at gamete numbers 46, 106 and 107.

can potentially be exploited to re-isolate a mutation that is lost in a G1 experiment; however, the data in *Figure 8* show that clusters are not always re-isolated, even when large numbers of gametes are sampled. The isolation of clusters in experiments that involve two- or three-generation pedigrees should be avoided, because of the time and space required for these breeding schemes.

Sampling too few gametes from a single male may lower the observed mutation rate in the experiment, whereas sampling too many gametes will probably isolate repeat mutations. Since mutagenesis is a random event, one male may carry many mutations, but none in the desired locus. Isolating a mutation from a desired locus may require mutagenizing and sampling many males. However, the observed mutation rate may be affected by under-sampling from a given male. After ENU treatment, sperm in the testis belong to one of two classes, wild type or mutagenized. However, a sperm mutagenized at a single locus is rare in the population, and a Poisson distribution would apply. The probability of the event occurring (mutation isolation) is given as $P(m)$ in the Poisson formula:

$$P(m) = ((\mu)^n e^{-\mu})/n!$$

where n is the number of trials and $\mu = np$, where p is the proportion of sperm carrying the mutation.

In a population of 100 sperm from one male, only one sperm may carry a mutation at a locus of interest. The chances of isolating that single mutation are good if 50 or 100 gametes are sampled, but low if only ten gametes from that male are sampled. In the examples below, $P(0)$ is the probability of not isolating the mutation, $P(1)$ is the chance of isolating it once, and $P(2)$ is the probability of isolating it twice.

Case 1: 100 trials, probability of mutation = 0.01 (1 in 100)

$$\mu = 100 \times 0.01 = 1$$
$$P(0) = (1)^0 e^{-1} = 0.368$$
$$P(1) = (1)^1 e^{-1} = 0.368$$
$$P(2) = \frac{(1)^2 e^{-1}}{2} = 0.184$$

Case 2: 50 trials

$$\mu = 50 \times 0.01 = 0.5$$
$$P(0) = (0.5)^0 e^{-0.5} = 0.6065$$
$$P(1) = (0.5)^1 e^{-0.5} = 0.303$$
$$P(2) = \frac{(0.5)^2 e^{-0.5}}{2} = 0.076$$

Case 3: 10 trials

$$\mu = 10 \times 0.01 = 0.1$$
$$P(0) = (0.1)^0 e^{-0.1} = 0.905$$
$$P(1) = (0.1)^1 e^{-0.1} = 0.0905$$
$$P(2) = \frac{(0.1)^2 e^{-0.1}}{2} = 0.0045$$

One question is whether sampling ten gametes from each of ten males (100 gametes in total) would increase the chances of recovering the mutation, since the same number of gametes is sampled. More than one male may carry an allele of the same mutation, so the probabilities are not mutually exclusive, yielding a probability $P = (0.0905)^{10} = 3.7 \times 10^{-11}$. Sampling 50 gametes each from two males gives $P = (0.303)^2 = 0.092$. Thus, undersampling spermatozoa from each male would decrease the chances of isolating the mutation, and lower the observed mutation rate. *Figure 8* shows real data that illustrate the point. Of the 18 new mutations isolated in this experiment, none would have been isolated if only ten gametes were sampled from each male. If 50 gametes had been sampled from each male, five of the 18 mutations would have been isolated. Fourteen of the 18 mutations would have been isolated if 100 gametes had been sampled from each male. The dose used in the experiment was a relatively low dose of ENU (single 250 mg/kg dose). Therefore, increasing the number of males to sample an adequate number of gametes may not yield the same mutation rate as a properly sampled experiment.

A higher dose of ENU would reduce the population of spermatogonial stem cells, so fewer gametes would be required for adequate sampling. After treatment with a high dose of ENU (total dose $\geq$ 300 mg/kg), sampling 30–50 gametes per male should provide an adequate sampling size while avoiding the isolation of repeat mutations. After treatment with a lower dose of ENU (total dose $<$ 300 mg/kg), approximately 100 gametes per male should be sampled (*Figure 8*). In practice, if one were performing a specific locus experiment to isolate multiple alleles of a given locus, and wanted to sample 3000 gametes, one should administer a high dose of ENU and sample 50 gametes per male from 60 males.

4.2 Strain background effects

4.2.1 Inbred strains to use for mutagenesis

A number of inbred strains have been tested for their ability to tolerate ENU. As a rule, F1 hybrid animals will tolerate higher doses, regain fertility faster and live longer after treatment. However, a number of inbred strains have also proven useful for ENU mutagenesis. In many experiments involving sensitized pathways and behaviour analysis, an inbred background will be desirable. One inbred strain, BTBR/N, is particularly resistant to the toxic effects of ENU and yields a very high mutation rate after treatment (37). This strain is reported to yield the same mutation rate after a single dose of 250 mg/kg as with a fractionated dose of 4 $\times$ 100 mg/kg (A. Shedlovsky and W. F. Dove, personal communication). Other inbred strains used for mutagenesis successfully are C57BL/6J and BALB/cJ, and several C3H substrains. FVB/N, which is commonly used to create transgenic mice, does not tolerate ENU well, but can recover from a low dose of ENU to produce mutant offspring (71). Inbred strains vary widely in their tolerance of ENU, so a dose protocol should be determined for each strain (*Protocol 5*).

Protocol 5. Optimizing ENU dose for inbred strains

Equipment and reagents

- ENU (see *Protocol 1*)
- Fertile female mice
- Male mice of appropriate inbred strain, 8–12 weeks old

Method

1. Inject five sets of ten males as in *Protocol 2*, each with a different dose of ENU. The dose regimes are:
   - dose 1: 4 × 100 mg/kg
   - dose 2: 3 × 100 mg/kg
   - dose 3: 4 × 85 mg/kg
   - dose 4: 3 × 90 mg/kg
   - dose 5: 1 × 250 mg/kg
2. After 8 weeks, set up males with fertile females for fertility testing, two females per male.
3. Males should regain fertility at 11–22 weeks. Record date of fertility recovery (calculated as time after final injection).
4. Signs of ENU toxicity are scruffy fur and death. Some males look healthy, but do not recover fertility. If males die, record date of death.
5. If none of the males at the above doses recover fertility, perform a second round of doses. These doses are not performed routinely, because mutation recovery at low doses has not been documented, except for FVB/N mice. Again, inject another five sets of ten males for each dose:
   - dose 6: 1 × 200 mg/kg
   - dose 7: 1 × 150 mg/kg
   - dose 8: 1 × 120 mg/kg
   - dose 9: 4 × 50 mg/kg
   - dose 10: 3 × 60 mg/kg
6. Repeat step 2.
7. If males do not become sterile, they are not mutagenized. If males become sterile, and then regain fertility, perform an SLT or other mutation assay to determine optimum dose rate for your experiment.

4.2.2 F1 hybrids

No F1 hybrid animal tested to date for its ability to tolerate ENU treatment has failed. Most F1 hybrid animals can tolerate a single dose of 250 mg/kg or a fractionated dose of 4 × 100 mg/kg. One exception to this rule is (101 × C3H)F1 hybrid animals, which tolerate a dose of 4 × 90 mg/kg, but not 4 × 100 mg/kg. (SEC/J × C57Br/6J)F1 males can tolerate a dose of >4 × 100 mg/kg (J. Weber and M. J. Justice, unpublished results). Because of these slight variations, it is recommended that a modified dose protocol be developed for each new F1 hybrid to be injected with ENU. This modified dose protocol should include doses of 4 × 120 mg/kg, 4 × 100 mg/kg, 4 × 90 mg/kg, 4 × 85 mg/kg, 3 × 100 mg/kg, 250 mg/kg and 200 mg/kg.

4.3 Other observations

4.3.1 Rate of recovery to fertility

Each animal that has been treated with ENU should become sterile after treatment. To test fertility, treated males should be placed with fertile females 8 weeks after the final ENU injection. Females should not become pregnant immediately. The fertility recovery rate can vary widely among different strains and individual mice. For example, C57BL/6J mice become fertile about 11–12 weeks after the final injection for a 3 × 100 mg/kg dose. In contrast, BALB/cJ mice become fertile 17–23 weeks after injection for the same dose. If mice are not fertile 27 weeks after the final injection, they should be discarded. Some mice that have regained fertility as late as 20–22 weeks have yielded extremely high mutation rates (M. J. Justice, unpublished results).

4.3.2 Mutation rates for different loci

The specific locus rate in *Table 1* is an average mutation rate for the seven loci of the SLT. This rate is not meant to extrapolate to all loci in the genome; however, in practice, observed mutation rates in a variety of experiments have mimicked this average rate. A variety of factors can affect the observed per-locus mutation rate. These include the gene's physical size, functional redundancy, functional aspects of the protein, phenotype screens and screens for dominance or recessivity. To compare two loci in the SLT, *agouti* is a very small target of 800 base pairs, and ENU-induced mutations at *agouti* are rare: only eight have been recovered in 40 years of experiments at Oak Ridge National Laboratory. In contrast, 69 mutations have been recovered at *dilute*, a very large target that produces at least three RNA isoforms, the largest of which is 12 kb (79). The target size is not the only factor that can influence mutation rate. For example, phenylalanine hydroxylase (PAH) is encoded by a transcript of only 1.2 kb, yet the per-locus mutation rate is one in every 175 gametes screened (37). A lesion in any part of the human *PAH* gene can produce an impaired protein, possibly explaining the high mutation rate for the relatively small target size (62).

Very few loci in the mouse have been explored thoroughly by point mutagenesis. Therefore, more information regarding induced mutation rates is likely to accumulate as additional mutagenesis experiments are carried out.

4.3.3 Cancer susceptibility and lifespan

ENU is a carcinogen as well as a germline mutagen, and many mice will succumb to various types of cancer after treatment. Some inbred strains whose genetic backgrounds predispose to cancer will die after treatment. For example, C57BL/6J mice are susceptible to T-cell leukaemia, and many of these mice die of disease shortly after regaining fertility. It is recommended that matings be performed as soon as the males are fertile, so that proper gamete sampling can be achieved.

ENU will often shorten the lifespan of the mouse. Because ENU is a stem-cell mutagen, it affects various stem cells, including haematopoietic stem cells. Thus, males are immunosuppressed and highly susceptible to infection during and after injection. Pathogen load in the colony has an effect on survival of treated males, and could possibly have an effect on mutation rate (M. J. Justice, unpublished results). Males will live an average of 1 year after treatment in a clean colony, rather than the normal 18 months to 2 years. Very few treated males live longer than 18 months.

5. Other mutagens

5.1 Radiation mutagenesis

5.1.1 X-rays: treatment and mutations recovered

Historically, the primary interest in mouse irradiation was to estimate the genetic risk for humans of exposure to various doses and types of radiation. In a seminal manuscript, W. L. Russell and colleagues suggested that acute doses of radiation were more hazardous than chronic doses (80). Since a basic tenet of radiation genetics from *Drosophila* experiments was that dose rate does not affect mutation rate, Russell's results allowed the scientific community to re-evaluate the hazards of irradiation, and validated the mouse as a model of risk estimates to mammals. Subsequent manuscripts evaluated the factors that affected radiation-induced mutations. A thorough review is presented by Searle (81), which describes the mutations recovered using different dose regimens for X-rays.

The male germline is affected by X-irradiation at two stages: spermatogonial stem cells and post-meiotic spermatids, and the stage sampled after treatment affects mutation recovery. Unlike ENU, X-rays are effective at inducing mutations in post-meiotic stages sampled 6–7 weeks after treatment, where mutations are derived from irradiated spermatozoa and spermatids. At a dose of 300 rads, mutations are recovered at a rate of 33×10^{-5} (81). At this stage, the mean mutation rate per locus per roentgen is twice that for spermatogonial irradiation. The disadvantage of mutation recovery at this stage is that offspring can be obtained for only 2 weeks after treatment. Many of the new mutations obtained from post-meiotic stages are large deficiencies or other chromosome aberrations (82). Many of the X-ray-induced SLT mutations recovered in these dose experiments have been maintained as stocks and are useful for fine-structure functional studies of the mammalian genome (5,77).

The female germline is also affected by treatment with X-irradiation, although the rate varies by oocyte stage. The estimate of spontaneous mutation rate in oocytes is significantly lower than in males (70). A characteristic of specific-locus mutations induced in oocytes is that they include a considerably higher percentage of large lesions that affect multiple loci than do mutations induced in spermatogonia. The stage of oocyte maturation affects female mutagenesis. Mitosis in the female mouse embryo occurs during the migration of the primordial germ cells and continues after the cells have reached the germinal ridge. Pre-meiotic DNA synthesis begins about 12–13 days after conception and is completed by day 16. Oogonia that do not enter meiotic prophase degenerate, so mouse females are born with their total supply of oocytes. By the time of birth, most oocytes are in late pachytene or early diplotene; 4–5 days after birth, only dictyate oocytes are found. The oocytes continue to mature for 6 weeks to the time of ovulation (83). In a typical adult female irradiation experiment, females are treated with radiation and mated to males any a given time from week 1 through week 7 after treatment, sampling irradiated dictyate oocytes in various stages of maturation. Most of the mutations recovered in female experiments were collected in the first 6 weeks following treatment (see ref. 70, for review); however, females treated with X-irradiation tend to become infertile after bearing one or two litters. Therefore, female mutagenesis is not recommended for mutation isolation.

Earlier stages of oocyte development can be tested by treating the females as newborns or as embryos by exposure of the mother. These germ cells would be in pre-dictyate stages of meiotic prophase. Radiation-induced mutation rates in embryonic, fetal and newborn females are lower than those in the mature and maturing oocytes of adults (70).

5.1.2 Other types of radiation

Limited studies have been carried out with neutrons and gamma rays. High LET radiation in the form of fission neutrons has proved very effective for the induction of mutations in spermatogonia and oocytes (70,81). Fission neutrons are more effective at inducing mutations in the female germline than radiation, and female sterility is not a problem after neutron irradiation. Gamma sources are not effective at inducing mutations in mice (81).

5.2 Chlorambucil

Chlorambucil is a chemotherapeutic agent that was the second chemical found to induce mutations at a high frequency in the mouse. In initial experiments, mutations were recovered at frequencies as high as 1.3×10^{-3} per locus per gamete (69). The highest frequency of mutations is found in progeny sired during the third week after chlorambucil injection, suggesting that mutations occur most frequently in sperm derived from chlorambucil-exposed early spermatids. Subsequent analysis of the mutations recovered from these experiments showed that the lesions induced by chlorambucil were large deletions and translocations (84). These early results suggest that chlorambucil might be a powerful mutagen for obtaining large deletions useful for functional studies. However, since mutated gametes will only be sampled during the third week after injection, many treated males are needed to recover mutant offspring.

6. Future prospects

6.1 DNA repair

DNA-repair enzymes provide checks on the many insults received from the environment. Studies of the devastating effects of the functional loss of repair enzymes show that DNA repair plays a crucial role in mutagenesis (see refs 85 and 86, for example). Often, cell lines mutant in DNA repair enzymes are analysed for their ability to tolerate mutagens, showing that the cellular response to mutagens in a repair-deficient background can either worsen the cell's condition or allow it to escape killing (see ref. 87, for example). Future studies will probably incorporate this knowledge into studies of germline mutagens in mice. For example, a mouse deficient in DNA alkyltransferase will probably respond to ENU treatment differently from a wild-type mouse. Further, inbred strains of mice differ in their response to mutagen treatment, possibly because of allelic variation at DNA repair enzymes (71) (D. A. Carpenter and M. J. Justice, unpublished results). Studies carried out in DNA repair-deficient backgrounds may identify conditions in which mutagenesis is more effective, or in which mutations other than point mutations are recovered at a high efficiency.

6.2 Sequence-based screening for lesions

An alternative approach to screening for desired phenotypes is to screen for DNA lesions in a gene of interest and subsequently determine a phenotypic effect. This approach requires a sequence-based hunt for mutations. Such studies have been attempted by a few laboratories using single-strand conformational polymorphism (SSCP) analysis (unpublished). Although very few mutations have been detected, these studies have the added benefit that they will determine which changes in the gene are neutral and fail to cause any noticeable phenotype in the whole organism. To reliably and efficiently detect the single-base changes induced by ENU in DNA, high-efficiency DNA sequencing techniques must be used. Innovative new technologies, including DNA sequencing by hybridization to DNA chips, are already becoming available for large-scale resequencing (88). New technologies for the detection of single nucleotide polymorphisms (SNPs) may also be used for mutation detection (89).

6.3 The future of mutagenesis

Mutagenesis experiments will soon produce large numbers of mouse mutations that will be a valuable resource for deciphering the functional meaning of genes and their contextual interactions. The generation of large numbers of mutations will create new ethical and scientific issues within the mouse community: new databases and phenotype assays will be required, cost-effective mutation recovery from cryopreserved or freeze-dried sperm will become essential, and animal care and cost issues will arise. A current campaign of the Humane Society of the United States to reduce pain and distress in animals may affect animal models by requiring potentially controversial protocols for treating or killing mutant animals (90). The utility and necessity of disease models may be debated, in part, because of this campaign. A few ENU mutagenesis experiments have isolated a large number of mutations that represent only a fraction of the mutational potential of the mammalian genome. Many large groups and small labs are now attempting ENU experiments that will generate new human disease models and reveal new mammalian developmental pathways. Our view of gene function in mammals will be dramatically and permanently changed by these experiments. The impact on drug development and human health will be enormous.

References

1. Woychik, R.P., Klebig, M.L., Justice, M.J., Magnusson, T., and Avner E. D. (1998) *Mutat. Res.* **400**, 3.
2. Peters, J. and Brown, S.D.M. (1996) *Trends Genet.* **12**, 433.
3. Davis, A.P. and Justice, M.J. (1998) *Mammal. Genome* **9**, 325.
4. Capecchi, M.R. (1989) *Science* **244**, 1288.
5. Rinchik, E.M. (1991) *Trends Genet.* **7**, 15.
6. Meisler, M.H. (1992) *Trends Genet.* **8**, 341.
7. Ramirez-Solis, R., Liu, P., and Bradley, A. (1995) *Nature* **378**, 720.
8. Zambrowicz, B.P. (1998) *Nature* **392**, 608.
9. Davis, A.P. and Justice, M.J. (1998) *Trends Genet.* **14**, 438.
10. Loveland, K.L. and Schlatt, S. (1997) *J. Endocrinol.* **153**, 337.
11. Broudy, V.C. (1997) *Blood* **4**, 1345.
12. Mouse Genome Database, v.2.1. (1998) Mouse Genome Informatics, Bar Harbor, Maine (http://www.informatics.jax.org/).
13. Schlager, G. and Dickie, M.M. (1967) *Genetics* **51**, 319.
14. Fletcher, C.F., Lutz, C.M., O'Sullivan, T.N., Shaughnessy, J.D., Hawkes, R., Frankel, W.N., Copeland, N.G., and Jenkins, N.A. (1996) *Cell* **87**, 607.
15. Ebersole, T.A., Chen, Q., Justice, M.J., and Artzt, K. (1996) *Nature Genet.* **12**, 260.
16. Michaud, E.J., van Vugt, M.J., Bultman, S.J., Sweet, H.O., Davisson, M.T., and Woychik, R.P. (1994) *Genes Dev.* **8**, 1463.
17. Cachon-Gonzalez, M.B., Fenner, S., Coffin, J.M., Moran, C., Best, S., and Stoye, J.P. (1994) *Proc. Natl Acad. Sci. USA* **91**, 7717.
18. Kuster, J.E., Guarnieri, M.H., Ault, J.G., Flaherty, L., and Swiatek, P.J. (1997) *Mammal. Genome* **8**, 673.
19. Ashby, J. and Tinwell, H. (1994) *Mutagenesis* **9**, 179.
20. Ashby, J., Gorelick, N.J., and Shelby, M.D. (1997) *Mutat. Res.* **388**, 111.
21. Marker, P.C., Seung, K., Bland, A.E., Russell, L.B., and Kingsley, D.M. (1997) *Genetics* **145**, 435.
22. Kohler, S.W., Provost, G.S., Fieck, A., Kretz, P.L., Bullock, W.O., Sorge, J.A., Putman, D.L., and Short, J.M. (1991) *Proc. Natl Acad. Sci. USA* **88**, 7958.

23. Gossen, J.A., de Leeuw, W.J.F., Tan, C.H.T., Zwarthoff, E.C., Berends, F., Lohman, P.H.M., Knook, D.L., and Vijg, J. (1989) *Proc. Natl Acad. Sci. USA* **86**, 7971.
24. Ashby, J., Gorelick, N.J., and Shelby, M.D. (1997) *Mutat. Res.* **388**, 111.
25. Davis, A.P. and Justice, M.J. (1998) *Genetics* **148**, 7.
26. Bode, V.C. (1984) *Genetics* **108**, 457.
27. Ehling, U.H., Favor, J., Kratochvilova, J., and Neuhauser-Klaus, A. (1982) *Mutat. Res.* **92**, 181.
28. Ehling, U.H. (1983) In *Utilization of Mammalian Specific Locus Studies in Hazard Evaluation and Estimation of Genetic Risk* (ed. F.J. de Serres and W. Sheridan), p. 169. Plenum Press: New York.
29. Selby, P.B. and Selby, P.R. (1977) *Mutat. Res.* **43**, 357.
30. Selby, P.B. and Selby, P.R. (1978) *Mutat. Res.* **51**, 199.
31. Snell, G.D. and Aebersold, P.C. (1937) *Proc. Natl Acad. Sci. USA* **23**, 374.
32. Generoso, W.M., Bishop, J.B., Goslee, D.G., Newell, G.W., Sheu, C.J., and von Halle, E. (1980) *Mutat. Res.* **76**, 191.
33. Haldane, J.B.S. (1937) *Amer. Nat.* **71**, 337.
34. Justice, M.J. and Bode, V.C. (1986) *Genet. Res.* **47**, 187.
35. Shedlovsky, A., Guenet, J.-L., Johnson, L.L., and Dove, W.F. (1986) *Genet. Res.* **47**, 135.
36. Shedlovsky, A., King, T.R., and Dove, W.F. (1988) *Proc. Natl Acad. Sci. USA* **85**, 180.
37. Shedlovsky, A., McDonald, J.D., Symula, D., and Dove, W.F. (1993) *Genetics* **134**, 1205.
38. Rinchik, E.M., Carpenter, D.A., and Selby, P.B. (1990) *Proc. Natl Acad. Sci. USA* **87**, 896.
39. Rinchik, E.M., Carpenter, D.A., and Handel, M.A. (1995) *Genetics* **92**, 6394.
40. Kasarkis, A., Manova, K., and Anderson, K.V. (1998) *Proc. Natl Acad. Sci. USA* **95**, 7485.
41. Brandon, E.P., Idzerda, R.L., and McKnight, G.S. (1995) *Curr. Biol.* **5**, 625.
42. Brandon, E.P., Idzerda, R.L., and McKnight, G.S. (1995) *Curr. Biol.* **5**, 758.
43. Brandon, E.P., Idzerda, R.L., and McKnight, G.S. (1995) *Curr. Biol.* **5**, 873.
44. Copeland, N.G. *et al.* (1993) *Science* **262**, 57.
45. Maddox, J. (1995) *Nature* **376**, 459.
46. Dujon, B. (1996) *Trends Genet.* **12**, 263.
47. The *C. elegans* Sequencing Consortium. (1998) *Science* **282**, 2012.
48. Oliver, S.G. (1996) *Nature* **379**, 597.
49. Chiang, C., Litingtung, Y., Lee, E., Young, K.E., Corden, J.L., Westphal, H., and Beachy, P.A. (1996) *Nature* **383**, 407.
50. Roessler, E., Belloni, E., Gaudenz, K., Jay, P., Berta, P., Scherer, S.W., Tsui, L.C., and Muenke, M. (1996) *Nature Genet.* **14**, 357.
51. Geisterfer-Lowrance, A.A.T., Christe, M., Conner, D.A., Ingwall, J.S., Schoen, F.J., Seidman, C.E., and Seidman, J.G. (1996) *Science* **272**, 731.
52. Cordes, S.P. and Barsh, G.S. (1994) *Cell* **79**, 1025.
53. Schumacher, A., Faust, C., and Magnuson, T. (1996) *Nature* **383**, 250.
54. McDonald, J.D., Bode, V.C., Dove, W.F., and Shedlovsky, A. (1990) *Proc. Natl Acad. Sci. USA* **87**, 1965.
55. Symula, D.J., Shedlovsky, A., Guillery, E.N., and Dove, W.F. (1997) *Mammal. Genome* **8**, 102.

56. Singer, B. and Dosahjh, M.K. (1990) *Mutat. Res.* **233**, 45.
57. Vogel, E. and Natarajan, A.T. (1979) *Mutat. Res.* **62**, 51.
58. Morimoto, K., Tanaka, A., and Yamaha, T. (1979) *Gann* **70**, 693.
59. Kohler, S.W., Provost, G.S., Fieck, A., Kretz, P.L., Bullock, W.O., Sorge, J.A., Putman, D.L., and Short, J.M. (1991) *Proc. Natl Acad. Sci. USA* **88**, 7958.
60. Shibuya, T. and Morimoto, K. (1993) *Mutat. Res.* **297**, 3.
61. King, D.P. *et al.* (1997) *Cell* **89**, 641.
62. McDonald, J.D. and Charlton, C.K. (1997) *Genomics* **39**, 402.
63. Huang, J.-D., Cope, M.J.T.V., Mermall, V., Strobel, M.C., Kendrick-Jones, J., Russell, L.B., Mooseker, M.S., Copeland, N.G., and Jenkins, N.A. (1998) *Genetics* **148**, 1951.
64. Huang, J.-D., Mermall, V., Strobel, M.C., Russell, L.B., Mooseker, M.S., Copeland, N.G., and Jenkins, N.A. (1998) *Genetics* **148**, 1963.
65. Russell, W.L., Kelly, E.M., Hunsicker, P.R., Bangham, J.W., Maddux, S.C., and Phipps, E.L. (1979) *Proc. Natl Acad. Sci. USA* **76**, 5818.
66. Hitotsumachi, S., Carpenter, D.A., and Russell, W.L. (1985) *Proc. Natl Acad. Sci. USA* **82**, 6619.
67. Justice, M.J. and Bode, V.C. (1988) *Genet. Res.* **51**, 95.
68. Cox, R.D., Hugill, A., Shedlovsky, A., Noveroske, J.K., Best, S., Justice, M.J., Lehrach, H., and Dove, W.F. (1998) *Genomics*, **57**, 333.
69. Russell, L.B., Hunsicker, P.R., Cacheiro, N.L.A., Bangham, J.W., Russell, W.L., and Shelby, M.D. (1989) *Proc. Natl Acad. Sci. USA* **86**, 3704.
70. Russell, L.B. and Russell, W.L. (1992) *Mutat. Res.* **296**, 107.
71. Davis, A.P., Woychik, R.P., and Justice, M.J. (1998) *Mammal. Genome*, **10**, 308.
72. Russell, W.L., Hunsicker, P.R., Raymer, G.D., Steele, M.H., Stelzner, K.F., and Thompson, H.M. (1982) *Proc. Natl Acad. Sci. USA* **79**, 3589.
73. Russell, W.L., Hunsicker, P.R., Carpenter, D.A., Cornett, C.V., and Guinn, G.M. (1982) *Proc. Natl Acad. Sci. USA* **79**, 3592.
74. Moser, A.R., Pitot, H.C., and Dove, W.F. (1990) *Science* **247**, 322.
75. Vitaterna, M.H., King, D.P., Chang, A.M., Kornhauser, J.M., Lowrey, P.L., McDonald, J.D., Dove, W.F., Pinto, L.H., Turek, F.W., and Takahashi, J.S. (1994) *Science* **264**, 719.
76. Rogers, D.C., Fisher, E.M., Brown, S.D., Peters, J., Hunter, A.J., and Martin, J.E. (1997) *Mammal. Genome* **8**, 711.
77. Justice, M.J., Zheng, B., Woychik, R.P., and Bradley, A. (1997) *Methods* **13**, 423.
78. Karim, F.D., Chang, H.C., Therrien, M., Wassarman, D.A., Laverty, T., and Rubin, G.M. (1996) *Genetics* **143**, 315.
79. Russell, L.B., Russell, W.L., Rinchik, E.M., and Hunsicker, P.R. (1990) In *Banbury Report*. (ed. Allen, J.W., Bridges, B.A., Lyon, M.F., Moses, M.J., and Russell, L.B.). **34**, pp. 271–289. Cold Spring Harbor Laboratory Press, NY.
80. Russell, W.L., Russell, L.B., and Kelly, E.M. (1958) *Science* **128**, 1546.
81. Searle, A.G. (1974) *Adv. Radiat. Biol.* **4**, 131.
82. Russell, W.L. and Russell, L.B. (1959) *Radiat. Res.* **1** (Suppl.), 296.
83. Oakberg, E.F. (1979) *Mutat. Res.* **59**, 39.
84. Rinchik, E.M., Bangham, J.W., Hunsicker, P.R., Cacheiro, N.L.A., Kwon, B.S., Jackson, I.J., and Russell, L.B. (1990) *Proc. Natl Acad. Sci. USA* **87**, 1416.
85. Fishel, R., Lescoe, M.K., Rao, M.R.S., Copeland, N.G., Jenkins, N.A., Garber, J., Kane, M., and Kolodner, R. (1993) *Cell* **75**, 1027.

86. Baker, S.M. *et al.* (1995) *Cell* **82**, 309.
87. Liu, L., Markowitz, S., and Gerson, S.L. (1996) *Cancer Res.* **56**, 5375.
88. Southern, E.M. (1996) *Trends Genet.* **12**, 110.
89. Ross, P., Hall, L., Smirnov, I., and Haff, L. (1998) *Nature Biotechnol.* **16**, 1347.
90. The Humane Society of the United States (1998) *The Pain and Distress Campaign Brochure.* The Humane Society of the United States, Washington, DC.

9

Generation of transgenic mice from plasmids, BACs and YACs

ANNETTE HAMMES and ANDREAS SCHEDL

1. Introduction

Transgenesis, or the stable integration of foreign DNA into a host genome, has developed into one of the most powerful techniques for analysing gene function and regulation (1). Applications of this technology are numerous and range from investigating the basic mechanisms of gene regulation to the generation of models for human diseases, which can be used for pathophysiological and therapeutic studies. The list of over-expression studies includes the use of cDNA constructs encoding transcription factors, cell surface receptors and structural proteins. In addition constructs encoding antisense RNA, to inhibit endogenous gene expression, can be used in transgenic experiments. Before describing the technical steps involved in the generation of transgenic mice, we will briefly highlight a number of important points to consider in designing a transgenic experiment. Due to space limitations this section cannot be comprehensive and the interested reader is referred to original articles where transgenic technology has been applied.

1.1 Principles and general considerations

The most direct way to generate transgenic mice is by microinjection of a DNA solution into the pronucleus of fertilized oocytes. After successful delivery of DNA, the transgene integrates into the mouse genome and replicates with the endogenous chromosome, thus becoming part of the genome of the newly forming embryo. As the integration usually occurs only on one chromosome, the resulting founder mouse will be hemizygous for the transgene. It should be noted that in approximately 10–30% of transgenic animals integration of the transgene occurs not during the one-cell stage but later in development. The founder mouse will then be mosaic for the transgene and germline transmission is not guaranteed.

In most cases the transgene integration occurs randomly at a single site, although integration by homologous recombination has also been observed in very rare cases. The actual mechanism of integration is still unknown, but some investigators believe that it coincides with the occurrence of double-strand breaks of the nuclear DNA during the injection procedure. For a standard transgene, integration of 1–50 copies is normal, but integration of up to 1000 copies has been observed (2). The individual copies are usually found in a tandem array as head-to-tail fusions. Relatively rare are integrations at several sites in the mouse genome. In these cases the offspring of the transgenic founder are likely to carry only a subset of the integrated copies in the founder animal, due to segregation of the chromosomes.

1.2 Difficulties and limitations

While the construct design can be readily optimized, there are other factors which cannot be controlled for and which can influence the expression of the transgene. Probably the most important factor is the so-called position effect, which in many cases affects the expression of the transgene; rather than being dependent on the number of copies of the transgene, the level of expression often is influenced by the genomic sequences flanking the site at which the transgene is integrated. Position effects can be of various types. They may reduce or completely abolish transgene expression, either through the action of specific silencing elements or simply because the transgene has inserted into a transcriptionally inactive region of the genome. Alternatively, flanking sequences may contain regulatory elements of nearby genes that act on the promoter of the transgene as an enhancer, which can lead to ectopic expression of the transgene. As position effects can influence the way in which the gene is expressed, it is important to generate several independent transgenic lines from one construct, check the level and the sites of expression and compare the phenotypes of different lines.

A second aspect of random integration is that insertion of the transgene can occur within a transcriptional region of an endogenous gene, creating an insertional mutation and disrupting the locus. As most genes do not show an obvious phenotype when heterozygously mutated, this usually does not pose a problem for the founder mice. However, when such mice are crossed to achieve homozygosity, a proportion of the resultant transgenic lines show a phenotype caused by disruption of a genetic locus (3). Interpretation of transgenic phenotypes therefore has to be done cautiously, and it may be best to look at the hemizygous phenotype. About 5–10% of transgenic lines show an insertional mutation resulting in a pathological phenotype in mice homozygous for the transgene. These animals can be of great value as they provide a route for cloning the corresponding affected gene.

Other factors, such as differences in genetic background and environmental and dietary variability, may also contribute to unexpected and ambiguous results from transgenic experiments. Gene dosage effects and compensatory alterations due to gene over-expression may also cause problems in analysing the animal model and interpreting the results.

In addition, the presence of multiple copies integrated at a single site has been shown to reduce the efficiency of expression independent of the integration site (4). The exact reason for this phenomenon is unclear, but it has been speculated that the repeat structure is recognized by the host cell, leading to a change in chromatin structure and a concomitant inactivation of the transgene.

1.3 Construct design

Transgenic mice have two major applications: expression of a gene under control of a specific promoter (either its own or a heterologous one), and the analysis of the regulatory regions of a locus. In the first case, transgenes are usually composed of a tissue-specific or ubiquitous promoter, e.g. the enhancer/ promoter region of the immediate-early gene of the human cytomegalovirus (CMV) linked to the cDNA of interest. Plasmid sequences flanking the transgene have been shown to negatively influence transgene expression and it is therefore important to design the construct in such a way that it can be easily released from the vector by restriction digestion (5–7). It has also been demonstrated that efficient expression of a transgene requires splicing (8) and polyadenylation of the transcribed product. Thus when a high level of expression is desired it is advisable to use genomic sequences or to include an intron (endogenous or heterologous; 9,10) downstream of the transcription start site, as well as an SV40 or polyoma polyadenylation signal at its 3′ end. Finally, to ensure an accurate and efficient start of translation the start codon (ATG) should be preceded by a Kozak consensus sequence (11).

Complementation (rescue) experiments of mouse mutants usually require expression of a gene under the control of its own promoter. However, regulatory domains can be very complex and cover several hundred kilobases. In these cases one can make use of larger constructs such as bacterial artificial chromosomes (BACs; 12) or yeast artificial chromosomes (YACs; 13,14). The large cloning capacity of these vectors allows the insertion of an entire locus including all elements required for transcriptional and post-transcriptional regulation. Although more demanding to generate, transgenic mice produced with YACs usually show copy-number-dependent and position-independent expression of the encoded gene(s). Fewer independent lines carrying the same construct have therefore to be established to confirm the phenotype.

Several reporter genes have been successfully employed in transgenic mice to analyse regulatory elements. Methods for reporter gene detection can be found in Chapter 3. Probably the most widely used reporter is the *lacZ* gene, which codes for the bacterial enzyme β-galactosidase. It allows rapid and sensitive detection of expression by whole-mount staining of embryos. It can also be used to carefully follow the expression pattern of a gene during development, due to its sensitivity. If detection in the living animal is required, one of the enhanced versions of the green fluorescent protein (e.g. EGFP from Clontech) can be used as a marker (15). Sensitivity is, however, lower than with β-galactosidase, allowing detection only using a very strong promoter. Finally, for quantitative analysis the chloramphenicol acetyl transferase (CAT) and firefly luciferase genes are used. Expression is detected in tissue extracts of the transgenic animals using the enzymatic activity of the CAT protein or firefly luciferase, respectively.

1.4 Perspectives

Besides the above-mentioned advantages of using tissue-specific promoters, new approaches provide not only spatial but also temporal control over the onset of transgene expression. This is often useful to overcome compensatory, lethal or other unexpected secondary effects due to aberrant (ectopic) expression and/or the gain of function during early embryonic development. So far only a few inducible systems have been established. Probably the most powerful approach, developed by Gossen and Bujard (16), is based on the inducibility of the tetracycline-responsive element coupled to a minimal promoter. To use this system for temporal control of transgene expression it is necessary to establish two different transgenic mouse lines. The first line carries the cDNA driven by a minimal promoter and the *tet* operon, which is controlled by a tetracyclin-dependent transactivator. The second mouse line is transgenic for the tetracycline-controlled transactivator driven by a tissue-specific or ubiquitous promoter. Induction of gene expression is accomplished by supplying tetracycline to the food or water of the mice.

1.5 Factors influencing the efficiency of transgenesis

One of the most critical factors for the successful generation of transgenic mice is the quality of DNA. Slight contamination with toxic substances such as phenol or too high concentrations of EDTA will kill the oocytes. Furthermore, dust particles will block the microinjection needle. Thus the greatest care should be taken when preparing DNA solutions for injections.

The topology of the construct used for microinjection also influences the success of transgenesis. With linear DNA up to 30% of successfully injected and transferred embryos will be transgenic, whereas efficiencies drop to 5% when injecting circular (supercoiled) plasmids. DNA concentrations ranging from 0.01 to 50 ng/μl can be used, but the efficiency varies dramatically. With plasmid constructs highest efficiencies are achieved at a concentration of ~2 ng/μl (17). Excessive amounts of DNA have been shown to be toxic to oocytes, reducing the number which survive. We therefore strongly advise never exceeding a concentration of 10 ng/μl. The amount of DNA transferred into the pronucleus is variable and depends on the investigator. An average injection corresponds to about 1–2 pl of DNA. Obviously the number of transferred molecules depends also on the size of the transgene. For example, with a 5 kb construct ~350–700 molecules are transferred, whereas with a 500 kb construct only three to seven copies are injected. To increase the chance of integration we therefore suggest using a slightly higher concentration (5 ng/μl) for constructs larger than 100 kb.

1.6 Choice of mouse strains

Before beginning a transgenic experiment one of the first considerations should be the mouse strain into which the transgene will be inserted. In general it is advisable to generate mice on a defined genetic background rather than with outbred strains. This is particularly important when subtle phenotypes are expected, which could be complicated by genetic background effects. In addition, some mouse strains are more suitable than others for pronuclear injections simply because of the size and visibility of the pronucleus and the resistance of the cell and nuclear membranes. Finally, several mouse strains allow a large number of oocytes to be recovered after superovulation and their eggs are relatively resistant to *in vitro* manipulations. Mouse strains that are well suited for the generation of transgenic mice include the inbred C57BL/6J and FVB/N strains and the outbred NMRI strain. Similarly, the use of the F1 generation of CBA × C57BL/6J crosses yields a high number of high-quality oocytes.

2. DNA isolation

2.1 Plasmid DNA

The desired fragment (the transgene) for microinjection is released by restriction digestion, separated from the vector by agarose gel electrophoresis, and isolated from the agarose as described below. At this point we would like to emphasize again that the success rate in producing transgenic mice depends very much on the purity of the DNA solution and that isolation procedures should therefore be performed with the greatest possible care.

In general, we recommend that the DNA restriction fragment to be used for microinjection is recovered and purified by electroelution from agarose gels. A very good method for recovering pure DNA is the Biotrap BT 1000 electroseparation system (Schleicher and Schuell). DNA fragments between 14 and 15 000 bp can be quantitatively eluted with this system. The gel is put into the Biotrap chamber, and when an electric field is applied, the DNA runs from gel through a membrane, which is permeable to DNA but impermeable to larger particles such as small pieces of agarose. The DNA is concentrated in a small trap, which is limited by the DNA-permeable membrane on one side and a DNA-impermeable membrane on the other side.

Protocol 1. Isolation of plasmid DNA

Equipment and reagents

- DNA preparation kit (e.g. from Qiagen)
- Agarose, e.g. Seakem (FMC Bioproducts)
- 1× TAE buffer, prepared from a 50x stock (50× TAE buffer is 2 M Tris–acetate, 0.05 M EDTA pH 8.0)
- Ethidium bromide solution (10 mg/ml)
- Absolute and 70% ethanol
- 3 M sodium acetate, pH 5.2
- 10 mM Tris–HCl pH 7.4, 0.1 mM EDTA
- DNA electroelution system, e.g. Biotrap BT 1000 (Schleicher and Schuell)
- 0.22 μm sterile filter, e.g. Millex-GV_4 (Millipore)

Method

1. Prepare high-quality plasmid DNA according to standard protocols (19). DNA preparation kits can be used, e.g. from Qiagen. Digest 10–20 μg DNA with appropriate restriction enzyme(s) to release the insert.
2. Load the digested sample onto a standard agarose gel in 1× TAE buffer.
3. Stain the gel in ethidium bromide solution after electrophoresis.
4. Cut out the desired fragment from the gel under long-wave UV light.
5. Put the gel piece into the Biotrap electroelution chamber, which is set up according to the manufacturer's instructions, and cover it with 1× TAE buffer.
6. Place the Biotrap system in a horizontal electrophoresis tank filled with pre-cooled 1× TAE buffer.
7. Run the electrophoresis for 90 min at 10 V/cm. Within that time the DNA should be collected in the trap limited by the two membranes. The gel piece can be stained again in ethidium bromide solution and checked under UV light to ensure no DNA remains.
8. Reverse the polarity of the electrodes for 30–60 sec to detach any DNA that has stuck to the DNA-impermeable membrane on the exit side of the trap.
9. Transfer the DNA solution from the trap into a 2 ml microcentrifuge tube and precipitate the DNA by adding 2.5 vols of absolute ethanol and 0.1 vol. of 3 M sodium acetate (pH 5.2) and placing at –20 °C for at least 30 min.
10. Pellet DNA by centrifugation, wash in 70% ethanol, and air dry.
11. Resuspend in the desired amount of injection buffer containing 10 mM Tris-HCl (pH 7.4) and 0.1 mM EDTA.

12. Check the DNA concentration as described below (*Protocol 4*) and dilute the DNA solution with injection buffer to a final concentration of 2 ng/μl for injection.
13. To remove any particles, spin the DNA through a 0.22 μm sterile filter. Store at 4°C until use.

2.2 YAC DNA

The size of YAC DNA renders its isolation at high concentrations rather difficult. Furthermore, there are no simple methods for separating the YAC from other yeast chromosomes apart from gel electrophoresis. To obtain a large quantity of YAC DNA it is important to prepare agarose plugs of high concentration (e.g. see ref. 18). We find the following protocol the most reliable for the isolation of high-quality YAC DNA. It involves two gel runs. Firstly, pulsed field gel electrophoresis (PFGE) serves to separate the YAC from the endogenous yeast chromosomes. In the second gel run the DNA is concentrated at the border of a 4% agarose gel. Due to its large size, YAC DNA is highly susceptible to breakage and care should be taken in handling the DNA (use pipette tips with the end cut off to widen the tip diameter). Be sure to use water of highest quality, e.g. Milli-Q (Millipore). To protect the DNA from breakage after isolation from the agarose gels, buffers containing a high concentration of NaCl and low concentrations of polyamines (spermine and spermidine) should be used. NaCl has been shown to stabilize DNA, in particular during heating. Polyamines are polycations which form intra-molecular bridges with the DNA, thus compacting the molecules into globular structures. Although DNA molecules can be stabilized using polyamines alone, we find that NaCl is required to counteract DNA precipitation caused by polyamines.

Protocol 2. Isolation of YAC DNA

Equipment and reagents

- Agarose plugs containing YAC DNA at a high concentration (see protocols described elsewhere, e.g. in ref. 18)
- Pulsed field gel electrophoresis system (e.g. from Bio-Rad)
- Low melting point (LMP) agarose, e.g. NuSieve (FMC)
- High quality agarose, e.g. SeaKem (FMC)
- TENPA buffer: 10 mM Tris pH 7.4, 1 mM EDTA, 100 mM NaCl, 70 μM spermidine, 30 μM spermine[a]
- Injection buffer: 10 mM Tris pH 7.4, 0.1 mM EDTA, 100 mM NaCl, 70 μM spermidine, 30 μM spermine[a]
- Agarase, e.g. GELase (Epicentre)
- Dialysis membrane (VMWP 02500, 0.025 μm pore size, from Millipore)

Method

1. Cast a 0.25× TAE gel using high-quality agarose, e.g. Seakem (FMC) at a concentration of 1% with a preparative lane of about 5 cm (tape together several teeth of the comb).

Protocol 2. *Continued*

2. Wash the agarose plugs containing the yeast DNA extensively (4 × 15 min in TE, and 2 × 15 min in water).
3. Load the plugs next to one another in the preparative lane and seal with 1% LMP agarose in 0.25× TAE.
4. Run the gel in cooled 0.25× TAE under conditions optimized for the separation of the YAC from the endogenous chromosomes.
5. Cut off each side of the gel including 0.5 cm of the preparative lane (see *Figure 1*) and stain for 30 min in 0.25× TAE buffer containing ethidium bromide (0.5 μg/ml).
6. Mark the position of the YAC lane under UV light using a scalpel.
7. Reposition the stained parts of the gel next to the preparative lane and excise the gel slice containing the YAC DNA using the marked parts as a guidance.
8. Cut out a slice containing endogenous yeast chromosomes above and below the YAC to serve as controls for the second gel run.
9. Place the excised gel slices on a second gel tray at an angle of exactly 90° to the PFGE run, with the YAC slice in the middle flanked by the two marker slices (see *Figure 1*). Cast a 4% Nusieve (FMC) gel in 0.25× TAE around it.
10. Run overnight at a voltage of 6 V/cm with circulating buffer at 4°C.
11. Cut off the two marker slices and stain with ethidium bromide solution to locate the DNA. All of the YAC DNA should be concentrated within the LMP gel. If electrophoresis has not proceeded far enough, continue the gel run.
12. Excise the equivalent position of the YAC DNA.
13. Equilibrate the gel slice in 20 ml of TENPA buffer for at least 1.5 h.
14. Transfer the slice to a 1.5 ml microcentrifuge tube, remove any additional buffer that has been carried over and incubate for 3 min at 68°C.
15. Spin for 5 sec in a microcentrifuge to bring down all pieces of agarose.
16. Incubate for further 5 min at 68°C.
17. Spin in a microcentrifuge for 5 sec, place the tube containing the molten agarose at 42°C and incubate for 5 min.
18. Meanwhile, load 4 U of Gelase (Epicentre) per 100 mg melted agarose into a Gilson pipette and keep at room temperature to allow the temperature to adjust. Do not add the Gelase straight from the –20°C freezer, otherwise parts of the agarose may set.

19. Slowly release the Gelase while stirring the solution gently with the pipette. Incubate at 42°C for 2–3 h.
20. Transfer the resulting DNA solution onto a dialysis membrane (Millipore, 0.025 μm pore size, VMWP 02500) and dialyse for 30 min against injection buffer.
21. Store the DNA at 4°C. Do not use the same batch for microinjection for longer than 2 weeks. Storage in solution for longer than this may result in fragmentation of the DNA.

[a] When isolating YACs <200 kb, spermidine and spermine can be omitted from both the TENPA and injection buffer.

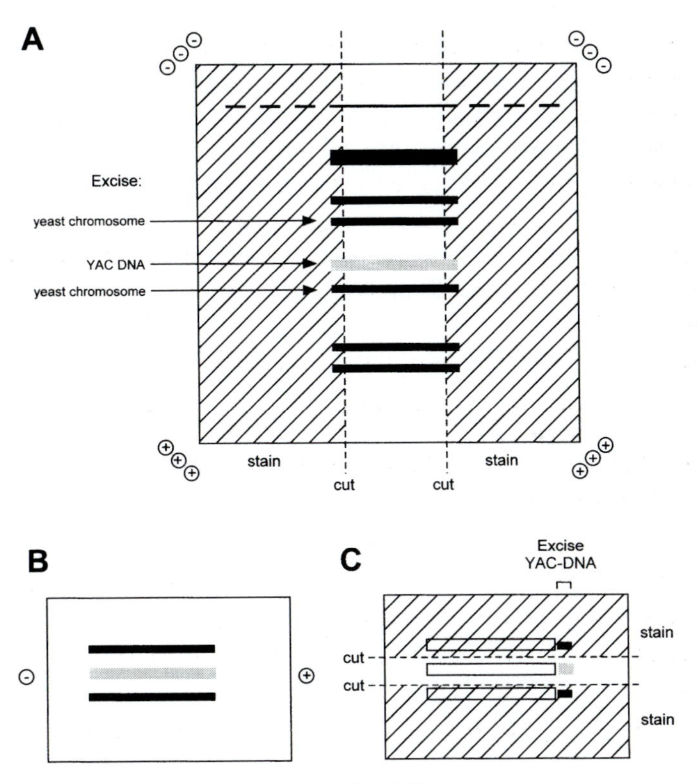

Figure 1. YAC DNA isolation procedure. (A) YAC DNA is separated by preparative pulsed field gel electrophoresis. Cut off both sides of the gel, including about 0.5 cm of the preparative lane, stain in buffer containing ethidium bromide and mark the position of the YAC under UV light. Reassemble the gel and cut out the gel slice containing the YAC DNA using the stained parts for guidance. Cut out two of the endogenous chromosomes as markers for the second gel run. (B) Position the markers and the YAC DNA on a minigel tray, pour a high concentration (4%) low melting point agarose gel around it and run at an angle of 90° to the PFGE run. (C) Stain the marker lanes and use them as guidance to cut out as small as possible a slice containing the YAC DNA.

2.3 BAC DNA

To obtain a large quantity of BAC DNA, the construct is amplified in *Escherichia coli* in a similar way to normal plasmid DNA and prepared using a plasmid purification kit suitable for the isolation of large DNA constructs (e.g. KB-100 Magnum; Genome Systems Inc.). BAC DNA is circular; it may be preferable to linearize the construct before injecting, otherwise breakage within the gene of interest may occur. This could lead to a non-functional transgene or a protein with unwanted dominant effects. Most BAC libraries are constructed as partially *Sau*3A-digested genomic DNA cloned into the *Bam*HI site of the pBeLoBAC vector. The cloning site is flanked by *Not*I sites which in many cases will be sufficient to excise the insert as a single fragment. If *Not*I sites are present within the insert, isolation will be difficult and a change of the vector may be necessary.

The BAC insert can be isolated by preparative PFGE as described for YAC DNA. An alternative method for the isolation of BAC DNA, based on DNA elution through a Sepharose (CL4b) column, is available (12) and is described below.

Protocol 3. Isolation of BAC DNA

Equipment and reagents

- Plasmid purification kit appropriate for the isolation of large DNA constructs (e.g. KB-100 Magnum from Genome Systems Inc.).
- Sepharose matrix CL4b (Pharmacia)
- 0.25% Bromophenol Blue
- 5 ml disposable plastic pipette with cotton-wool plug
- Injection buffer: 10 mM Tris–HCl pH 7.4, 0.1 mM EDTA, 100 mM NaCl

Method

1. Digest about 50 μg of high-quality BAC DNA with the restriction enzyme *Not*I, or another suitable enzyme.
2. To prepare a Sepharose column take a 5 ml plastic pipette and block it by blowing the cotton-wool plug into the tip of the pipette. Clamp it onto a stand and fill it gradually with Sepharose CL4b (Pharmacia) up to 1 cm below the top of the pipette (*Figure 2*).
3. Equilibrate the column with 30 ml of injection buffer. You can make a reservoir for the buffer by attaching a 10 ml syringe to the top of the pipette. **Note**: Never let the column run dry!
4. Add 5 μl of 0.25% Bromophenol Blue to the digested BAC DNA and load the sample onto the column.
5. Wait until the DNA has just entered the column, then add 0.5 ml of injection buffer.

6. Reattach the syringe and load with 10 ml of injection buffer.
7. Collect 0.5 ml aliquots until the blue dye has reached the bottom of the pipette.
8. There will always be a slight contamination with vector DNA. To identify the fraction containing the highest concentration of high-quality insert DNA with the lowest amount of vector, load 50 μl of each sample onto an agarose gel and run overnight under optimized conditions.
9. Check the DNA concentration (see *Protocol 4*) and, if necessary, dilute to a final concentration of 2 ng/μl with injection buffer.

2.4 Testing the DNA concentration and quality

We have found measurements of DNA concentration by spectrophotometry to be unreliable, as contamination with other UV-absorbing substances may

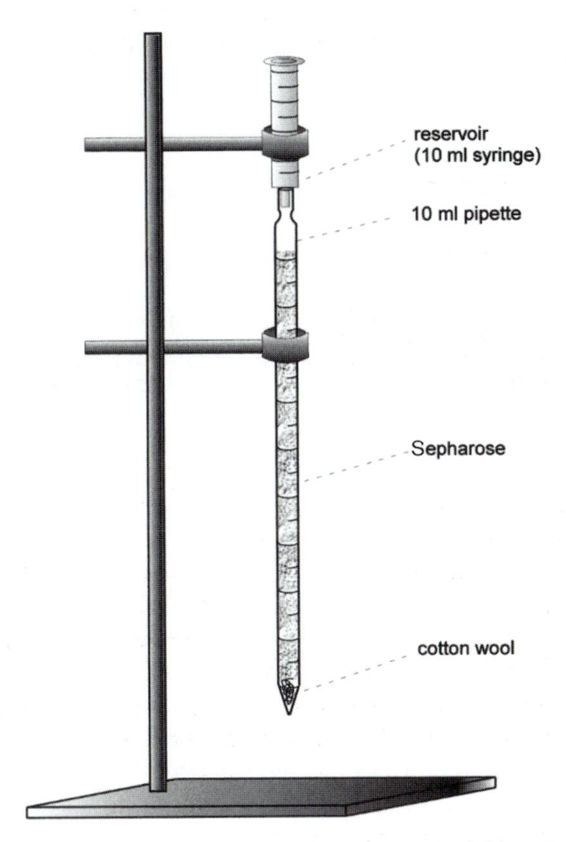

Figure 2. BAC DNA isolation. Clamp a 5 ml pipette onto a stand, blow the cotton wool up to its tip and fill it with Sepharose (CL4b). A 10 ml syringe can be attached at the top of the pipette as a reservoir.

lead to an over-estimate of DNA concentration. To determine the DNA concentration it is therefore advisable to compare it with a plasmid/bacteriophage λ DNA standard of known concentration. We usually use bacteriophage λ DNA (e.g. from New England Biolabs) diluted to a concentration of 1 ng/μl and load 1, 2, 5, 10 and 20 μl onto a thin 0.8% agarose gel. Samples (1 and 5 μl) of the isolated transgenic construct are run next to the standard and the intensities of the ethidium bromide-stained bands are compared. The band should be sharp: any sign of smearing below it indicates the early stages of degradation. The same method can be used to check the concentration of YAC and BAC DNA.

Protocol 4. Testing the DNA concentration

Equipment and reagents

- Agarose
- Bacteriophage λ DNA
- DNA to be tested
- Ethidium bromide
- Gel electrophoresis apparatus

Method

1. Pour a thin 0.8% agarose minigel.
2. Dilute bacteriophage λ DNA to a final concentration of 1 ng/μl and load 1, 2, 5, 10 and 20 μl onto the gel.
3. Load 1 and 5 μl of the isolated DNA next to the standard and run a standard gel electrophoresis.
4. Visualize the ethidium bromide-stained DNA under UV light and compare the intensity of the isolated DNA with that of the DNA standard.

3. The microinjection set-up

An overview of the microinjection set-up is given in *Figure 3*.
The following specialized equipment is needed for microinjection:

- vibration-damped table (e.g. Series 63; Spindler & Hoyer GmbH)
- microscopes with inverted optics (e.g. Zeiss Axiovert or Leica DM IRB); the microscopes should be equipped with 10x eyepieces, a 32× or 40× objective and a 5× or 10× objective.
- micromanipulators: mechanical micromanipulators (e.g. from Leitz, Zeiss-Jena or Narashige) or electronically controlled micromanipulator (e.g. Eppendorf PatchMan/Micromanipulator 5171)
- De Fonbrune microforge (Micro Instruments)
- micro-needle puller: vertical (e.g. needle puller model 750, Kopf TM Instruments) or horizontal (e.g. Sutter P-97, Sutter Instruments)

Figure 3. Micromanipulator set-up. 1, Stereo-microscope; 2 and 2b, electronic micromanipulator for the holding pipette (Eppendorf Patchman); 3 and 3b, electronic micromanipulator for the injection pipette (Eppendorf micromanipulator 5171); 4, microinjection needle holder; 5, holding pipette holder; 6, Celltram Air (Eppendorf) for controlling the oocytes; 7 and 7b, transjector with control unit.

- holding pipettes (e.g from W-P Instruments TW-150) made of borosilicate glass tubing without an internal filament; length 10 cm, outer diameter 1.0 mm, inner diameter 0.75 mm, thickness of glass wall = 0.125 mm
- injection needles (e.g from W-P Instruments TW-150), made of thin-walled borosilicate glass tubing, with a thin internal filament: length 12 cm, outer diameter 1.0 mm, inner diameter 0.58 mm, filament diameter 0.133 mm
- injector: 50 ml syringe or Eppendorf transjector

3.1 Location and design of the injection table

The microinjection procedure is performed with very fine-tipped needles and the slightest vibration of the microscope or the micromanipulators will make injections impossible. To avoid vibrations being transferred to the manipulators it is essential to position the microscope on a very stable table, ideally within the basement where the vibrations of the building are usually minimal. Usually a table with a heavy base-plate is sufficient, but if the building transfers vibration easily it may be necessary to obtain a special vibration-damped table (e.g. Spindler & Hoyer GmbH, Göttingen, Series 63). As the plate of this table rests on air, and therefore has no direct connection to the ground, the microscope is particularly well shielded against low-frequency vibrations. Even then, in rare cases complete elimination of the problem may be impossible and the set-up may have to be moved to a different room or building.

3.2 Microscope

Ideally, an inverse stereo-microscope with image-erected optics and a fixed stage is used for microinjections. This allows easier control of the oocyte and injection needle than the standard inverted optics. Excellent microscopes are the Zeiss Axiovert and the Leica DM IRB. Both microscopes can be equipped with a fine adjustable stage, which allows the oocytes to be moved around much more rapidly (see later).

Injections are usually performed at 300–400× magnification. The microscopes should therefore be equipped with 10× eyepieces and either a 32× or 40× objective. As some injection chambers have a rather thick base, objectives with a long working distance should be chosen. In addition, a 5× or 10× objective for use when adding and removing the eggs to the injection chamber and positioning the holding and injection needles is required. To locate the pronuclei within the fertilized oocytes a contrasting method is desirable, at least when working under high magnification. Nomarski optics, Hoffmann modulation and interference contrast all work very well. The condenser of the microscope should have a working distance of at least 2 cm to allow enough space for the manipulation with the microinjection needle.

3.3 Micromanipulators

Micromanipulators are required for the precise manipulation of both the oocyte and the injection needle. A joystick is highly advisable (ideally one for each side), which allows the simultaneous movement along two axes. The most commonly used are from Leitz, Zeiss-Jena and Narashige. The movement of the joystick is proportionally transmitted to the microinjection needle through oil-filled tubing. The sensitivity can be adjusted by a precision screw.

A quite recent development is the electronically controlled Eppendorf PatchMan/Micromanipulator 5171. Instead of a mechanical transmission, the Eppendorf system works with three electric motors, which allows a dynamic control over the injection needle. The speed of movement of the injection needle depends on the extent to which the joystick is moved from its zero position. This allows both fine and rapid movement of the injection needle. In contrast to the Patchman, the Micromanipulator 5171 accommodates a fire button allowing the automatic injection of an oocyte. Adapters for most microscopes are provided by the companies producing the manipulators.

3.4 The holding pipette

A good holding pipette is essential for efficient injections. Holding pipettes can be pulled either by hand or using a micro-needle puller (e.g. Sutter P-97). Use 10–15 cm long borosilicate glass tubing without an internal filament (see above). Hold the centre of the tubing above a small gas flame until it is soft (a good indication is when the flame turns yellow). Remove the tube from the flame and at the same time pull strongly at both ends. If you pull too early, while the glass is still within the flame, the fine ends will burn immediately. If you wait for too long, the glass will cool down, with the result that the diameter of the needle will be too large. The outer diameter of the holding pipette should, ideally not be much bigger than the oocyte itself (80–100 μm), and certainly not more than 150 μm. The glass capillary is marked with a diamond pen about 1.5 cm below the neck and broken by flicking the glass tube. Check the end under a stereo-microscope. Discard all needles with a ragged end, as only a blunt and clean break will lead to a good holding pipette. Clamp the needle into a De Fonbrune microforge and position the needle just above the heating filament. Turn on the heating until the filament is bright red. Move the holding needle closer to the heat until you see the end melting. Keep it in this position until the inner diameter of the opening is about one-third the diameter of an oocyte (20 μm). This should take no longer than 30 sec. Use only needles with a smooth and straight end. The correct shape of a holding pipette is shown in *Figure 5*. The preparation of a good holding pipette may take several attempts, but it is worth it, as you will save time during the microinjection process.

For controlling the oocyte through the holding pipette several methods have been employed. Probably the simplest way is by mouth control. Connect silicon tubing with a mouthpiece to the end of the needle holder. To pick up an oocyte simply suck; to release slowly blow into the mouthpiece. As an alternative a micrometer syringe can be connected via silicon tubing to the needle holder and filled with light paraffin oil. Avoid any air bubbles within the system or fine control of the movement of oocytes will be difficult. The suction is controlled by slowly turning the micrometer screws. Finally, the commercially available CellTram Air system (Eppendorf) provides excellent control over the oocytes.

3.5 Microinjection needles

Microinjection needles are prepared on a puller. Both vertical (e.g. needle puller model 750, Kopf TM Instruments) and horizontal designs can be used, though we have obtained more reproducible results with horizontal pullers (e.g. Sutter P-97, Sutter Instruments). Use thin-walled borosilicate glass tubing with a thin internal filament (see above).

The tip of the microinjection needle should be so fine that you cannot see the opening under a light microscope. The various settings of the micro-injection needle puller, such as heating, pull force, etc., will have to be varied to obtain tips with an optimal shape. Some settings will produce needles with a fused end. In this case, choose a different setting or break the needle by touching the holding pipette carefully (under the microscope). The size of the opening will depend also on the construct that is injected. A standard plasmid construct requires a diameter of <1 μm. For larger constructs such as YACs or BACs we suggest using a bigger opening. We usually obtain this by breaking a standard injection needle on the holding pipette. The opening should be big enough so that you can just see it under 400× magnification. The larger opening not only reduces the risk of shearing the DNA during the injection process, but also allows easier cleaning of the injection needle if it becomes blocked.

4. The microinjection experiment

A microinjection experiment will take a total of 4–5 days. For a quick over-view of the procedure for generating transgenic mice we have included a flow chart depicting all important steps of the experiment (*Figure 4*).

4.1 Superovulation and isolation of fertilized oocytes

Immature females are hormonally superovulated to obtain sufficient embryos for the microinjection. The procedure of superovulation has already been described in detail (Chapter 1, *Protocol 5*). We will therefore describe only the isolation of superovulated embryos.

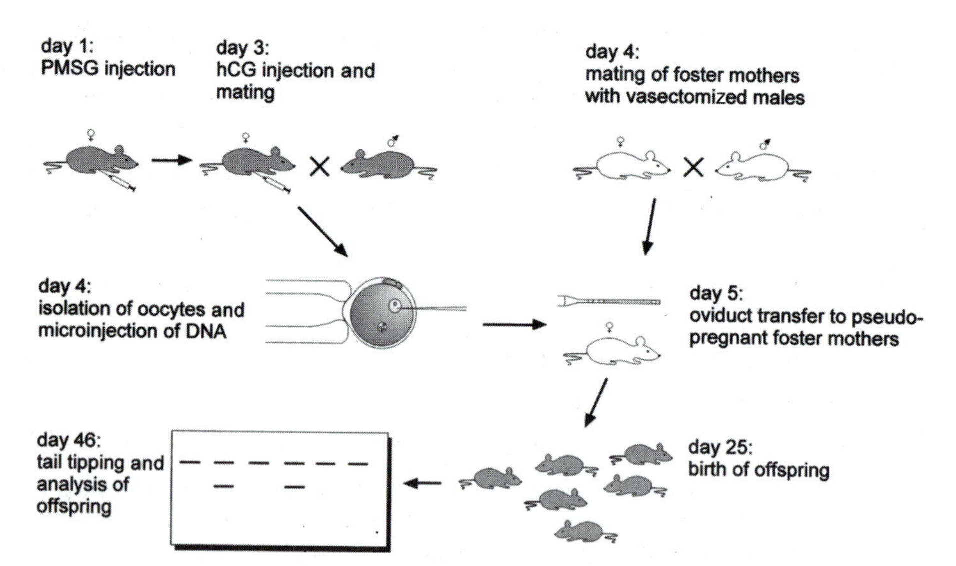

Figure 4. Flow chart for the generation of transgenic mice. See text for details.

The ideal age of mice for superovulation varies depending on the mouse strain, but is usually in the range of 3–4 weeks. For the strain C57BL/6J, hormone induction (administration of PMSG) on day 25 seems to give the highest number of oocytes per female. The timing of hormone administration is also critical and has to be adjusted to the light–dark cycle. The following parameters are optimized for the cycle set from 6.00 a.m. to 6.00 p.m.

Protocol 5. Superovulation and isolation of fertilized oocytes

Equipment and reagents

- Pregnant mare serum gonadotropin (PMSG), e.g. Intergonan (Intervet)[a]
- Human chorionic gonadotropin (hCG), e.g. Ovogest (Intervet)[a]
- Surgical instruments (e.g. from Fine Scientific Tools): blunt-tipped forceps, fine scissors, fine, blunt-tipped, curved forceps, watchmaker's forceps
- M2 medium (e.g. from Sigma)[b]
- M16 medium (e.g. from Sigma)
- Penicillin/streptomycin solution, 100× (10 000 U/ml penicillin and 10 mg/ml streptomycin) (e.g. from Sigma)
- Hyaluronidase (e.g. from Merck)
- Paraffin oil, embryo-tested (e.g. from Sigma)
- CO_2 incubator (e.g. from Heraeus Instruments)

Method (see also Chapter 1, Protocol 5)

1. At 2.00 p.m. on day 1, inject each donor mouse intraperitoneally with 5 U of PMSG.
2. At midday on day 3, inject each donor mouse intraperitoneally with 5 U of hCG. In the evening set up matings with proven stud males (aged 8 weeks to 1 year).
3. In the morning of day 4, check for successful matings by looking for the presence of vaginal plugs. Use these animals for the isolation of fertilised oocytes (Chapter 1, *Protocol 5*). Fertilization is usually complete by 6 h after mating.
4. Sacrifice the mouse by cervical dislocation.
5. Position the animal on its back and clean the abdomen with 70% ethanol.
6. Make a horizontal incision in the mid-abdomen and pull the skin off the abdomen.
7. Open the peritoneum and locate the ovaries.
8. Dissect the ovary, oviduct and uterus.
9. Cut off the oviduct and place it in a depression slide with M2 medium.[b]
10. Put the slide under a stereo-microscope and locate the ampulla among the loops of the oviduct. The ampulla contains the embryos and is clearly visible as the most swollen upper part of the oviduct.
11. Open the ampulla with a pair of watchmaker's forceps and push

Protocol 5. *Continued*

out the fertilized oocytes which are surrounded by follicle (cumulus) cells.

12. Transfer the embryos to M2 medium with hyaluronidase (300 mg/ml) and incubate them until the follicle cells are detached from the oocytes. This should take between 30 sec and 3 min depending on the batch of hyaluronidase. **Note**: hyaluronidase may damage the oocytes, so prolonged treatment should be avoided.
13. Transfer the embryos into fresh M2 medium with a transfer pipette and repeat this washing step three or four times.
14. Incubate the fertilized oocytes in M16 medium at 37 °C and 5% CO_2. The most convenient way is to keep the oocytes in micro-drop cultures. Pipette several 200 μl drops of M16 medium in a 15 cm^2 Petri dish and overlay them with paraffin oil (embryo-tested). Transfer the oocytes from each donor into a separate drop.

[a] PMSG and hCG should be diluted to a concentration of 50 U/ml and aliquots frozen at –20°C until use. Defrost each aliquot only once.
[b] While M16 medium is optimized for culturing oocytes, its pH changes quickly when kept outside the CO_2 incubator. Manipulation of oocytes (e.g. washing steps and injection) should therefore always be performed in M2 medium.

4.2 Microinjections

4.2.1 Timing of injections

After fertilization, pronuclei form and start to increase in size until the nuclear membranes break up and the pronuclei fuse. The ideal time for injections is the period 3–5 h before this fusion, as then the pronuclei are clearly outlined and rather large.

4.2.2 Injections

There are several alternative methods for injecting DNA into the pronucleus of an oocyte. The simplest is a 50 ml syringe connected via silicon tubing to the back of the injection needle holder. After insertion of the needle into the pronucleus the DNA is injected simply by applying strong pressure to the back of the syringe until the desired increase in the pronuclear size is achieved. Alternatively, an automatic injection machine (e.g. Eppendorf) can be used, which allows two pressures to be set. A low compensation pressure is permanently applied to the end of the needle to counteract the capillary force. Once the injection needle has been inserted into the pronucleus, a higher injection pressure can be applied (e.g. via a foot pedal). The compensation pressure seems to be of great importance when injecting YAC DNA.

Protocol 6. Microinjection

Equipment and reagents

- Microloader tips (e.g. from Eppendorf)
- Embryo-tested paraffin oil (Sigma)
- Depression slides (e.g. from Roth)
- M2 medium (see *Protocol 5*)
- Injection set-up (see Section 3)
- Embryos from *Protocol 5*
- DNA for injection from *Protocol 1, 2* or *3*

Method

1. Prepare a flat drop of about 40 μl of M2 medium (about 0.8 cm diameter) on a depression slide and overlay it with paraffin oil (embryo-tested).
2. Transfer 10–20 fertilized oocytes into the medium and mount the slide onto the microscope stage. The oocytes should be kept no longer than 30 min outside the incubator. If they start to show any sign of shrinking, transfer them back to the M16 micro-drop culture and place them into the incubator for recovery. They can be injected at a later stage.
3. Using an Eppendorf microloader transfer about 1–2 μl of DNA into the back of the injection needle and insert it into the needle holder. Try to avoid introducing air bubbles into the needle. The capillary force will pull the liquid to the tip of the needle.
4. Slowly lower the injection needle and the holding pipette to the plane of the oocytes, taking care not to break them at the base of the depression slide.
5. For efficient injections we suggest that a routine should be developed for arranging the oocytes on the injection slide. Uninjected eggs can be kept at the top, injected eggs at the bottom and lysed or unfertilized eggs at the left of the depression slide. Initially, you will find it necessary to switch from high to lower magnification when moving the oocytes around. However, after some practice you will get used to the distances between the three locations of oocytes and you may want to try moving the stage under the oocyte instead of moving the holding pipette. This routine increases the speed and also reduces the risk of accidentally breaking the injection needle with your holding pipette.
6. Under low-power magnification (10× objective) pick up a fertilized oocyte by applying suction to the holding needle and return the needle to the centre of the field of vision (alternatively move the stage to position the oocytes next to the holding pipette).
7. Switch to high-power magnification. Position the embryo in such a way that the male pronucleus (usually the bigger one containing only

Protocol 6. *Continued*

a single nucleolus) is in the equatorial plane of the oocyte and relatively close to the injection needle (see *Figure 5*). If the pronucleus is not in the equatorial plane the injections will be difficult as the oocyte will turn away when you try to enter with the injection needle.

8. Slowly insert the injection needle into the pronucleus. In the ideal situation the pronucleus will start swelling as soon as you apply pressure to the injection pipette. Try to avoid the nucleolus: it is extremely sticky and once touched you will pull out a substantial amount of DNA (visible as a string). If you find a small bubble arising at the end of the microinjection needle and clearly distinguishable from the pronucleus, you have not entered the pronucleus properly. The outer membrane of the oocyte can be extremely flexible and form a thin layer along the injection needle. To disrupt this you can either try to continue the pressure until the bubble bursts or insert the needle further into the inner part of the oocyte. Sometimes it helps to move the needle beyond the pronucleus. Once the membrane is disrupted retract the needle to the pronucleus to carry out the injection.
9. As soon as the diameter of the pronucleus has increased by a factor of ~1.5–2, pull the needle out off the oocyte. We have found that the survival rate of the oocytes improves dramatically if this retraction is performed rapidly.
10. If the cell has survived, transfer it to the bottom of the injection slide. If the cell has started to lyse (granules of the cytoplasm will start leaking out of the cell) keep the cells to the left or right side of the slide.
11. Pick up the next oocyte to inject as described.

4.3 Oviduct transfer

Transfer of injected embryos is performed either the same day or the day after microinjections (culturing overnight to the two-cell stage in M16 medium). If no pseudopregnant females are available, it is also possible to culture the embryos for another 2 days and transfer them to the uterus at the morula stage. It should be pointed out that certain mouse strains show a block at the two-cell stage when cultured *in vitro*. In this case transfer at the morula stage is not an option.

4.3.1 Pseudopregnant females

Female mice at the age of about 8 weeks to 6 months are mated with vasectomized males the evening before the transfer (see Chapter 1,

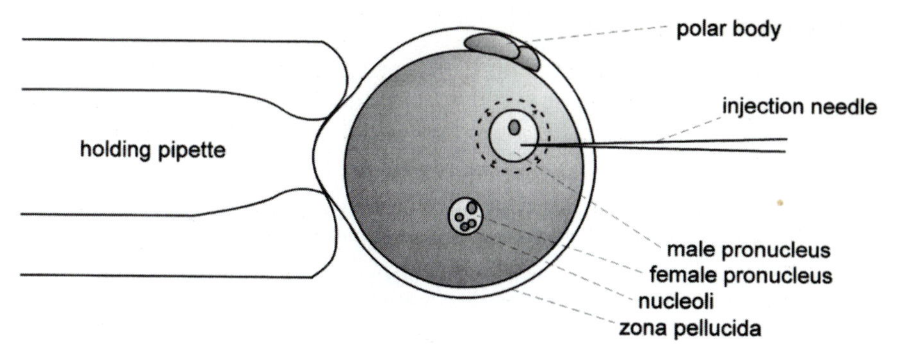

Figure 5. Pronuclear injection. The injection needle is inserted into the male pronucleus (which is larger than the female pronucleus and usually has only one nucleolus). Successful injection results in a clear swelling of the pronucleus.

Section 6.2). The next morning the females are checked for the presence of a copulation plug (vaginal plug). Usually, 10–20 % of the females are positive for the plug and can be used for the transfer. Note that female mice tend to synchronize their oestrus when kept for a long time in the same cage. This can lead to some days with a low plug rate and others with a very high one.

4.3.2 Preparation of the transfer pipette

To ensure that the transfer of embryos is performed in the smallest possible volume of medium, the diameter of the transfer pipette should be only slightly bigger than that of the oocytes.

Heat the tip of a Pasteur pipette in a flame and pull it to an internal diameter of 120–150 μm. Score with a diamond pen and break the tip of the transfer pipette carefully at this point. The break has to be clean. Heat the tip in a flame to polish it. A diagram of an ideal transfer pipette is shown in *Figure 6*.

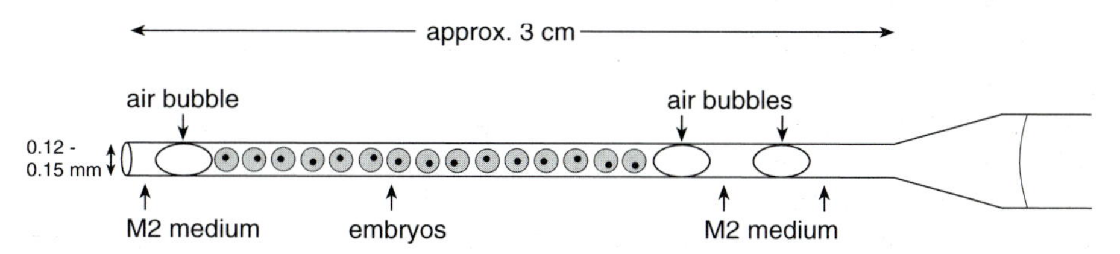

Figure 6. Diagram of a transfer pipette containing embryos. First fill the transfer pipette with M2 medium. Take up an air bubble, then medium and a second small air bubble. Collect the embryos and line them up next to each other with as little medium in between as possible. Take up a third air bubble and then some medium. This arrangement of medium, air bubbles, and embryos is suitable for an optimal control when collecting and blowing out the embryos.

4.3.3 Oviduct transfer

Protocol 7. Oviduct transfer

Equipment and reagents

- Anaesthetic: 100 mg/ml ketamine hydrochloride solution, e.g. Ketavet (Pharmacia–Upjohn) and 2% xylazine hydrochloride solution (5,6-dihydro-2-(2,6-xylidino)-4*H*-1,3 thiazinhydrochloride), e.g. Rompun (Bayer)
- Eye drops, e.g. Liquifilm (Pharm Allergan)
- Epinephrine hydrochloride, 1.2 mg/ml, e.g. Suprarenin (Hoechst AG)
- Surgical silk with needle (e.g. from Ethicon)
- Transfer glass pipette
- Sterilized surgical instruments (e.g. from Fine Scientific Tools): blunt-tipped forceps, fine sharp scissors, watchmaker's forceps, fine, blunt-tipped, curved forceps, Serrefine clamp, wound clips, wound clip applicator (special forceps to close the wound clip)
- Stereo-microscope (e.g. Leica MZ12 or Zeiss SV11)
- Neomycin/bacitracin ointment or powder, e.g. Nebacetin (Yamanouchi Pharma).

Method

1. Sterilize all instruments used for the transfer, either by autoclaving or by dipping them in ethanol and flaming them.
2. Weigh the mouse and anaesthetize it by intraperitoneal injection of 1 μl per gram body weight of an anaesthetic mixture containing 0.9 μl of a 100 mg/ml ketamine hydrochloride solution, e.g. Ketavet (Pharmacia–Upjohn) and 0.1 μl of a 2% xylazine hydrochloride solution, e.g. Rompun (Bayer).
3. Before starting the surgery carefully check whether the animal is deeply anaesthetized. Reflexes can be checked, e.g. by pinching the tail. The reflex of the eyelid can be checked by applying eye drops. Use of eye drops is recommended anyway, to prevent the eyes from drying during the operation.
4. Place the mouse in the prone position onto a lid of a Petri dish (*Figure 7*) and clean its back with a tissue soaked in 70% ethanol.
5. Load the embryos into the transfer pipette: transfer the embryos from M16 into M2 medium. Collect the embryos according to the scheme depicted in *Figure 6*. Between 15 and 30 embryos are usually transferred into each oviduct. Transfer of fewer than ~15 embryos leads to very small litter size and in this case the foster mother often does not take care of the newborn mice.
6. Place the pipette carefully next to the microscope.
7. To expose the reproductive tract of the recipient mouse, the skin and the body wall have to be cut at the right position. Make the incision ~1 cm lateral to the dorsal midline in the lumbar region at the level where the abdomen bulges maximally from the midline (*Figure 7*). Alternatively, when embryos are to be transferred into both oviducts, the incision can be placed at the dorsomedial position. Use blunt-

ended forceps to lift the skin and fine dissection scissors to cut it (*Figure 8A*). Wipe off any loose hair from the site of incision with 70% ethanol.

8. Slide the incision laterally until it is over the reproductive tract. The ovary and the attached fat pad are visible through the body wall as an orange and white spot, respectively. Pick up the body wall with a pair of forceps. Make a short incision into the peritoneal wall, ideally directly above the periovarian fat pad. Try to avoid damaging nerves and major blood vessels. Be careful not to cut too deeply, to prevent injury to the ovary, oviduct or uterus.
9. Pull a surgical silk through the body wall as shown in *Figures 8B* and *9*. This will make it easier to stitch the incision after the transfer.
10. Grasp the fat pad with fine blunt-tipped forceps and pull it carefully through the opened body wall towards the midline. The ovary, oviduct and uterus are attached to the fat pad. As soon as the reproductive tract is exposed, clip a Serrefine clamp onto the fat pad to keep the ovary and oviduct in a position outside the body wall on the back of the mouse (*Figures 7* and *8C*).
11. Carry out the following steps under the stereo-microscope. Locate the opening of the oviduct (infundibulum), which is hidden among the coils of the oviduct and located close to the ovary. Usually, it faces towards the tail of the mouse. If you handle the transfer pipette with your right hand it is convenient to put the mouse under the microscope with its head facing to the left.
12. To get access to the infundibulum, tear the ovarian bursa (a thin transparent tissue layer penetrated by blood vessels) using a pair of watchmaker's forceps (*Figure 9*). Take care not to damage any major blood vessels. If required, a few microlitres of the vasoconstrictor epinephrine hydrochloride (1.2 mg/ml) can be dripped onto the bursa before tearing it open to avoid heavier bleeding which could complicate the transfer.
13. Carefully insert the end of watchmaker's forceps into the infundibulum to widen it and to remove any mucus present at the opening (*Figure 8E*). Use a small sterile swab to remove fluid or blood, take the transfer pipette in the other hand and insert the tip into the infundibulum (*Figure 8F*). Blow the embryos along with the air bubbles into the oviduct. The air bubbles should be visible through the slightly transparent oviduct.
14. Check the transfer pipette to see that all embryos have been transferred.
15. Take off the clamp from the fat pad, grasp the body wall with forceps and lift it carefully so that the reproductive tract can slide back more easily into the peritoneum (*Figure 8G*). The fat pad can be picked up but try to avoid touching the oviduct or the uterus.

Protocol 7. *Continued*

16. Sew the incision in the body wall with one or two surgical knots (*Figure 8H*).
17. Close the skin incision with two small wound clips (*Figure 8J*) and put neomycin/bacitracin ointment or powder onto the wound.
18. Put the mouse back into its cage and leave it in a quiet and warm place. It should wake up ~30–40 min after injection of the anaesthetic.

Note: Before performing an oviduct transfer on an anaesthetized animal, we highly recommend practising on a dead animal using a transfer pipette loaded with a solution of Bromophenol Blue. If the 'transfer' is successful, the blue colour should spread along the coils of the oviduct.

Success rate: Usually about 10–20% of the injected and transferred embryos survive and are born 19–20 days after the transfer. On average 10–20% of these newborn mice are positive for the transgene and will pass it on to the following generations.

5. Analysis of transgenic founders

Potential transgenic founder mice which develop from the injected zygotes are screened for the insertion of the transgene by PCR or Southern blot analysis. The genomic DNA for this analysis is usually isolated from tail

Figure 7. Preparation of the recipient mouse for embryo transfer. The picture shows the surgical exposure of the reproductive tract of the pseudopregnant mouse. Place the anaesthetized mouse on the lid of a petri dish. Expose the ovary, oviduct and uterus. Fix the position of the oviduct outside the body on the back of the mouse by clipping a Serrefine clamp on the fat pad connected with the reproductive tract. The surgical silk pulled through the body wall helps in getting hold of the incision and sewing it with one or two stitches after the transfer.

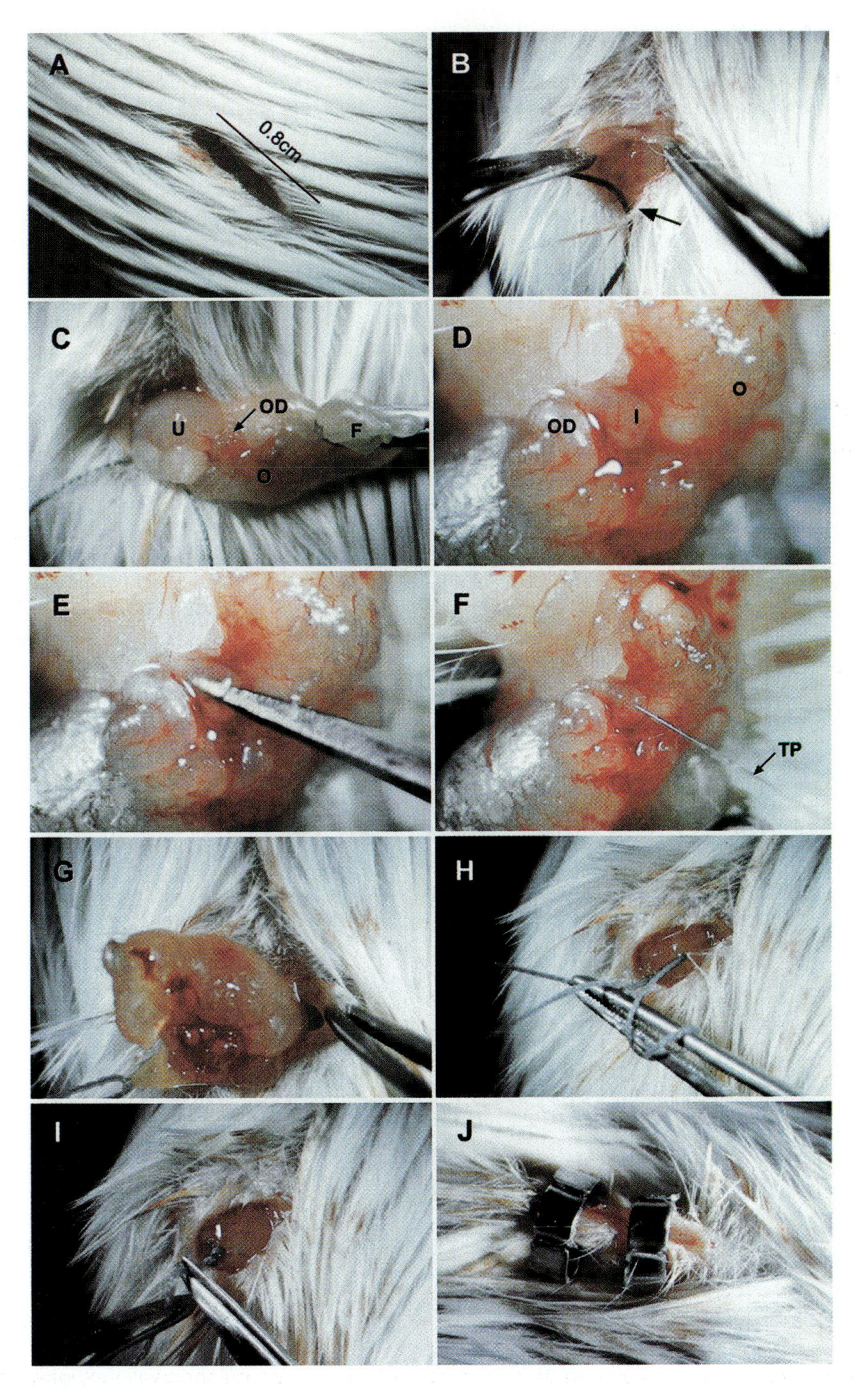

Figure 8. Caption See overleaf

Figure 8. Detailed documentation of the various steps of embryo transfer into a pseudo-pregnant mouse. U, uterus; OD, oviduct; O, ovary; F, fat pad; I, infundibulum; TP, transfer pipette. (A) Make an incision of about 1 cm in the skin approximately 1 cm lateral to the dorsal midline in the lumbar region at the level where the abdomen bulges maximally from the midline. Alternatively, if embryos are to be transferred into both oviducts, the incision can be placed at the dorsomedial position. Wipe off cut hair from the incision with 70% ethanol. (B) Lift the peritoneal wall with fine forceps and make a short incision, ideally right above the periovarian fat pad, avoiding damage to nerves and blood vessels. Pull a surgical silk through the body wall as shown schematically in *Figure 5*. (C) Grasp the fat pad with fine blunt-tipped forceps and pull it carefully through the opened body wall towards the midline. The ovary, oviduct and uterus are attached to the fat pad. (D) Higher magnification of the reproductive tract with the ovary, oviduct and the upper part of the uterus after tearing open the ovarian bursa. The opening of the oviduct (infundibulum) is clearly visible. The infundibulum is usually located close to the ovary and faces toward the tail of the mouse. (E) To be sure that the infundibulum is accessible and not still partly covered by the bursa, one can try to carefully enter the infundibulum with the tip of watchmaker's forceps as shown in this figure. (F) Transfer of embryos into the oviduct. Insert the tip of the transfer pipette into the infundibulum and blow the embryos along with the air bubbles into the oviduct. Be careful not to insert the tip of the transfer pipette too far into the infundibulum or to touch the oviduct wall. (G) Grab the body wall with forceps and lift it carefully so that the reproductive tract can slide back more easily into peritoneum. The fat pad can be picked up but try to avoid touching the oviduct or the uterus. (H) Sew the incision in the body wall with one or two stitches and close with a surgical knot. (I) Cut the surgical silk about 3 mm above the knot. (J) Close the skin incision with two small wound clips.

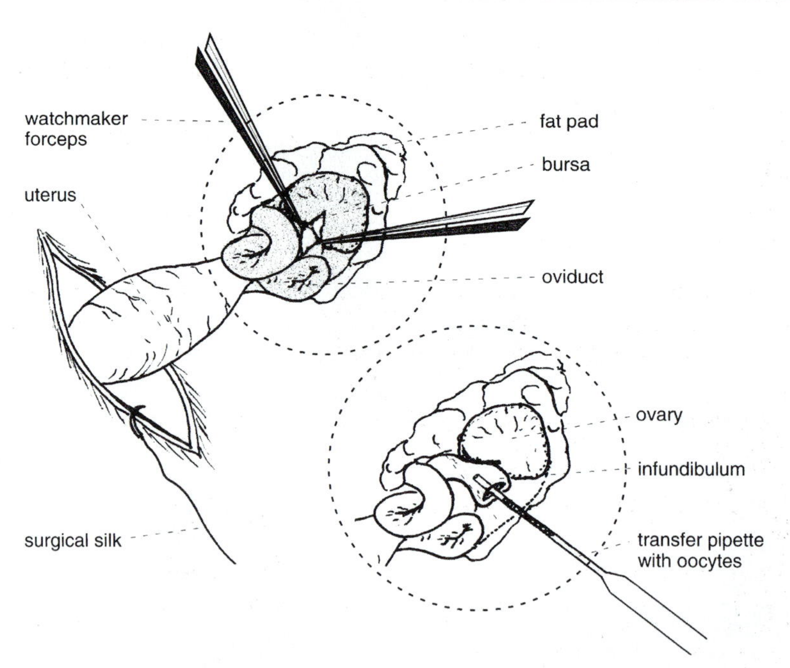

Figure 9. Schematic drawing of the reproductive tract of a mouse surgically exposed for the embryo transfer.

biopsies. Alternatively, the tissue from an ear punch, which is performed to label the animals, yields a sufficient amount of DNA for PCR analysis. Tail tipping and ear punching are performed at weaning age (3 weeks).

The founder mice should also be checked for the number of integration sites of the transgene. Usually, there is only one integration site but in rare cases the transgene integrates into more than one chromosome or chromosomal area. In this case, genotyping by quantitative Southern blot analysis is difficult to interpret. Further breeding is necessary to segregate the chromosomes containing transgenes and obtain animals with a single integration site. The best method for examining the number of integration sites is by fluorescence *in situ* hybridization (FISH) analysis (see Chapter 6). Less reliable, but easier, is Southern blot analysis. Set up the restriction digestion and hybridization in such a way that in addition to the transgene (usually integrated as multiple tandem repeats), its 5′ or 3′ flanking genomic sequence can be detected. If there is only one integration site, only one band representing the flanking genomic area is expected. As the length of this fragment is unknown it is sometimes difficult to set up optimal conditions for the Southern blot analysis.

5.1 Tail tipping and ear punching

Tail tipping is performed as follows:

(a) To anaesthetize the animals, pipette a few millilitres of ether on paper towels in a 1 L beaker. Put the mouse into the beaker and take it out as soon as it is anaesthetized.
(b) Cut off about 1 cm of the tail tip with a sterile razor blade. To stop the bleeding, apply a fibrin-containing ointment.
(c) Place the tail biopsy in a microcentrifuge tube and store at –20°C until required for isolation of DNA.

To identify the mice, the ears can be quickly clipped according to an identification scheme usually used to number the animals. Alternatively, animals can be marked by tattooing their toes or tails (see Chapter 1)

5.2 Isolation of genomic DNA from tail biopsies

Protocol 8. Isolation of genomic DNA

Equipment and reagents

- Lysis buffer (0.3 M sodium acetate pH 5.5, 10 mM Tris pH 7.0, 1 mM EDTA, 1% SDS)
- Proteinase K (10 mg/ml), e.g. from Boehringer
- RNase A solution (4 mg/ml), e.g. from Sigma
- Phenol–chloroform–isoamylalcohol (25:24:1 by volume) equilibrated with Tris–HCl pH 8.0
- Isopropanol
- TE buffer (10 mM Tris pH 8.0, 1 mM EDTA)
- Saturated NaCl (6 M)

Protocol 8. *Continued*

A Preparation of genome DNA

1. Add 500 μl of lysis buffer and 10 μl of proteinase K (10 mg/ml) to each tail biopsy.
2. Place the tubes on a rocking platform and incubate at 55 °C overnight.
3. Incubate the samples on ice for 5 min.
4. Add 200 μl of a saturated NaCl (6 M), mix well and incubate on ice for another 5 min.
5. Centrifuge the tube for 20 min at 15 000 r.p.m. to pellet the cell debris and transfer the supernatant to a fresh tube.
6. Add 2 μl of a 4 mg/ml RNase A solution and incubate at 37 C for 15 min.
7. Add 1 volume of phenol–choroform–isoamylalcohol (25:24:1 by volume) equilibrated with Tris–HCl pH 8.0 and mix the phases well.[a]
8. Centrifuge for 10 min at 12 000 r.p.m. and transfer the aqueous phase to a fresh tube.[a]
9. Precipitate the DNA with 1 vol. of isopropanol at room temperature. The DNA should be immediately visible as a white stringy precipitate. Either pull out the DNA using a pipette tip and place it in a new tube, or pellet it by centrifugation and discard the supernatant.
10. Wash the DNA once in 70% ethanol, shake gently and centrifuge again.
11. Dry the DNA pellet at room temperature.
12. Add 100 μl of TE buffer to the pellet and leave the sample at 37 °C for 30 min to dissolve. Vortexing of the genomic DNA should be avoided in all steps.
13. Quantify the DNA and use about 10 ng for PCR and 10 μg for Southern blot analysis.
14. Store the DNA at 4 °C.

B. *Alternative method for very crude DNA preparations*

1. Incubate the tail biopsy overnight at 55 °C in 500 μl lysis buffer (see above).
2. Freeze the tubes at –20 °C overnight.
3. Centrifuge the frozen sample at 14 000 r.p.m. at 4 °C until melted (this takes about 10 min). This pellets most of the SDS.
4. Transfer the supernatant containing the genomic DNA into a fresh tube.
5. Use 0.2 μl of the supernatant for a 20 μl PCR reaction.

[a] Steps 7 and 8 are optional. Alternatively, proceed directly with step 9 after the RNase treatment.

5.3 Southern blot analysis

An alternative to the PCR analysis of transgenic founder mice is to detect the transgene by Southern blot analysis using standard procedures (19).

Founder and F1 generation mice can be checked by qualitative Southern blot analysis whereas quantitative Southern blot analysis should be applied for genotyping the F2 generation to distinguish between heterozygous and homozygous animals. For quantitative Southern blot analysis it is important to load equal concentrations of DNA for each sample and, after detecting the transgene, to rehybridize the genomic DNA with a probe for an endogenous gene as a standard.

References

1. Gordon, J.W., Scangos, G.A., Plotkin, D.J., Barbosa, J.A., and Ruddle, F.H. (1980) *Proc. Natl Acad. Sci USA* **77**, 7380.
2. Lo, C.W., Coulling, M., and Kirby, C. (1987) *Differentiation* **35**, 37.
3. Costantini, F., Radice, G., Lee, J.L., Chada, K., Perry, W., and Son, H.J. (1989) *Prog. Nucleic Acids Res. Mol. Biol.* **36**, 159.
4. Garrick, D., Fiering, S., Martin, D.I., and Whitelaw, E. (1998). *Nature Genet* **18**, 56.
5. Chada, K., Magram, J., Raphael, G., Radice, E., Lacy, E. and Constantini, F. (1985). *Nature* **314**, 388.
6. Krumlauf, R., Hammer, R.E., Tilghman, S.M., and Brinster, R.L. (1985). *Mol. Cell. Biol.* **5**, 1639.
7. Townes, T.M., Lingrel, J.B., Brinster, R.L., and Palmiter, R.D. (1985). *EMBO J.* **4**, 1715.
8. Brinster, R.L., Allen, J.M., Behringer, R.R., Gelinas, R.E., and Palmiter, R.D. (1988) *Proc. Natl Acad. Sci. USA* **85**, 836.
9. Choi, T., Huang, M., Gorman, C., and Jaenisch, R. (1991) *Mol. Cell. Biol.* **11**, 3070.
10. Palmiter, R.D., Sandgren, E.P., Avarbock, M.R., Allen, D.D., and Brinster, R.L. (1991). *Proc. Natl Acad. Sci. USA* **88**, 478.
11. Kozak, M. (1987) *Nucleic Acids Res.* **26**, 8125.
12. Yang, X.W., Model, P., and Heintz, N. (1997). *Nature Biotechnol.* **15**, 859.
13. Schedl, A., Montoliu, L., Kelsey, G., and Schütz, G. (1993). *Nature* **362**, 258.
14. Peterson, K.R., Clegg, C.H., Huxley, C., Josephson,, B.M., Haugen, H.S., Furukawa, T., and Stamatoyannopoulos, G. (1993). *Proc. Natl Acad. Sci. USA* **90**, 7593.
15. Chiocchetti, A., Tolosano, E., Hirsch, E., Silengo, L., and Altruda, F. (1997). *Biochim. Biophys. Acta* **1352**, 193.
16. Gossen, M. and Bujard, H. (1992). *Proc. Natl Acad. Sci. USA* **91**, 9302.
17. Brinster, R.L., Chen, H., Brinster, R.L., Chen, H.., Trumbauer, M.E., Yagle, M.K., and Palmiter, R.D. (1985). *Proc. Natl Acad. Sci. USA* **82**, 4438.
18. Schedl, A., Larin, Z., Montoliu, L., Thies, E., Kelsey, G., Lehrach, H., and Schütz, G. (1993). *Nucleic Acids Res.* **21**, 4783.
19. Sambrook, J., Fritsch, E.F., and Maniatis, T. (1989). *Molecular Cloning: A Laboratory Manual*, 2nd edn. Cold Spring Harbor Laboratory Press, Cold Spring Harbor, NY.

10

Directed mutagenesis in embryonic stem cells

ANTONIUS PLAGGE, GAVIN KELSEY and NICHOLAS D. ALLEN

1. Introduction

Since the first description of homologous recombination in mouse embryonic stem (ES) cells (1,2), gene targeting has become a routine technology in mouse molecular and developmental genetics. At its simplest it allows the introduction of disabling mutations into a gene of interest in ES cells and passage of the disrupted allele through the mouse germ line in order to assess gene function from the phenotypic consequences. Mutations in some hundreds of genes have now been made by this route (see the transgenic/targeted mutation database (TBASE; http://tbase.jax.org) and Mouse Knockout and Mutation Database (MKMD); http://BioMedNet.com/db/mkmd) (Chapter 7). Over the past few years, approaches have become more sophisticated and the potential for more refined sequence manipulation is being realized (see *Table 1*). Amongst these refinements are, for example, conditional mutations which can be induced at defined times in development, or within specific lineages, to assess aspects of function of a gene for which a constitutively null mutation is lethal. In addition, targeted alleles can be 'cleaned up' by removing selectable markers to avoid potentially confounding effects arising from the presence of heterologous sequences. Finally, it is feasible to introduce subtle mutations, even single nucleotide exchanges, into genes. Many of these extra possibilities have come about through the exploitation of site-specific recombinases (reviewed in ref. 3). The potential of such technology should be carefully considered before embarking on a simple targeting experiment. Although generation of a more elaborate construct may be more time-consuming, in comparison with the time involved in targeting ES cells and obtaining germline transmission it is insignificant and the additional time investment in designing a more versatile construct might have significant longer-term benefits. In this chapter, we provide an overview of targeting strategies and the factors to be taken into account in the design of a targeting experiment, from initial construct preparation to the generation and analysis of mice carrying the desired mutation. We give particular emphasis to the potential

Table 1. Gene targeting strategies

Strategy	Description	Type of genetic modification	Comments
Gene replacement	Inactivation of gene function by replacing coding exons by a selectable marker	Suitable for the generation of a non-functional allele	The presence of the selectable marker may alter gene expression at/around targeted locus.
Gene insertion	Inactivation of gene function by inserting a selectable marker into the gene, accompanied by insertion of vector sequences and duplication of genomic sequences	Usually results in the generation of a non-functional allele.	Duplication at the targeting site may allow splicing around insertion and 'leakiness' of mutation. Insertion is the first step of a hit-and-run strategy.
'Hit-and-run'or 'in–out'	Two-step targeting strategy based on a single insertion type vector. Hit: targeting leads to integration of the vector and duplication of genomic sequences at the target locus. Run: the duplication created is resolved by an intrachromosomal recombination	Small non-selectable mutations	Allows removal of selectable markers from targeted allele. Negative selection required to recover rare reversion events.
'Double replacement' or 'tag-and-exchange'	Two independent replacement vectors in sequential targeting steps: (1) introduction of positive and negative selectable markers into target locus; (2) replacement of selectable markers and introduction of a non-selectable mutation by the second targeting vector	Small non-selectable mutations	Allows removal of selectable markers from the targeted allele. Multiple replacements at targeted locus possible.
Recombinase systems (Cre-*loxP*, Flp-*frt*)	Site-specific recombination between *loxP* or *frt* sites. Recombination products depend upon orientation of sites. Intra- and intermolecular events are possible, the former being favoured.	Wide range: removal of heterologous sequences; temporally or spatially regulated knock-outs; targeted integrations; small non-selectable mutations; large deletions; translocations; inversions; duplications	*LoxP* or *frt* sites remain in the genome after site-specific recombination.

use of site-specific recombinases in targeting. Because of the breadth and complexity of the techniques involved in targeting to cells, we only provide protocols in this chapter for the embryo manipulations required to make chimeric animals carrying the ES cells. We refer readers to another volume in this series, *Gene Targeting: a Practical Approach* (4), for more extensive coverage of ES cell methods and fuller details of other topics introduced in this chapter.

2. Basic elements of construct design

DNA transfected into mammalian cells integrates into the genome at a low frequency, predominantly at random locations. A small proportion of integration events (typically 0.1–10%) may occur by homologous recombination precisely at the endogenous counterpart of the transfected DNA: the basis of gene targeting. A successful targeting strategy must be capable of clearly discriminating random from targeted events. In addition, construct design may enhance the probability of homologous recombination. Some variables are clearly within the control of the investigator, but others may vary on a locus-to-locus basis. For example, although genes thought to be quiescent in ES cells can be targeted, the effect of transcriptional activity or chromatin organization at the locus on the probability of homologous recombination is not fully understood. Below, we discuss some of the principles to be considered in relation to a basic gene replacement construct.

2.1 Replacement versus insertion

Homologous recombination between a linear DNA molecule and the genome can occur in two ways, depending on the arrangement of homologous sequences in the targeting construct. These are illustrated in *Figure 1*. Recombination to produce a replacement of the host genome sequence with vector occurs via two crossover events at sequences flanking a selectable marker by a mechanism sometimes referred to as 'omega-type' recombination. By contrast, recombination of 'O-type' requires only a single crossover and produces an insertion of the vector into the genome and a duplication of the regions of homology present in the vector. This type of construct is less frequently used for simple knock-outs, in part because there is no deletion of any part of the target gene, and the duplication may result in the mutation being leaky by splicing around the insertion. In addition, if the insertion is targeted near the 5′ end of the gene, as illustrated in *Figure 1*, then it is possible that the duplication at the targeted locus will regenerate a fully or partially functional promoter. However, the efficiency of targeting by insertion at the *Hprt* locus is nine times higher than by targeting by replacement at the same locus (5). The duplication produced by the insertion method, furthermore, is the basis of 'hit-and-run' targeting, which can engineer single base changes into the genome (see Section 4.2).

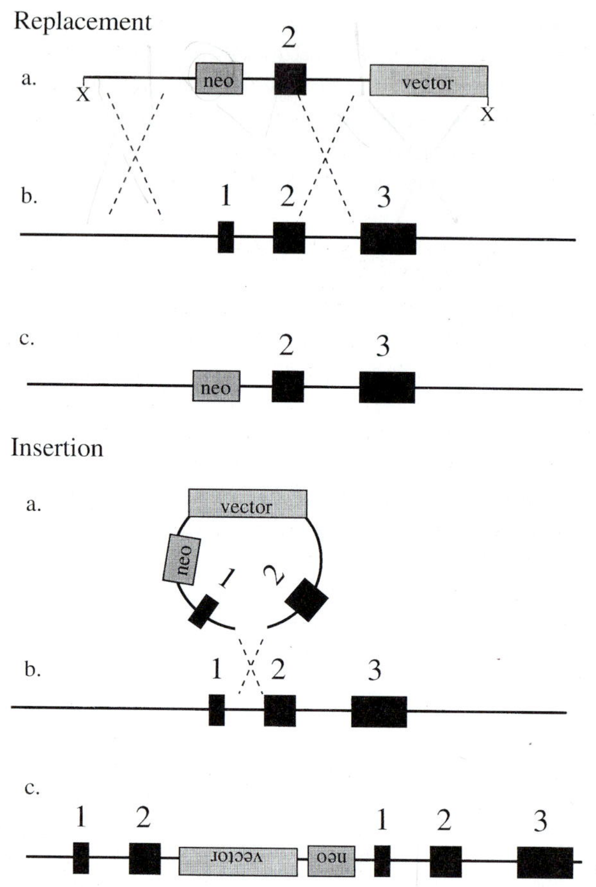

Figure 1. Basic gene targeting by gene replacement and by insertion. (1) Replacement. (a) The targeting construct comprises two regions of homology flanking a selectable marker cassette (*neo*) inserted in place of exon 1 (exons are represented by dark boxes). X indicates a unique site at which the construct is linearized. (b) Recombination (dashed lines) between regions of homology in the targeting construct and the genomic locus leads to a replacement at the wild-type locus. (c) The targeted locus has exon 1 replaced by the neomycin resistance gene, *neo*. (2) Insertion. (a) The targeting constuct has a single region of homology cloned adjacent to the *neo* selectable marker. The construct is linearized at a unique site in the region of homology. (b) Recombination (dashed lines) occurs at the ends of the linear molecule. (c) The targeted locus has the construct inserted into the gene flanked by a duplication of the region of homology.

2.2 Regions of homology

The length of the regions of homology, both individually and combined, has a profound effect on the frequency of targeting. Increasing the total homology in an *Hprt* construct from 1.7 to 6.8 kb, for example, was found to result in at least a 20-fold increase in targeted events (6). A total of 5–10 kb is a useful size at which to aim. It is possible that targeting frequency increases further with greater length of homology; however, it may become unwieldy to generate larger constructs by conventional restriction enzyme cloning methods. The distribution of the regions of homology around the selectable marker also has an important impact. Reducing the shorter homology arm to much less than ~1 kb results not only in diminshed frequency, but also reduced fidelity of recombination (6,7). On the other hand, one short arm facilitates identification of homologous recombination by PCR (Section 2.6).

Sequence identity within the homology arms should be maintained. In practice this means that, to avoid sequence differences between mouse strains, the DNA from which the homology arms are made should originate from the same mouse strain as that of the ES cell. It has been shown that targeting the *Rb* locus in ES cells derived from 129 mice with a 129 derived construct was 20-fold more efficient than using a construct from BALB/c DNA (8). Sequence comparison of the two construct DNAs revealed sequence divergence, including exchanges or deletion/insertion variants, at a frequency of about 1 per 160 bp. Genomic libraries of 129 DNA, isogenic with respect to the commonly used ES cells, are available from genome resource centres (for example, a PAC library is available from HGMP Hinxton, UK (http://www.hgmp.mrc.ac.uk) and PAC and cosmid libraries from RZPD, Berlin (http://www.rzpd.de)) or commercial sources.

2.3 The mutation

A few precautions are appropriate before induction of a null mutation. Detailed knowledge of the gene structure is advantageous, in terms of the location of exons encoding functionally important domains of the protein, and possible alternative start sites and splice variants. The generation of a null allele would require complete deletion of the gene or deletion of a critical functional domain(s). It is important to consider the likelihood that alleles may be produced that have residual activity, or have gained additional functions. Targeting events that leave part of the coding region upstream or downstream of the insertion intact might have the potential to produce a truncated gene product. In addition, splicing can occur around an exon containing the selection cassette. It is also possible that the presence of heterologous sequences in the targeted allele, i.e. selectable marker cassette, could alter the regulation or expression of closely neighbouring genes, which could complicate a phenotype. Removal of such extraneous sequences after targeting will be appropriate in some circumstances and can be achieved by site-specific recombinases (Section 3.1) or the use of a 'hit-and-run' strategy (Section 4.2).

2.4 Components of the targeting cassette

Since the frequency of integration of transfected DNA is low, a selectable marker is required as part of the mutation cassette. The most routinely used markers are encoded by bacterial genes and offer resistance, for example, to neomycin, hygromycin or puromycin. Some ES cell lines (e.g. E14TG) are HPRT deficient, permitting the use of constructs encoding HPRT to be selected in the presence of HAT medium. This same marker also allows negative selection, against its presence, by the purine analogue 6-thioguanine. Negative selection can be used as the basis of enrichment strategies (Section 2.5), to select for excision after deletion catalysed by site-specific recombinases (Section 3.1), and in 'hit-and-run' and 'double replacement' approaches (Section 4). A widely used negative selectable marker is the herpes simplex virus thymidine kinase gene (*HSVtk*). The wild-type *HSVtk* gene causes male sterility and mutant forms should be used if a targeted allele containing this marker is to be passed through the male germline (9). Sophisticated targeting strategies involving more than one transfection step may require multiple selection markers and it should be borne in mind that exposure to several drugs and multiple rounds of selection may compromise ES cell pluripotency.

The selectable marker requires a promoter that is active in ES cells and relatively position-independent. The phosphoglycerate kinase-1 (*Pgk1*) promoter or the polyoma enhancer/*HSVtk* promoter combination (pMC1) are frequently used. The sequence context around the translation start site of prokaryote-derived selectable markers should conform to the eukaryotic Kozak consensus to ensure efficient translation in ES cells. (For a more extensive discussion of marker cassette promoters, see ref. 10).

In addition to the insertion of a selectable marker, targeting offers the possibility of introducing novel genetic elements into the gene, such as expression reporter markers, for example, β-galactosidase (LacZ), green fluorescent protein (GFP) or placental alkaline phosphatase (PAP). These could be useful to gain information on the expression of the locus at cellular resolution, where this is not already known, and open the possibility of following the fate of cells carrying the targeted mutation in chimeras (Section 6.5).

2.5 Enrichment for targeted events

Several methods exist to enhance the frequency of homologous recombination, either by suppressing random insertions, or by providing a selective advantage for correctly targeted events. In positive–negative selection (11), the construct is provided with a negative selection cassette, for example the *HSVtk* gene, adjacent to one of the regions of homology. The rationale is that many random events will lead to integration of the entire targeting vector including the negative selection cassette. If expression ensues, the colony is killed in the presence of the negative selective drug. Positive–negative selection can enrich for targeted events by up to 20-fold. Enrichment may be increased by flanking the construct at both ends with negative selection markers. Positive–negative selection is used routinely in some laboratories, although there is some concern that exposure of ES cells to a more complex selection regime can compromise totipotency.

Where a targeting construct has the selectable marker driven by its own promoter, random integration into the genome will allow expression of the selectable marker. However, if the locus to be targeted is active in ES cells, then a promoterless marker will be inactive at many random locations, and only become expressed when appropriately inserted into an active locus, including the gene to be targeted. Similarly, by omitting a translational start site from the selectable marker, activity could be made dependent upon insertion in-frame into the locus (and generation of a fusion protein). An analogous enrichment strategy with a marker lacking polyadenylation signals could be considered.

2.6 Screening strategies

A rapid and reliable screening strategy is an essential component of targeting and should be developed during construct design. Typically, 100–300 ES clones may need to be analysed to identify a handful of targeted events. Screening has two phases: a first screen (which might involve PCR or Southern analysis) for the rapid detection of candidate targeted colonies, and a second screen to verify correct targeting, necessitating more detailed Southern analysis.

A PCR screen ideally employs a primer adjacent to the region of homology (logically outside the shorter arm), and a second within the novel sequence within the targeting construct, e.g. the selectable marker. In this way, a correctly sized PCR product is only generated when the two primer sites are juxtaposed by homologous recombination. A PCR screen has the advantage that it can be conducted with minimal expansion of ES clones, from cells in 24- or 96-well plates (12), saving culture time and helping to preserve the totipotency of the ES cells.

A primary Southern screen uses a restriction enzyme that cuts once outside one region of homology and a second time within (or the other side of) the construct, in combination with a single-copy probe outside the region of homology. Homologous recombination should lead to the generation of a diagnostic restriction fragment, readily distinguishable from the endogenous copy, while random integrations do not alter the wild-type pattern detected by the external probe. It is possible to use a probe internal to the construct to detect a diagnostic restriction fragment, but a proportion of random events will lead to a similar-sized restriction fragment, so the discriminative power is not high. Sufficient DNA can be obtained for Southern analysis from colonies of 24-well plates (13). Factors to be considered in the choice of restriction enzyme are:

- a novel, targeted, fragment that is smaller than the endogenous fragment avoids the possibility of false positives due to partial digestion of DNA;
- some restriction enzymes having the CpG dinucleotide in their recognition sequence are inhibited by cytosine methylation, and may lead to unreliable digestion.

The second phase, verification of correct targeting, requires detailed Southern analysis of the structure of the locus, typically with probes both 5′ and 3′ and possibly internal to the construct, and digests with a number of restriction enzymes. This is necessary to check that precise recombination has occurred in both homology arms, and to identify possible integration problems, such as integration of the entire targeting vector, tandem duplication of the construct, or rearrangements such as interstitial deletions or larger deletions of the targeted locus (for details, see ref. 14). Some such rearrangements may be favoured at particular loci if they serve to increase expression of the selectable marker and provide a selective advantage to the colony.

2.7 Alternative approaches for construct design

Targeting constructs are generated conventionally by a reiterative process of ligation and transformation into *Escherichia coli*. Such cloning can be time-consuming and it requires a reasonably detailed restriction map of the locus, or complete sequence, to design an appropriate cloning strategy. It is possible to circumvent some of the restriction digestion and ligation steps by alternative cloning approaches. Libraries have been constructed so that clones of the locus of interest already have some of the modifications necessary for making a targeting construct and can significantly reduce the steps required to construct the vector (15). Another method uses random insertion of the *Ty* transposon *in vitro* using commercially available enzymes and following transfection into *E. coli* produces clones that need only addition of a selection cassette to be ready for targeting (16). Homologous recombination is the predominant fate of transfected DNA in yeast, and this has been exploited in a method to generate vectors for homologous recombination using fewer ligation steps (17).

3. The use of site-specific recombinases in targeting

Site-specific recombinases promise to broaden radically the scope of manipulation of the mouse genome. The recombinases Cre and Flp belong to the λ integrase family; Cre is derived from the bacteriophage P1 and Flp is encoded by the *Saccharomyces cerevisiae* 2 μm plasmid. Both recognize 34 bp inverted repeat sequences separated by an asymmetric 8 bp spacer, known as *loxP* (in the case of Cre) and *frt* (in the case of Flp) sites; the latter contains a third repeat at one end (reviewed in refs 3 and 18).

```
                          spacer
loxP site:  ATAACTTCGTATA ATGTATGC TATACGAAGTTAT
            ------------>          <------------
                          spacer
frt site:   GAAGTTCCTATAC TTTCTAGA GAATAGGAACTTCGGAATAGGAACTTC
            ------------>          <-----------  <-----------
```

loxP and *frt* sites are available in a number of plasmids in combination with versatile polylinkers and selectable marker cassettes (17,23–25).

Cutting and rejoining of the DNA strands takes place within the 8 bp spacer

(19). The type of recombination between *loxP* or *frt* sites depends on the location and orientation of the recognition sequences, as determined by the 8 bp spacer. If the two target sites are located on the same DNA molecule in the same orientation, the intervening DNA is excised, resulting in two DNA molecules each retaining a single site (see *Figure 2*). If the two sites are oppositely orientated, inversion of the intervening DNA occurs. Intermolecular recombination can also occur. In the case of sites on separate circular and linear DNA molecules, an insertion takes place; where two linear molecules are involved, a translocation event. Each of these reactions is reversible, the intramolecular reactions being favoured kinetically. Neither Cre nor Flp requires co-factors, and both enzymes have been shown to work *in vitro* as well as in various cells and organisms (3). These features make the two recombinases ideal tools for defined genetic manipulations and the pioneering work of Sauer and O'Gorman, who established the Cre and Flp systems for use in eukaryotic cell culture (20–23), has opened the way to new approaches in gene targeting technology in ES cells. Here, we consider the applications for which the recombinase systems are employed (see *Table 1*).

3.1 Excision of heterologous DNA from a targeted locus

Heterologous DNA sequences, such as selection marker genes and their promoters in targeted genes, can influence the expression of neighbouring

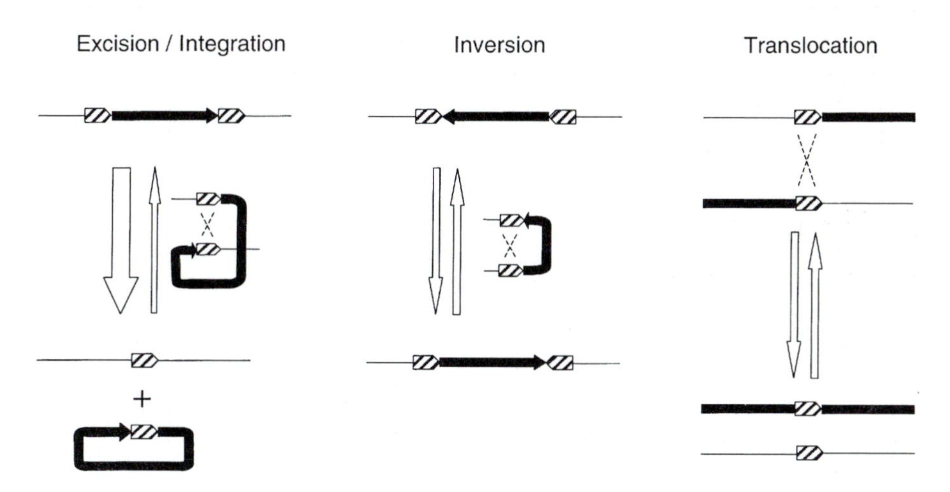

Figure 2. Types of recombination between two *loxP* or *frt* sites dependent on their location and orientation. Excision of intervening DNA takes place between directly repeated sites on the same molecule. The reverse reaction, integration of a circular molecule into a linear DNA strand, is kinetically less favoured. Inversion of the intervening sequence occurs if the two sites are oppositely orientated. A translocation is the result of recombination between two sites on different linear molecules.

Table 2. Ways of supplying recombinase

Source	Purpose	Examples/references
Recombinase expression plasmid	Transient expression of recombinase in transfected ES cells and injected oocytes	pMC-Cre (29), pBS185 (22), pCAGGS-Cre (39), pOG 44 (23), phACTB::Flp (78), pANMerCreMer[a] (43)
Recombinase RNA	Transient expression of recombinase in injected oocytes	Plasmid T7IC[b] (40)
Recombinase-expressing transgenic mice	Ubiquitous or restricted supply of recombinase *in vivo* for general or conditional recombination	General 'deleter' strains: hCMV-Cre (35), EIIa-Cre (36), Prm1-Cre[c] (37), ZP3-Cre[d] (38), hACTB::Flp (78) Strains expressing recombinase in a restricted way[e]
Recombinase-adenovirus	Spatially restricted supply of recombinase *in vivo* via adenovirus infection	AdCre (52)

[a]Expression plasmid containing a tamoxifen-inducible form of Cre.
[b]Plasmid designed for *in vitro* transcription of Cre RNA.
[c]Strain expressing Cre in the male germline.
[d]Strain expressing Cre in the female germline.
[e]For a complete list see website at http://www.mshri.on.ca/nagy/Cre.htm and compare text.

loci, and could have an unpredictable influence on the knock-out phenotype (26–28). A 'cleaner' mutation can be generated by flanking the heterologous sequences by *loxP* or *frt* sites ('floxed' or 'flrted' in the jargon) and subsequently using recombinase to excise the heterologous sequence from the targeted locus, leaving behind a single *loxP* or *frt* site. Excision can be mediated by recombinase delivered directly to targeted ES cells or to mice carrying the targeted allele (see *Table 2*).

3.1.1 Excision from ES-cell clones with recombinase expression plasmids

A straightforward approach to produce excision is to transfect targeted ES-cell clones with a recombinase expression plasmid. It is possible to select for the recombination event by including a negative selectable marker, e.g. *HSVtk*, in the targeting construct between the *loxP* or *frt* sites, which becomes lost by excision (29–31). A less direct enrichment strategy utilizes a marker in the recombinase expression plasmid to select for cells that have taken up the plasmid (26,32,33). Alternatively, non-selective PCR-based screening approaches can been used (34); in our experience, the frequency of transfection and excision is sufficiently high to dispense with counter-selection. The frequency of the re-insertion event at the residual *loxP/frt* site appears to be negligible. A second electroporation and cloning results in increased passage number of the ES cells before injection into blastocysts, with the potential to reduce the pluripotency of the cells.

3.1.2 Excision in mice using transgenic mice expressing recombinase

To avoid this increase in passage number, excision can be performed by producing mice with a targeted allele and crossing them with transgenic mice expressing the recombinase at early stages of embryonic development or in the germline ('deleter' strains). There are several transgenic strains in which expression of recombinase is regulated by a variety of promoters. For example, transgenic mice expressing Cre driven by the ubiquitously active hCMV minimal promoter or the adenovirus EIIa promoter have been reported to lead to Cre activity in pre-implantation embryos, and efficient excision of floxed DNA from all tissues including the germline has been obtained (35,36). By crossing chimeric mice containing recombination target sites with hCMV Cre mice, recombination was reported to be complete in all cells, such that constitutive excision had already occurred in the progeny (35). Highly specific Cre expression in the male and female germlines has been obtained through the use of the protamine 1 (37) and ZP3 (38) promoters, respectively. In these lines recombination occurs in the germline of double transgenics (with both Cre and 'floxed' genes) and it is their offspring that have a complete excision.

3.1.3 Oocyte microinjection of recombinase

Site-specific recombination can be catalysed by microinjecting recombinase, either as an expression plasmid or as mRNA, directly into fertilized eggs of mice carrying the floxed allele. Excision appears to be complete and non-mosaic in the resulting mice, saving the generation time required to produce second-generation offspring. The expression plasmid pCAGGS-Cre used for microinjection by Araki *et al.* (39) contains the strongly active chicken β-actin promoter regulating Cre recombinase. Excision was generally complete before the morula stage. Integration into the mouse genome of the Cre expression plasmid can be avoided by using circular plasmid for microinjection. Alternatively, *in vitro* transcribed Cre RNA has been used for cytoplasmic microinjection of fertilized eggs (40). Although recombination was mosaic at early stages, older embryos were shown to be completely recombinant.

3.2 Use of recombinases to generate conditional somatic knock-outs

The Cre/*loxP* and Flp/*frt* recombination systems allow the generation of conditional knock-outs, i.e. animals in which a gene knock-out is restricted to certain cells or occurs in response to an exogenous induction signal. This is achieved by spatial and/or temporal control of recombinase expression. These are made by introducing *loxP* or *frt* sites into flanking or intronic regions of a gene by targeting so that the gene function is left unperturbed. Mice generated from these targeted ES cells will express the gene normally from the 'floxed' or 'flrted' allele(s) until regulated expression of recombinase induces excision (see *Figure 3*). This can be achieved by crossing to transgenic mice

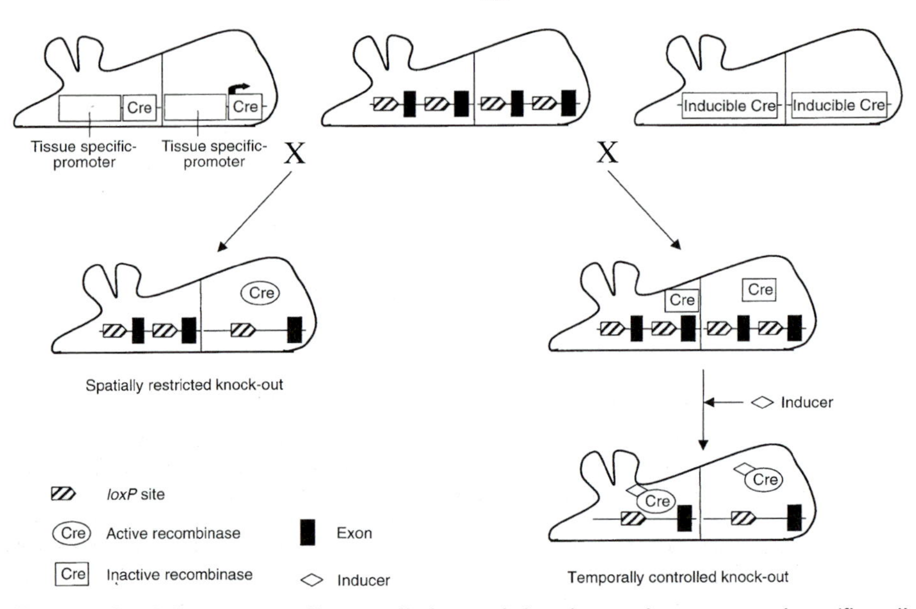

Figure 3. Spatially or temporally controlled somatic knock-outs. A mouse carrying a 'floxed' gene (middle) crossed with a mouse transgenic for a tissue-specifically expressed Cre (left) results in offspring carrying a Cre-mediated deletion of the gene in the corresponding tissue. To control the timing of a general deletion a mouse carrying a 'floxed' gene can be crossed to a transgenic mouse expressing an inducible form of recombinase ubiquitously (right).

expressing the recombinase under the control of a tissue-specific promoter (spatially restricted knock-outs) or as an inducible form, e.g., inducible Cre fusion proteins, or expression from inducible promoters (temporally controlled knock-outs). These conditional knock-outs open new opportunities to study the function of widely expressed genes in specific tissues. Embryonic or post-natally lethal phenotypes accompanying constitutive knock-outs can be circumvented to allow regionally restricted knock-outs at a later stage or a global knock-out at a chosen time-point. The possibilities are limited only by the availability of recombinase constructs with tightly regulated expression. The generation of transgenic lines with highly restricted and efficient recombinase expression to expand the application of these methods is a collective effort which can be followed at the web site: http://www.mshri.on.ca/nagy/Cre.htm.

3.2.1 Spatially restricted knock-outs using tissue specific promoters to drive recombinase expression

We consider two examples of tissue-specific knock-outs. By targeting three *loxP* sites into the gene for the repair enzyme DNA polymerase B (*Polb*), one upstream of the promoter and two flanking a *neo*^{r}/*HSVtk* cassette in the first intron, a two-step strategy could be used to generate a 'floxed' allele (see

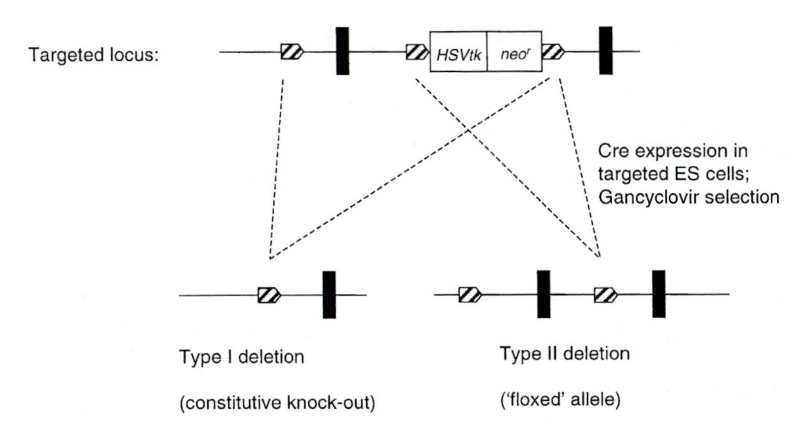

Figure 4. Generation of a 'floxed' allele in which selection markers have been deleted. Targeting is used to introduce positive and negative selectable markers flanked by *loxP* sites into an intron a third *loxP* site further upstream or downstream of the exons. After positive selection for homologous recombination (top) targeted ES cells are transiently transfected with a Cre expression plasmid and the following negative selection results in cell clones bearing type I and type II deletions. The negative selection excludes recombination between the left and middle sites which would leave the selection cassette intact (30).

Figure 4; 30). Targeted clones were electroporated with a Cre expression vector and selected for the loss of the *HSVtk* gene. Two types of recombination occurred:

- recombination between the first and third *loxP* sites, leading to deletion of the promoter and the first exon (see *Figure 4*, type I deletion), and
- recombination between the second and third *loxP* sites, resulting in the desired 'floxed' allele (see *Figure 4*, type II deletion).

Mice produced from cells with the type I deletion have a consitutive loss of *Polb*, which is lethal. Mice generated with the 'floxed' allele were crossed to mice carrying a Cre transgene driven by the *lck* promoter, which restricts expression to the T-lymphocyte lineage. The Cre-mediated knock-out occurred specifically in a proportion of T-cells.

Another experiment targeted two *loxP* sites flanking part of the NMDA receptor subunit 1(*Nmdar1*) gene in ES cells and subsequently in mice (41). The calcium/calmodulin-dependent kinase II α (*Camk2a*) promoter, active post-natally in the forebrain, was used to regulate Cre transgene expression. One transgenic line had a highly specific and efficient expression in the CA1 region of the hippocampus (42), which resulted in a complete absence of wild-type *Nmdar1* mRNA expression and impaired spatial memory (41).

3.2.2 Temporally regulated knock-outs using inducible recombinases

There are two broad approaches for temporally regulated knock-outs: regulatable recombinases comprising fusion proteins of Cre or Flp and the ligand-binding domains (LBDs) of steroid-hormone receptors; and the use of inducible promoters to drive recombinase expression. In a perfect system, there should be no background recombinase activity in the absence of the inducer, and the inducer should be capable of being delivered to all tissues and be devoid of non-specific side effects.

The fusion protein strategy relies on inactivation of the recombinase by binding of a heatshock protein 90 (HSP90)-containing complex to the LBD in the absence of ligand. Binding of ligand releases the fusion protein from the complex and allows recombinase activity. Mutated LBDs with reduced affinities for the endogenous hormone but increased affinity for an artificial ligand have been developed to enhance the specificity of the system. Cre recombinase fusions with mutated forms of the murine and human oestrogen receptor LBD have been tested in ES cells or transgenic mice(43, 44). The mutated oestrogen receptor LBDs have a ~1000-fold reduced affinity for the endogenous 17β-oestradiol but retain binding to the drugs tamoxifen and 4-OH-tamoxifen. A fusion protein based on the human oestrogen receptor LBD and the CMV promoter to drive expression has also been tested in transgenic mice (44). One line showed 4-OH-tamoxifen-dependent, though variable, excision of a 'floxed' DNA fragment in all tissues tested except in the thymus. Transient administration of 4-OH-tamoxifen had no overt deleterious effects on the mice. A different fusion protein based on a modified human progesterone receptor LBD, which does not bind endogenous hormones but retains affinity for the synthetic progestin RU486, has also been tested in ES cells in a *lacZ* gene activation assay (45). Recombination was strongly inducible to near completeness, although there was some background activity in the absence of RU486. The drug RU486 itself did not induce obvious morphological changes in the ES cells.

The second approach uses inducible promoters. A particularly promising and versatile inducible promoter system is the prokaryotic tetracyclin repressor (TetR) and operator elements (see *Figure 5*; 46–49). A fusion protein (tTA) of TetR and the HSV VP16 transcriptional activator domain has been generated which retains affinity for the *tetO* operator element and activates a coupled promoter. Binding of tetracyclin or doxycyclin to tTA inhibits association with *tetO* elements, leaving the promoter inactive. Inducibility therefore relies on withdrawal of the drug to enable the activator to bind the promoter. Transgenic mice expressing the tTA fusion protein from the CMV promoter/enhancer and the *tetO* minimal promoter coupled to reporter genes had induction or silencing of up to several thousand-fold, although variable between tissues and mouse lines (47).

An obvious extension is a combination of temporally and spatially restricted knock-out in which an inducible recombinase is expressed from a tissue-specific promoter. To this end a Cre–oestrogen-receptor fusion protein has been put under the control of a B-lymphocyte-specific promoter in transgenic mice (51). Some mosaicism was observed; 75% of lymphocytes expressed the

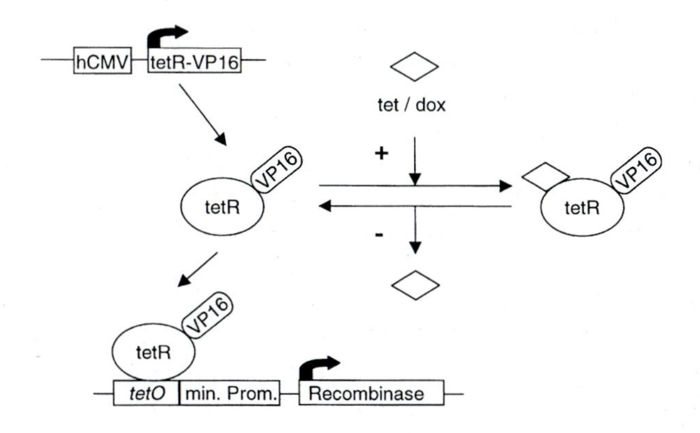

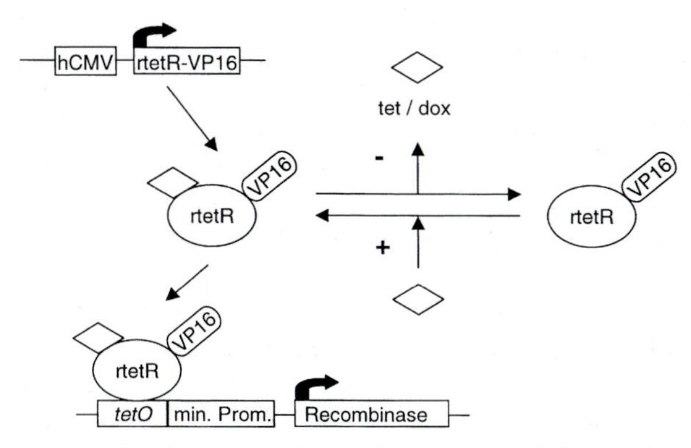

Figure 5. The tetracyclin-dependent inducible system. (a) In the 'Tet-OFF' version expression of the recombinase is inhibited upon administration of the inducer tetracyclin or doxycyclin. A fusion protein of the tetracyclin repressor and the HSV-VP 16 transcriptional activator domain (tTA) is expressed from a ubiquitously active promoter. Binding of the fusion protein to a minimal promoter containing *tetO* elements confers transcriptional activation of the associated recombinase. In the presence of tetracyclin the fusion protein is inhibited from binding to *tetO* elements. Induction (activation) of the recombinase requires withdrawal of the drug. (b) A mutated version of the fusion protein (rtTA) is used in the 'Tet-ON' system. Binding to *tetO* elements only occurs in the presence of tetracyclin. The system is silent unless the antibiotic is supplied.

A reverse system has been developed based on a mutated fusion protein (rtTA) which requires tetracyclins for binding to *tetO* elements (50). This system, used with the hCMV promoter, conferred an even higher range of reversible inducibility (up to 10^5-fold in some tissues) after administration of doxycyclin as an inducer (48). As a test of inducible recombination, a *tetO*–Cre transgene was driven by the *tetO* promoter and controlled by a CMV–tTA transgene. A *lacZ* gene was a marker for site-specific recombination (49). β-Galactosidase staining, and hence recombination, was observed after adminisatration of tetracycline, but with cellular heterogeneity. This mosaicism might be a property of the hCMV promoter and might be avoided by using different promoter elements.

Cre fusion protein and after induction ~80% of the these expressing cells showed *loxP* recombination.

A completely different route of administration is through infection by recombinant adenovirus (52–54). While the adenovirus can infect a variety of both actively dividing and quiescent cells, the range of tissues affected depends upon the route of infection. Intraveneous injection readily leads to strong infection in most tissues, e.g. liver and muscle, but the brain is hardly affected. By contrast, local administration of Cre–adenovirus to specific tissues resulted in infection and recombination in close proximity to the site of injection (54).

3.3 Knocking-in: recombinase mediated integration into chromosomally positioned target sites

The site-specific recombinase systems offer a straightforward means by which to introduce new sequences, or a series of mutated genes, into a locus of interest. This can also be accomplished by more conventional targeting, or by 'hit-and-run' or 'double replacement' approaches (Section 4). However, once a single *loxP* or *frt* site has been inserted into a locus, any number of new sequences could be introduced by exploiting site-specific recombination, more efficiently than by repeated targeting. Appropriate location of the *loxP* site will permit, for example, introduction of different genes into the genome under the control of the same endogenous promoter. Recombination between a circular DNA molecule containing a single *loxP* site in the presence of a Cre expression plasmid can mediate integration. However, since intermolecular recombination between target sites is kinetically less favourable than intramolecular recombination, the newly integrated DNA becomes flanked by *loxP* sites and is likely to undergo re-excision. In addition to restricting the expression of Cre to prevent re-excision, there are several other possibilities by which to maintain for correct knock-in events. Selection marker cassettes have been developed which result in the reconstruction of a functional resistance gene only after the site-specific integration event. One part of the selection marker is introduced first together with a single *loxP* site by homologous recombination. Recombinase mediated knock-in then delivers the missing part of the marker. Examples of such markers are the promoter and ATG-less neo^r cassette (55), and the *Hprt* minigene (56) (*Figure 6*). A second approach to reducing the frequency of re-excision is the use of mutated recombination sites. Different forms of *loxP* site can be used as the genomic target and in the knock-in plasmid: one site containing mutations in the left half of the inverted repeat, the other in the right half. After integration, a wild-type and a double-mutant *loxP* site are created, with the double-mutant site having reduced affinity for Cre recombinase, and the rate of intramolecular recombination between the mutant and wild-type sites being reduced. This results in an increased yield of stable knock-in colonies compared with the use of two wild-type *loxP* sites (57). Alternatively, the spacer region between the inverted repeats within *loxP* can be mutated to

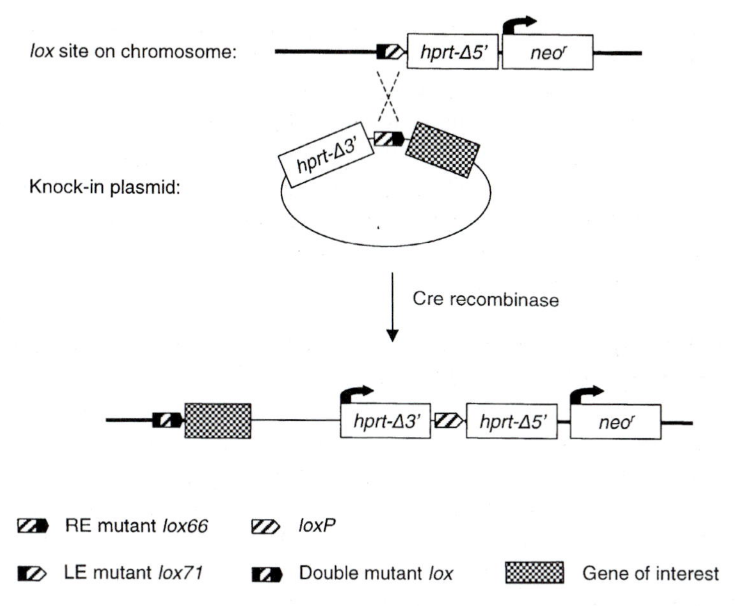

Figure 6. The use of a split selection marker cassette and mutant *lox* sites as a knock-in approach. A mutant *lox* site, the 3′-half of a split selection marker cassette (*Hprt*-Δ5′) and a different selectable marker (*neo*^r) are introduced via homologous recombination in a first step. Targeted ES cells are co-electroporated using a Cre expression vector and a knock-in plasmid containing the missing part of the split selectable marker (*Hprt*-Δ3′) and are selected for reconstitution of this marker. If the knock-in plasmid carrys a differently mutated *lox* site, a *loxP* site and a double-mutant *lox* site are created after site-specific recombination reducing the likelihood of re-excision. The sequences of the mutated sites are given below (nucleotide changes are underlined; ref. 57). Gene cassettes to be introduced into the genomic locus should be included in the knock-in vector downstream of the *lox* site. LE mutant *lox71* site: TACCGTTCGTATAGCATACATTATACGAAGTTAT; RE mutant *lox66* site: ATAACTTCGTATAGCATACATTATACGAACGGTA; double mutant *lox* site: TACCGTTCGTATAGCATACATTATACGAACGGTA.

create a heterospecific site (*lox511*) which does not cross-react with the wild-type site. This makes possible the parallel use of different *lox* sites in the targeting vector and the knock-in vector. In this case, independent Cre-mediated recombination between the heterospecific *lox* sites leads to an exchange of the DNA fragment in between those sites on the knock-in plasmid and the genomic locus. This kind of replacement was shown in 3T3 cells to be significantly more efficient than an insertion via a single *loxP* site (58).

4. Subtle mutations without site-specific recombinases

Although the development of site-specific recombinases Cre and Flp in gene-targeting technology has opened a whole range of new strategies, subtle genomic alterations can also be made by alternative methods (see *Table 1*).

4.1 Double replacement

In the double-replacement method a single targeted ES-cell clone can be used to generate a series of replacement alleles. In addition no selection marker or other heterologous sequence remains in the modified locus. The method consists of two rounds of homologous recombination (see *Figure 7*). First, a positive/negative selectable marker cassette is introduced into the locus of interest by a conventional replacement targeting to produce a 'tagged' allele. Suitable selection markers include a double *neo*^{r}/*HSVtk* cassette (59,60) or an *Hprt* minigene (61). In a second transfection, a targeting vector spanning the same genomic region and including the desired mutations, but no selection marker, is used. Selection against the *HSVtk* or *Hprt* gene leads to ES-cell clones in which the tagged allele has lost the selectable marker by replace-

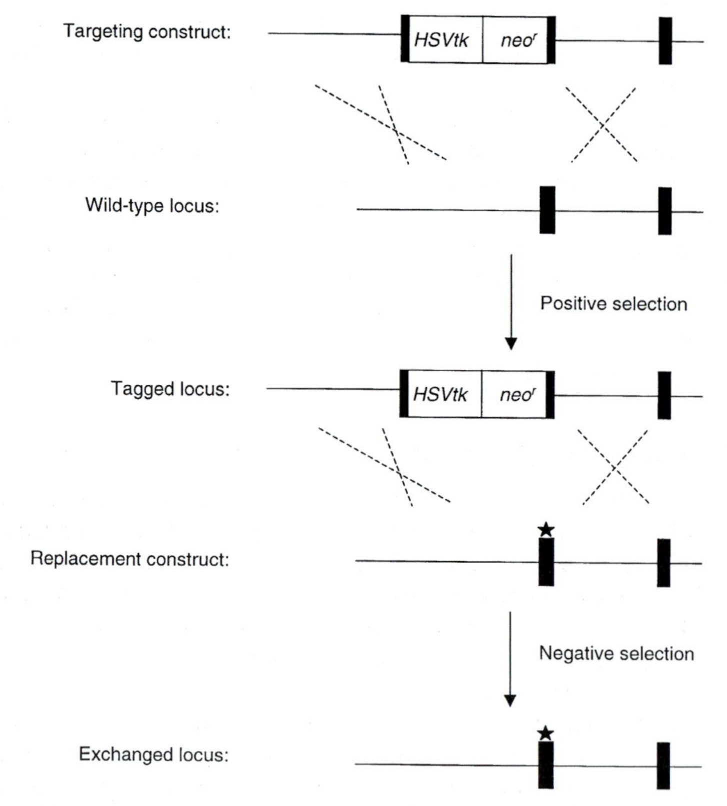

Figure 7. The double-replacement strategy. This consists of a two-step homologous recombination procedure in which a positive and negative selectable marker (*neo*^{r}/*HSVtk* or *Hprt*) are introduced in the first step. Targeted ('tagged') clones are used for the second round of homologous recombination employing a construct covering the same genomic region but including subtle mutations, such as point mutations in an exon (symbolized by a star). Selection is for the loss of the negative selectable marker.

ment with the mutated allele. Although random integration events and homologous recombination at the wild-type allele during the second transfection should not survive negative selection, in practice only a proportion of surviving ES-cell colonies undergo the desired replacement event: between one in 10 and one in 72 (59–61). The background comprises colonies in which the negative selectable marker is present, but presumably functionally inactivated, or is physically lost but not replaced by the second targeting vector. Nevertheless, the double-replacement strategy is a relatively efficient alternative to recombinase mediated 'knock-in'.

4.2 'Hit-and-run'

The 'hit-and-run' (62) or 'in–out' (63) procedure (see *Figure 8*) is based on an insertion type of targeting vector, in contrast to the replacement type vectors used in all the previously mentioned strategies. Recombination of the construct leads to partial duplication of the genomic sequences ('hit' or 'in' step) (see also *Figure 1*). Mutations introduced in the region of homology of the construct will be integrated into the genome as part of the duplication. The second step ('run' or 'out' step) relies on the spontaneous reversion of the duplicated sequences by an intrachromosomal recombination to remove the intervening sequences and selection for loss of the vector. Depending upon the position of the crossover, the locus either reverts to wild type, or is replaced by the mutated copy (63). Most colonies obtained after negative selection are accurate revertants of the duplicated region, but the portion of colonies that retain the introduced mutation varies (62,63). Although the frequency of the reversion step is low (8×10^{-7}), it is compensated for by the small amount of effort in plating.

5. Chromosome engineering in ES cells

Site-specific recombinase can be used to mediate gross rearrangements of chromosomes in ES cells and mice, including large deletions, duplications, inversions and even translocations, which can model human inherited or cancer-associated rearrangements (56,64). These manipulations rely on the use of pairs of *loxP* or *frt* sites and require two consecutive targeting events to insert the two sites, followed by transient expression of recombinase. Selection for rare recombination events through the use of split selection cassettes becomes imperative, and ES cells may be exposed to three separate drug-selection regimes during the experiment. The type of rearrangement depends on the relative orientation of the two sites, whether they are *in cis* or *in trans*, or located on non-homologous chromosomes.

5.1 Deletions, inversions and duplications

Conventional gene targeting produces deletions in the 10 kb range. In some cases larger deletions would be valuable: for example, gene clusters and

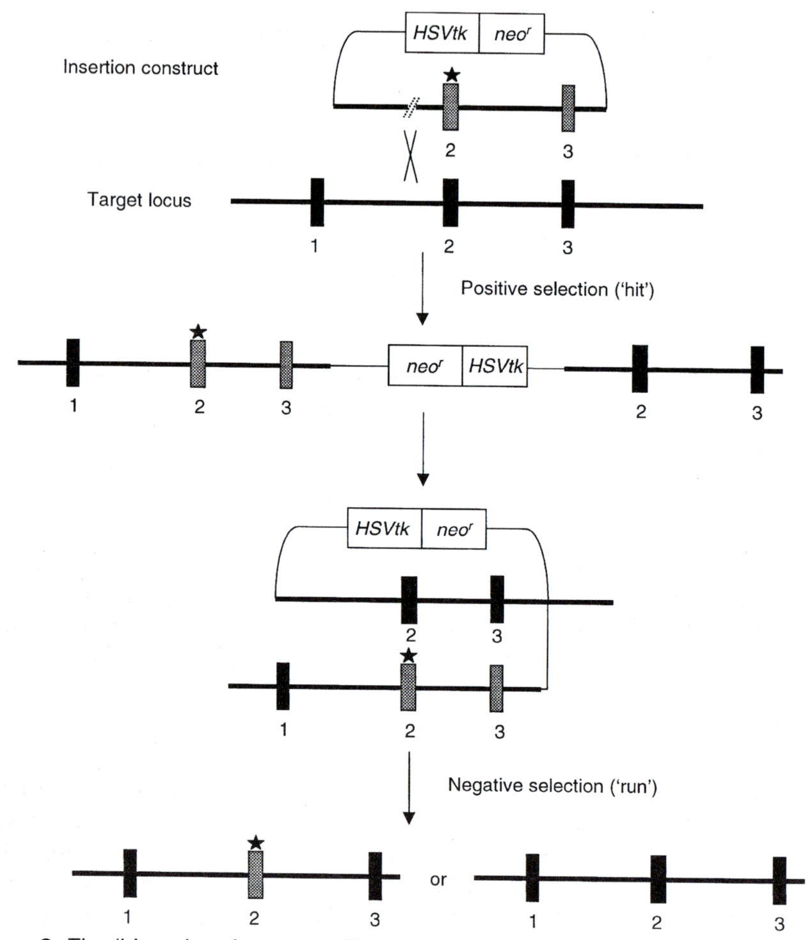

Figure 8. The 'hit-and-run' strategy. This procedure relies on insertion of the complete targeting vector, which leads to a duplication of the genomic sequences represented on the construct. The insertion type vector contains the selectable markers (*neo*^{r}/*HSVtk* or *Hprt*) within its backbone and is linearized for electroporation within the region of homology. Mutations to be introduced into the genome are included in the region of homology (symbolised by a star). The second step consists of a spontaneous recombination resulting in excision of the vector backbone and removal of duplicated sequences. Negative selection for the loss of the selection markers enriches for cells that have undergone the reversion. Mutations introduced in the first step can be retained or lost depending on the exact position of the second recombination.

regions corresponding to multigene syndromes. If the orientation of loci along the chromosome is known, the targeting strategy is conceptually simple: the targeting constructs are made so that the *loxP* sites are inserted at the boundaries of the region to be deleted, in the direct-repeat orientation. Two targetings are performed with constructs containing split selection cassettes and encoding resistance to different drugs, see *Figure 9* (56). Cre-mediated

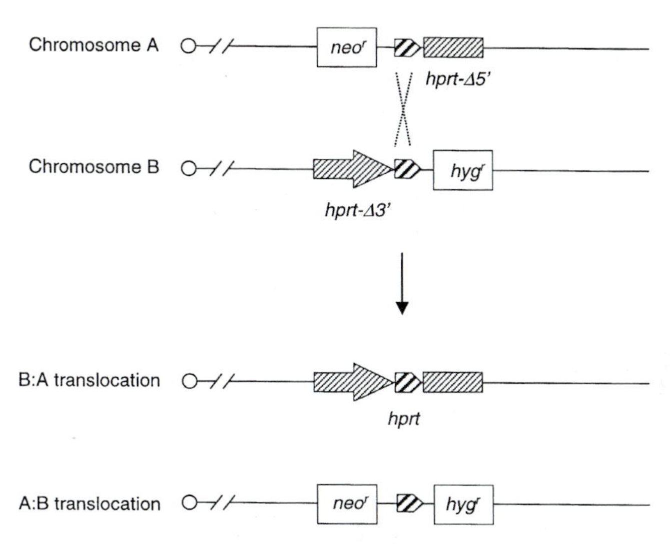

Figure 9. Chromosomal translocation induced by Cre recombinase between *loxP* sites (hatched arrowheads) residing on non-homologous chromosomes (A and B). The two genomic loci involved are targeted with distinct selectable marker cassettes (*neo*^{r} *loxP hprt*-Δ5′ and *hyg*^{r} *loxP hprt*-Δ3′). Site-specific recombination reconstructs a functional *hprt* allele for positive selection on the B:A translocation chromosome.

recombination leads to fusion of the *Hprt* half-sites, and deletion of the intervening DNA, including the other two selectable markers as a circular molecule. The two targeting steps will produce clones in which the *loxP* sites are *in cis* (on the same chromosome) and *in trans* (on each of the two chromosomal homologues). Other than FISH (see Chapter 6) there is no easy means by which to distinguish these alternatives, although the frequency of conversion to HAT resistance after Cre expression is indicative. The intra-chromosomal (*cis*) configuration has a higher recombination rate than the *trans* one and leads to loss of the two dominant markers (64). The outcome of *trans*-recombination is also more complex, with one chromosome sustaining a deletion, and one a duplication of the region. However, these can be segregated upon germline transmission of the manipulated ES cells.

If gene order or orientation of two loci is not known, the desired deletion or other rearrangement has to be generated empirically. This involves targeting multiple constructs to ensure that the appropriate relative order of *Hprt* half-genes and *loxP* sites is obtained. There are four possible combinations: >>, > <, < > and <<. *In cis* these lead respectively to deletion, inversion and duplication events; *in trans*, the recovery of recombinant clones is again much reduced and the outcomes can be more complex (64).

5.2 Translocations

Translocations can be produced by a conceptually similar procedure, except that two chromosomes are targeted (see *Figure 9*). Translocations have been engineered between the *Myc* and immunoglobulin heavy-chain loci on chromosomes 15 and 12, respectively, to model translocations common in plasmacytomas after sequential targeting was performed with *loxP/Hprt* split selection cassettes (56). Transfection with a Cre expression plasmid and selection for the intact *Hprt* marker led to the recovery of translocation events at a frequency of 5×10^{-8}.

6. Generating, analysing and maintaining knock-out mice

ES cells are introduced into developing embryos to generate chimeras either by blastocyst injection or by injection into or aggregation with eight-cell or morula-stage pre-implantation embryos (see *Figure 10*). These techniques are described in detail elsewhere (4), but here we discuss their relative merits and the critical parameters required for success and give basic protocols for producing chimeras by injection and by aggregation.

6.1 ES-cell culture and pluripotency

Whichever technique is used, it is critical to maintain cell cultures with a high degree of germline potency. by maintaining them under optimal conditions at all times (10,65). Although a variety of media are used to culture ES cells, it is

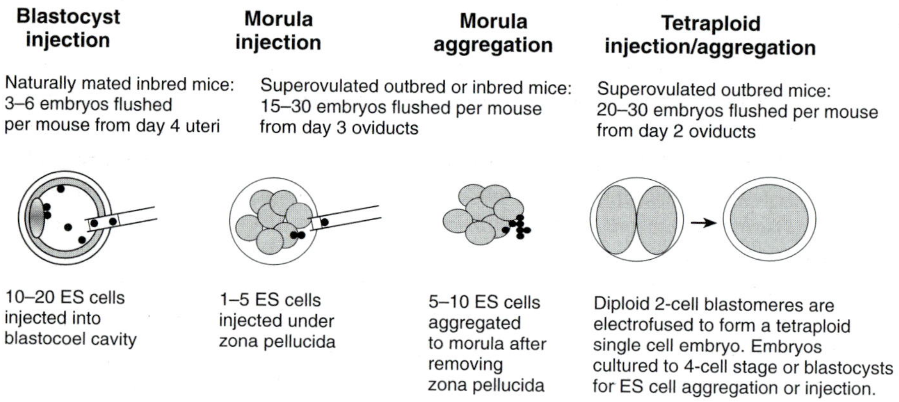

Blastocyst injection
Morula injection
Morula aggregation
Tetraploid injection/aggregation
Naturally mated inbred mice: 3–6 embryos flushed per mouse from day 4 uteri
Superovulated outbred or inbred mice: 15–30 embryos flushed per mouse from day 3 oviducts
Superovulated outbred mice: 20–30 embryos flushed per mouse from day 2 oviducts
10–20 ES cells injected into blastocoel cavity
1–5 ES cells injected under zona pellucida
5–10 ES cells aggregated to morula after removing zona pellucida
Diploid 2-cell blastomeres are electrofused to form a tetraploid single cell embryo. Embryos cultured to 4-cell stage or blastocysts for ES cell aggregation or injection.

Figure 10. Summary of methods used to generate ES cell/embryo chimeras. ES cells may be injected into blastocyst- or morula-stage embryos, or aggregated with morulae after removing the zona pellucida. Tetraploid embryos may also be used for injection or aggregation to make ES-cell-derived fetuses.

recommended that they be cultured as far as possible under the conditions in which the lines were derived. The most important component of the culture medium is the fetal calf serum. Since considerable variation is normally found in the quality of different serum batches, all serum used should be batch-tested for optimal ES-cell growth before embarking on a targeting experiment. The use of feeder cells is also recommended although ES cells can be successfully maintained in feeder-free cultures supplemented with the differentiation-inhibiting growth factor LIF (Leukemia Inhibitory Factor).

Despite the maintenance of optimal culture conditions, the pluripotency of ES cells will gradually wane with time. With extended culture, ES-cell lines will gradually accumulate mutations or epigenetic changes that affect gene regulation (66). With increasing passage number, mutations and chromosomal abnormalities that confer a growth advantage on ES cells will rapidly increase in the population. Gross genetic changes such as chromosome gain or loss and some translocations can be checked for by simple karyotyping (see Chapter 6). One common abnormality is trisomy for chromosome 8, which gives ES cells a growth advantage in culture, but prevents germline transmission (67). To minimize the effects of genetic and epigenetic changes it is important to use ES cells at the earliest passage possible. Unless ES-cell lines are derived in-house, most targeting experiments are performed with ES cells between passage 10 and 20.

Before starting gene-targeting experiments, the ES cell-line should ideally be cultured under the conditions to be used, and tested for its ability to make good chimeras with a high frequency of germline transmission. However, since a test for germline transmission can take a long time, other methods can be used to assess the potency of an ES-cell line.

6.1.1 Analysis of ES-cell potency in chimeras with tetraploid embryos

The most stringent test of ES-cell pluripotency is the ability of ES cells to form the entire embryo and develop to term and beyond. This can be tested by using the elegant trick of injecting or aggregating ES cells into tetraploid host embryos formed by simply fusing both blastomeres of a two-cell embryo back to one cell (68). The tetraploid cells can form an embryo poorly if at all but are capable of developing in the extra-embryonic membranes to form a fully functional placenta which can support the development of an ES-cell-derived fetus. Although very few cell lines pass this test by giving viable ES-cell-derived pups, good-quality ES-cell lines will nevertheless support development to late gestation (E15 or beyond), whereas embryos derived from poorer cell lines will give a high rate of resorptions at mid-gestation.

6.2 Equipment for injection of ES cells into embryos

Blastocyst and morula injection require the use of a micromanipulation rig that comprises two manipulators to hold and manoeuvre the embryo-holding and injection pipettes and, most commonly, an inverted microscope with Nomarski or differential interference optics. We will not describe the manipulation rigs in detail as these vary considerably between laboratories depending on which microscopes and manipulators have been purchased. In our laboratory we use standard Leitz micromanipulators and a Nikon Diaphot inverted microscope. In addition, equipment is required to make the glass holding and injection pipettes. This includes a glass capillary needle puller, a microforge and a needle-bevelling machine.

The holding pipette is made from glass capillary tube pulled to give an internal diameter of about 15–20 μm and the end polished using a microforge. Injection pipettes are pulled from thin-walled glass capillary tube using a pipette puller to give a taper of about 1 cm. The tapered glass is then broken to give a square end with a diameter of 15–20 μm. Although pipettes can be broken manually, they are broken more precisely and consistently with the aid of a microforge. Finally the injection pipettes are bevelled and given a sharp point to facilitate easy penetration through the zona pellucida and trophectoderm.

Protocol 1. Preparation of microinstruments

Equipment and reagents

- Micromanipulation rig with microscope (see Section 6.2.1 and Chapter 9, Section 3)
- Glass capillary needle puller (e.g. Sutter Instruments)
- Microforge (e.g. Beaudouin)
- Needle-bevelling machine (e.g. Narashige)
- Glass capillary tubes (Clark Electromedical Instruments GC 100-15)
- Thin-walled glass capillary tubes (Clark Electromedical Instruments GC 100T-15)
- Diamond pen

A. *Preparation of the holding pipette*

1. Pull the glass capillary with a pipette puller or by hand. Using a diamond pen, score the glass and break the capillary to give a clean (square) break with an internal diameter of approximately 15–20 μm. The size can be measured using a graticule in the eye-piece of the microforge.
2. Place the pipette in a needle holder from the micromanipulation rig and polish the break using a microforge.
3. For work in a depression slide, make a slight (5°) bend 1–2 mm from the tip of the pipette by turning it perpendicular to the microforge filament and bringing it close to the tip of the filament under moderate heat. Turn the heat off when a sufficient bend is made.

B. *Preparation of the injection pipette*

1. Pull a thin-walled glass capillary tube using a pipette puller to give a 1 cm taper to a diameter of approximately 15–20 μm.

2. To prepare the microforge, melt a glass bead onto the end of the microforge filament.
3. Mount the pipette on the microforge and move the pipette parallel to the filament to bring the glass bead adjacent to where the break will be made.
4. Gently heat the filament and bring the glass bead and pipette into contact. Heat the filament such that the pipette fuses to the glass bead, but does not bend (*Figure 11a*).
5. Turn the heat off; the filament will contract and snap the pipette where it makes contact with the glass bead (*Figure 11b*).
6. Attach a syringe to the back of the pipette holder and draw a small volume of water into the pipette tip.
7. Transfer the pipette to the grinding machine, and grind under droplets of water to give a bevel of 45°.
8. Attach the syringe to the back of the pipette holder again and wash ground glass from the needle tip by blowing any water within the pipette followed by air bubbles into a drop of distilled water.
9. Return the pipette to the microforge and gently bring the pipette tip up to the glass bead on the microforge filament; under a low heat, fuse the tip to the bead (*Figure 11c*). Immediately turn the heat off. As the filament contracts it will draw out a fine point from the tip of the bevelled pipette (*Figure 11d*).
10. For work in a depression slide, make a slight (5°) bend 1–2 mm from the tip of the pipette by turning it perpendicular to the microforge filament and bringing it close to the tip of the filament under moderate heat. Turn the heat off when a sufficient bend is made.

6.3 Host embryos

Following blastocyst injection or morula aggregation, ES cells become incorporated within the developing inner cell mass (ICM) of the embryo and contribute to the development of the different embryonic lineages, in competition with host cells. Chimeras with the highest ES-cell contributions will be obtained using ES-cell:host combinations in which competition by the ES-cell population is favoured. For this reason host embryos are used from inbred strains of mice rather than the more robust F1 hybrid or outbred strains. The most common choice of host strain is C57BL/6, followed by BALB/c. Hosts are also selected depending on the strain coat colours so that chimeras can be readily identified; for example an ES-cell contribution from the commonly used agouti 129 strains is readily detected in chimeras on the black background of C57BL/6 and certain 129 strains are also detected against the

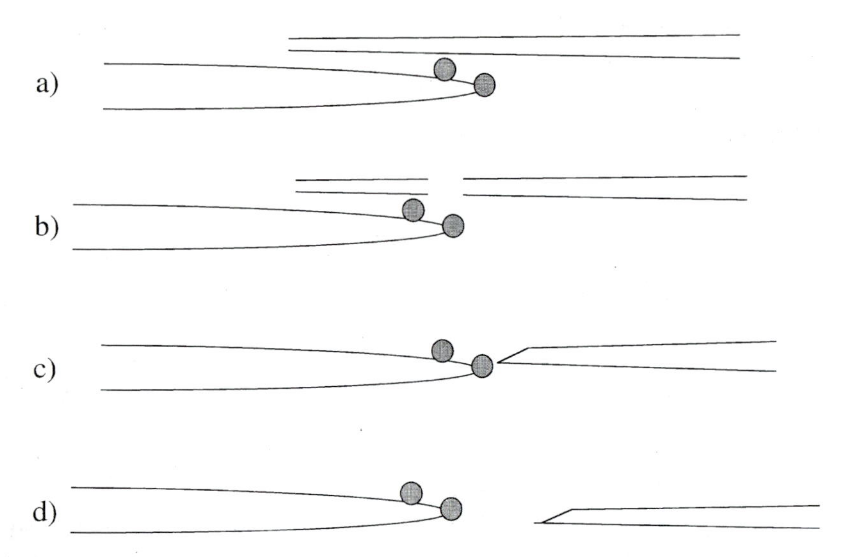

Figure 11. Injection pipettes. (a) Pulled glass capillary is aligned parallel to the heating filament of the microforge, which has been prepared by melting two glass beads at its tip as shown. The capillary is fused to the top bead under gentle heat, where the glass has an internal diameter of 15–20 µm. (b) As the heat is turned off the filament contracts, causing the capillary to break. (c) The pipette is bevelled on a grinding wheel and returned to the microforge, and the tip is fused to the bead at the tip of the filament. (d) As the heat is turned off, a fine/sharp point is drawn out from the tip of the pipette.

white background of BALB/c, whereas the (less used) ES cells of C57BL/6 origin need to be detected on a BALB/c host (69).

6.3.1 Blastocyst injection

The standard procedure is to inject 10–20 ES cells into the blastocoel cavity of recently cavitated blastocysts that have been recovered by flushing the uteri of day 4 pregnant mice (where day 1 is the day on which a vaginal plug appears) (see *Figure 10*). After injection embryos are cultured for a short period (2–3 h) to allow re-expansion of the blastocoel cavity, which collapses upon injection, and then transferred to the uterine horns of day 3 pseudopregnant mice. One shortcoming of using inbred strains of mice to provide host embryos is poor embryo recovery. From natural matings of C57BL/6 mice an expected blastocyst recovery is only about four per mouse, while for BALB/c mice the recovery is slightly lower. Greater numbers of embryos can be recovered by flushing them as morulae from the oviducts on day 3 and culturing them overnight. However, this procedure tends to delay the development of the blastocysts, so that they may only be cavitated sufficiently for injection by the evening of day 4 and the timing must be adjusted by changing the light cycles on which the mice are kept. Greater numbers of embryos can also be obtained by superovulating pre-pubertal females, which for C57BL/6 would be best performed at 3–4 weeks of age (70).

Protocol 2. Embryo preparation

Equipment and reagents

- M2 medium + 1 mg/ml BSA (Sigma, see Chapter 1, *Protocol 6*)
- M16 medium + 1 mg/ml BSA (Sigma, see Chapter 1, *Protocol 6*)
- Embryo-tested mineral oil (Sigma)
- Dissecting microsope
- Mouth pipette for handling embryos (see Chapter 2, *Protocol 1*)
- DMEM culture medium (Sigma)
- ES-cell culture medium (this will depend on the particular medium used for growth of the cells during targeting; see Section 6.1)
- 5 ml syringe with 27g needle
- 6 cm bacteriological dishes (e.g. Sterilin)
- Timed, superovulated mated females on day 3 (morulae) or timed, naturally mated females on day 4 (blastocysts) (see Chapter 1, *Protocol 5*)

A. *Eight-cell/morulae preparation*

1. Flush eight-cell/morula stage embryos in M2 handling medium + 1 mg/ml BSA, from the oviducts of superovulated mice on day 3 (day 1 is the day on which the vaginal plug appears) (see Chapter 9, *Protocol 5*).
2. Wash embryos by transferring them by mouth pipette through six drops of M16 medium + 1 mg/ml BSA and then culture in a fresh drop of M16 medium + BSA under embryo-tested mineral oil in a bacteriological dish (see Chapter 9, *Protocol 5*).

B. *Blastocyst preparation*

1. Dissect out, into a drop of DMEM, the uterine horns with attached oviducts of naturally mated mice on the morning of day 4.
2. Using a dissection microscope, flush the uteruses by piercing the uterine wall with a 27-gauge injection needle close to the isthmus (between the uterus and oviduct) and using a 5 ml syringe gently pass 0.5–1 ml of DMEM through the lumen of each uterine horn into a 6 cm bacterial culture dish.
3. Harvest all flushed embryos by mouth pipette, wash through four to six drops of culture medium and place in culture in a drop of embryo recovery medium (1:1 ES-cell culture medium:M16) under embryo-tested mineral oil until they are ready for use.

Protocol 3. ES cell preparation

Equipment and reagents

- 6 cm tissue culture dish
- ES-cell culture medium (see Section 6.1)
- ES cells
- Phosphate-buffered saline (PBS; Sigma)

Protocol 3. *Continued*

- Injection medium: Hepes-buffered DMEM + 10% fetal calf serum (ICN)
- Trypsin: 0.05% trypsin, 0.02% EDTA in modified Puck's saline A (Gibco/BRL)
- Depression slide (BDH)
- Embryo-tested mineral oil (Sigma)
- 50 ml polypropylene tubes (e.g. Falcon)
- Bench-top centrifuge

A. *Preparation of ES cells for injection*

1. Plate 10^4 cells for microinjection or aggregation 2 days before use on a 6 cm dish and culture in ES-cell culture medium.
2. One hour prior to use, harvest the cells. First, wash the cells with PBS, then add enough trypsin to cover the bottom of the dish (about 0.5 ml for a 6 cm dish). Watch the cells under a tissue-culture microscope until the colonies begin to detach, and then add 5 vols (2.5 ml) of ES-cell culture medium. Disperse the colonies to single cells by repeated pipetting.[a]
3. Harvest the cell suspension, by centrifugation at 1000 r.p.m. for 3 min; resuspend the ES cells in injection medium and place on ice.

B. *Microinjection chamber*

Note: Microinjection is performed in depression slides.

1. Wash the slides and rinse thoroughly with water followed by ethanol.
2. Place a small drop (50 µl) of injection medium in the depression and cover with mineral oil to prevent evaporation.
3. Transfer 10–20 embryos to this injection chamber and add enough ES cells to form a sparse lawn on the bottom of the chamber. If too many cells are added to the chamber they may clog the microinjection instruments. The ES cells will soon settle to the bottom; because they are small they will be in a slightly different focal plane from the embryos.

[a] If the ES-cell line used is grown on feeder cells, a significant proportion of feeders can be removed from the cell suspension by replating the cells onto a culture dish for up to 1 h. In this time feeder cells will adhere to the culture plate, leaving the majority of ES cells still in suspension.

Protocol 4. ES-cell injection into blastocysts

Equipment and reagents

- Injection apparatus and microinstruments as in *Protocol 1*
- Blastocyst-stage embryos (*Protocol 2A*)
- ES cells (*Protocol 3*)
- M16 medium + 1 mg/ml BSA (Sigma)

Method

1. Place 10–20 recently expanded blastocysts, in a cluster, in the micro-injection chamber with the ES cells (*Protocol 3*).

2. Orient each embryo to be picked up and hold it firmly on the holding pipette at the ICM end.
3. Collect 10–20 ES cells into the injection pipette.
4. Pierce through the zona pellucida and trophectoderm and deposit 10–20 cells in the blastocoel cavity.
5. Withdraw the pipette and release the injected embryo from the holding pipette to a separate part of the slide.
6. After injecting all the embryos in the chamber, transfer them into a culture drop of ES-cell culture medium, to allow them to re-expand.
7. Transfer re-expanded embryos to the uterine horns of day 3 pseudopregnant foster mothers (*Protocol 7*).

6.3.2 Morula injection

Chimeras can be made with a very high efficiency by combining ES cells with developing morulae by injection or aggregation. However, by introducing ES cells at this stage there is a tendency for the ES-cell contribution in the fetus to be excessively high. This can result in developmental problems associated with size regulation of the embryo. Furthermore, as with chimerism with tetraploid embryos, any genetic or epigenetic defects in the the cell line used would be exposed and could not be compensated for by the presence of wild-type cells. To limit this problem, the number of ES cells entering an aggregate must be restricted and the best quality cells should be used. In this respect morula injection is a preferred technique since the number of cells injected is precisely determined and the manipulation allows cells with the best morphology to be selected individually from a population. There are some advantages over blastocyst injection. Morulae can be recovered from the mice at a greater efficiency than blastocysts, and they are technically easier to inject, allowing many more embryos to be manipulated. Following injection, chimeric embryos can be cultured overnight to the blastocyst stage and transfered to the uterine horns on day 3 (*Protocol 7*), but better results are obtained if they are transferred to the oviducts of day 1 pseudopregnant foster mothers (see Chapter 9, *Protocol 7*). The frequency of obtaining germline transmitting chimeras by morula injection is comparable to that of blastocyst injection. When injecting morulae the number of ES cells must be restricted to between one and five; good germline-transmitting chimeras can be made by injecting morulae with a single ES cell. Studies on the fate of injected cells show that ES cells injected into morulae invariably become incorporated into the ICM (71,72); if excess cells (more than five) are injected the developing embryos will be predominantly ES-cell derived and consequently may not support development to term. This contrasts with blastocyst injection, where an excess of ES cells is generally injected into the blastocoel cavity. Although a number of cells become associated with the ICM, many other cells adhere to the trophectoderm or fail to enter the ICM due to the differentiation of the primitive endoderm, which may form a physical barrier.

Protocol 5. Injection of ES cells into morulae

Equipment and reagents

- Injection apparatus and microinstruments as in *Protocol 1*
- Morula-stage embryos (*Protocol 2A*)
- ES cells (*Protocol 3*)
- M16 medium + 1 mg/ml BSA (Sigma)

Method

1. Mount the holding and injection pipettes onto the micromanipulators.
2. Place 10–20 embryos into the depression chamber and cluster them into a single group (*Protocol 3*).
3. Pipette ES cells into the depression chamber to form a sparse lawn surrounding the embryos (*Protocol 3*).
4. Suck one embryo from the group onto the holding pipette and hold it firmly under gentle pressure.
5. Adjust the focal plane and suck a number of ES cells into the injection pipette. Take care to select the best, small, round ES cells, excluding larger differentiated cells or feeder cells.
6. Raise the injection pipette until the tip of the pipette and the zona pellucida of the embryo are in the same focal plane.
7. Gently pierce through the zona with the tip of the injection pipette and expel three to five cells under the zona pellucida.
8. Gently remove the injection pipette.
9. Release the injected embryo and move to a distinct part of the injection chamber to form a group of injected embryos, taking care not to mix injected and uninjected morulae as they are not easy to distinguish.
10. Repeat the injection procedure for all remaining embryos.
11. Collect the injected embryos from the injection chamber with a mouth pipette, wash through four to six drops of M16 medium + 1 mg/ml BSA and return them to the incubator until ready for transfer.
12. Manipulated morulae are best transferred directly to the oviducts of day 1 pseudopregnant recipient foster mothers (up to eight embryos per oviduct) (Chapter 9, *Protocol 7*). Alternatively the morulae may be cultured overnight in M16 medium + 1 mg/ml BSA and then transferred to the uterine horns of day 3 pseudopregnant recipient foster mothers (*Protocol 7*).

6.3.3 Morula aggregation

The third method of making chimeras is to aggregate a small number of ES cells to a morula. The zona pellucidae of eight-cell embryos are removed by brief exposure to acid Tyrode's solution and the embryos are then simply brought into contact with five to ten ES cells (73). The two cell types adhere, then upon culture through to the blastocyst stage the ES cells incorporate into the developing ICM. The chimeric embryos produced are most similar to those produced by morula injection. Aggregation methods have the advantage over injection of being cheap and requiring significantly less technical skill to perform; however, it is not possible to control the numbers or the quality of the cells. Since all embryos are recovered at the eight-cell stage by flushing the oviducts, donor mice can be superovulated and, as with morula injection, embryos from a wider range of host strains (including F1 hybrid and outbred strains) can be successfully used.

Protocol 6. Morula aggregation

Equipment and reagents

- 6 cm bacterial culture dish
- M2 medium + 1 mg/ml BSA (Sigma)
- M16 medium + 1 mg/ml BSA (Sigma)
- Blunt-ended darning needle
- Single-cell suspension of ES cells (*Protocol 2*)
- Mouth pipette for handling cells and embryos (see Chapter 2, *Protocol 1*)
- Eight-cell embryos from superovulated females (*Protocol 2A*)
- Tyrode's solution, acidic (Sigma)
- CO_2 incubator

A. *Preparation of aggregation wells*

1. Place six drops of M16 culture medium + BSA in a bacteriological-grade culture dish and cover with mineral oil.
2. Make a series of 12–15 small depressions beneath each drop by pressing a round or blunt-ended darning needle, washed with ethanol, onto the plastic surface, through the oil and medium.
3. Place the dish in the incubator to pre-equilibrate the culture medium with CO_2

B. *ES-cell preparation*

1. Plate the single-cell suspension of ES cells in ES-cell medium in a bacterial culture dish, and incubate for 1–2 h. During this time they will aggregate into small clumps of five to ten cells.
2. Using a hand-pulled, narrow-bore mouth pipette, place one clump (five to ten cells) of ES cells into each aggregation well.

Protocol 6. *Continued*

C. *Embryo preparation and aggregation*

1. Flush eight-cell-stage embryos from the oviducts of day 2.5 superovulated or naturally mated mice in M2 medium (*Protocol 2*).
2. Wash the embryos by transferring through three drops of M2 medium
3. Remove the zona pellucidae of the embryos by brief exposure to acid Tyrode's solution; using a mouth pipette release 10–20 embryos at the top of a drop of acid Tyrode's, keep the embryos suspended in the solution and observe them until the zona pellucida is seen to dissolve. Take care not to let the embryos fall to the bottom of the drop and touch the plastic, as they will tend to stick irretrievably.
4. Return the embryos to a drop of M2.
5. Wash the embryos through four drops of M16 + BSA medium.
6. Place one embryo in contact with each ES-cell clump in the pre-prepared aggregation wells.
7. Return the dish to the incubator and culture overnight.
8. Transfer compacted or early cavitating morulae to the uterine horns of day 3 pseudopregnant recipients (*Protocol 7*).

6.3.4 Transfer of embryos to the uterus

Morula- and blastocyst-stage embryos are transferred to the uterine horns of day 3 pseudopregnant recipient foster mothers. If bilateral transfers are to be made, six to ten embryos can be transferred per uterine horn. If a unilateral transfer is made, up to 12 embryos may be implanted. Embryos are transferred by mouth pipette using a transfer pipette as described for oviduct transfer (Chapter 9, *Protocol 7*). Pseudopregnant females are obtained by mating 6–8 week old oestrous females (20–24 g) with vasectomized males and checking for the presence of a vaginal plug (see Chapter 1, Section 6.2). We routinely use (C57Bl/6 × CBA)F1 or CD1 strain females as recipients. Preparation of the mice for uterine transfer is the same as for oviduct transfer (Chapter 9, *Protocol 7*).

Protocol 7. Transfer of morulae and blastocysts to pseudopregnant mothers

Equipment and reagents

- Mouth-transfer pipette for handling embryos (Chapter 2, *Protocol 1*)
- Chimeric morulae or blastocyts (*Protocols 4–6*)
- Pseudopregnant females (see Chapter 1, Section 6.2), 2.5 days after mating
- Anaesthetic (Chapter 9, *Protocol 7*)
- Surgical instruments (see Chapter 9, *Protocol 7*)
- 27g hypodermic needle
- Wound clips (Chapter 9, *Protocol 7*)

Method

1. Weigh and anaesthetize a day 3 pseudopregnant mouse. See Chapter 9, *Protocol 7* for details.
2. Collect chimeric embryos into the transfer pipette as in Chapter 9, *Protocol 7*.
3. Wet the back of the mouse with ethanol and make a 0.5–1 cm transverse dorsal incision at the level of the first lumbar vertebra.
4. Wipe the incision with a tissue to remove any cut hair.
5. Slide the skin incision over the ovarian fat pad, which can be seen through the peritoneal wall.
6. Hold the body wall over the fat pad using fine forceps and make a small (2 mm) incision using fine-point dissection scissors.
7. Using blunt-ended fine forceps, hold the fat pad and ease the fat pad, ovary, oviduct and uterus through the incision.
8. The uterus can be prevented from sliding back into the peritoneum by resting the oviduct fat pad on a square of tissue paper.
9. Using fine watchmaker's forceps, gently hold the uterine wall 1–2 mm below the oviduct and make a small incision through the uterine wall into the lumen with a 27-gauge hypodermic needle.
10. Still holding the uterus with the forceps, insert the transfer pipette through the incision into the uterine lumen and expel the embryos, watching movement of the marker bubble in the transfer pipette.
11. Withdraw the transfer pipette and release the uterus. Slide the uterus, oviduct, ovary and fat pad back into the peritoneum. If the peritoneal incision is small it may be left open, otherwise a single suture stitch may be necessary.
12. For a bilateral transfer repeat the procedure for the second uterine horn.
13. Close the incision in the skin using one or two wound clips.
14. Place the mouse on clean tissue paper in a clean cage, on a heat pad, and allow it to recover before returning it to the animal room.

6.4 Genetic background

Gene targeting and germline transmission of a mutation are the initial steps in the genetic analysis of gene function. The genetic background of mutant animals may influence the phenotypes observed to a greater or lesser extent. The issue of genetic background has come to the fore in neurosciences, particularly concerning the analysis of complex behavioural phenotypes. In the standard production of knock-out mice, genes are targeted in ES cells which are usually derived from one of several 129 strains. Germline transmission is often achieved by crossing chimeras to C57BL/6 mice which produce heterozygotes for the mutation in a (129 × C57BL/6)F1 background. However, if F1 progeny are then intercrossed to produce homozygous mutant animals, there will be segregation of all other 129 and C57BL/6 alleles in this F2 generation. If the mutated gene is influenced by the genetic background, variations in phenotype may occur. The problems associated with genetic background in the analysis of mutant phenotypes have been discussed extensively; one of the most critical issues is to define the appropriate genetic control group of animals against which mutant animals are assessed. A number of recommendations have been put forward in order to rationalize the problems and evolve guidelines to establish control groups for the effects of genetic background on mutant phenotypes (74).

6.4.1 Maintenance of mutations as co-isogenic and congenic lines and mutant analysis in a hybrid background

To avoid the problems of segregating backgrounds, there have been recommendations that the mutations are established on two backgrounds. The simplest method is to maintain the line as co-isogenic (or quasi-co-isogenic) by crossing the germline-transmitting chimeras to the strain of mice from which the ES cells were derived. This is usually one or other of the 129 series of strains. Although this seems straightforward, caution should nevertheless be taken to use appropriate 129 mice for breeding, as ES-cell lines in common use have been derived from a number of different 129 sub-strains (74).

A second approach is to make the mutant congenic with an inbred strain such as C57BL/6, by crossing to this strain initially and then backcrossing the mutation through successive generations to the inbred strain. Although a line would not be considered truly congenic before ten backcross generations had been achieved, the background would nevertheless be >90% congenic after five backcross generations and for the majority of experiments could be used at that stage.

Producing congenic lines has significant implications for animal husbandry and the time required to generate appropriate animals for a detailed study. With efficient breeding a five-generation backcross could be completed in 1 year. This time can be reduced using one of two breeding strategies. The first is 'speed breeding', in which female progeny from a backcross generation are typed for the presence of the mutation at 3–4 weeks of age, superovulated and mated to inbred males. Fertilized embryos are then collected from the heterozygous mutant females and transferred to pseudopregnant foster mothers. This strategy reduces the generation time to a minimum of 6–7 weeks. The second strategy is termed 'speed congenics'. In this case, progeny from each generation are typed against a panel of microsatellite markers that scan the genome for homozygosity at marker loci. Following a genome scan, animals with the greatest degree of homozygosity would be selected to breed to the next generation. Progeny from that generation then only need to be typed for markers that were previously heterozygous. This process of selective breeding significantly reduces the number of backcross generations required to reach a congenic-equivalent line (75).

In many instances phenotypic analysis may be better performed in more robust F1 hybrid animals than in inbred or congenic mice. Producing the mutation on two inbred (or congenic) backgrounds allows the production of hybrid mice by crossing the two strains. It is worth pointing out that all the offspring of a cross between two inbred lines are identical, but this is lost once the progeny are intercrossed or backcrossed.

6.5 Chimera analysis

Although germline transmission from chimeras is the end-point of most experiments, chimerism may also be used directly to analyse mutant phenotypes. In a number of cases, mutations may have dominant phenotypic effects that may compromise embryo development, in which case phenotypes may be analysed directly in chimeras. Alternatively, ES cells can be mutated at both alleles to allow phenotypic analysis of homozygous mutant cells in chimeras. In such cases, phenotypes can be observed in ES-cell-derived fetuses by chimerism with tetraploid embryos, or in normal diploid chimeras where the presence of wild-type cells from the host may rescue the development of mutant cells in particular cell lineages.

The first requirement of an experiment that uses chimeras is that the contribution of experimental cells can be unequivocally distinguished from the host cells. A number of markers are available, some of which can be built into the overall gene-targeting strategy (76). Markers fall into two categories: those that simply allow the overall level of chimerism in different tissues to be assessed, as a proportion of the cell population, and those that allow identification of cells *in situ*. In the first category, markers are generally DNA or biochemical polymorphisms that are readily detectable between the experimental cell and host strains. Glucose phosphate isomerase (GPI) is the most frequently used biochemical marker and allows cell proportions to be assessed down to a minimum contribution of 5%. For DNA analysis, microsatellite polymorphisms can be used by determining the relative abundance of the host and donor alleles by PCR. More commonly, targeting vectors are designed such that the targeted gene is substituted for a reporter gene, for example *lacZ*, GFP or PAP. The mutant cells may then be identified by their reporter gene expression.

A further approach is to make use of mutant mouse strains to provide host blastocysts to study gene functions through phenotypic complementation. For example, animals with a homozygous mutation in the *Rag2* gene (77) have a severe combined immunodeficiency due to early disruption in the development of both the B-cell and T-cell lymphocyte lineages, but this deficit can be rescued by embryo chimerism since donor cells will contribute to haematopoiesis and replace the mutant lineages. Thus *Rag2*$^{-/-}$ host embryos can be used to rescue other manipulated ES cells to be studied in the immune system. This is particularly important for mutations that have pleiotropic effects and may result in non-immune-related lethality. As the number of different mutant mice increases, it is likely that several other strains will become valuable as hosts to study gene functions in different lineages.

Acknowledgements

We thank Miguel Constancia for his valuable help and discussion of the manuscript. G.K. is a Senior MRC Fellow, N.D.A. is a BBSRC Advanced Fellow, A.P. is supported by a Link grant from the MRC and Merck Sharp and Dohme.

References

1. Thomas, K.R. and Capecchi, M.R. (1987). *Cell* **51**, 503.
2. Doetschman, T., Maeda, N. and Smithies, O. (1988). *Proc. Natl Acad. Sci. USA* **85**, 8583.
3. Kilby, N.J., Snaith, M.R. and Murray, J.A.H. (1993). *Trends Genet.* **9**, 413.
4. Joyner, A.L. (ed.) (1993). *Gene Targeting: A Practical Approach.* IRL Press, Oxford.
5. Hasty, P., Rivera-Pérez, J.R., Chang, C. and Bradley, A. (1991). *Mol. Cell. Biol.* **11**, 4509.
6. Hasty, P., Rivera-Pérez, J. and Bradley, A. (1991). *Mol. Cell. Biol.* **11**, 5586.
7. Thomas, K.R., Deng, C. and Capecchi, M.R. (1992). *Mol. Cell. Biol.* **12**, 2919.
8. te Riele, H., Maandag, E.R. and Berns, A. (1992). *Proc. Natl Acad. Sci. USA* **89**, 5128.
9. Salomon, B., Maury, S., Loubiere, L., Caruso, M., Onclercq, R. and Klatzmann, D. (1995). *Mol. Cell. Biol.* **15**, 5322.
10. Wurst, W. and Joyner, A.L. (1993). In *Gene Targeting: A Practical Approach* (ed. A.L. Joyner), pp. 33–61. IRL Press, Oxford.
11. Mansour, S.L., Thomas, K.R. and Capecchi, M.R. (1988). *Nature* **336**, 348.
12. Ramírez-Solis, R., Rivera-Pérez, J., Wallace, J.D., Wims, M., Zheng, H. and Bradley, A. (1992). *Anal. Biochem.* **201**, 331.
13. Laird, P.W., Zijderveld, A., Linders, K., Rudnicki, M.A., Jaenisch, R. and Berns, A. (1991). *Nucleic Acids Res.* **19**, 4293.
14. Hasty, P. and Bradley, A. (1993). In *Gene Targeting: A Practical Approach* (ed. A.L. Joyner), pp. 1–30. IRL Press, Oxford.
15. Nehls, M., Messerle, M., Sirulnik, A., Smith, A.J.H. and Boehm, T. (1994). *BioTechniques* **17**, 770
16. Westphal, C.H. and Leder, P. (1997). *Curr. Biol.* **7**, 530.
17. Storck, T., Krüth, U., Kolhekar, R., Sprengel, R. and Seeburg, P.H. (1996). *Nucleic Acids Res.* **24**, 4594.
18. Craig, N.L. (1988). *Annu. Rev. Genet.* **22**, 77.

19. Guo, F., Gopaul, D.N. and van Duyne, G.D. (1997). *Nature* **389**, 40.
20. Sauer, B. and Henderson, N. (1989). *Nucleic Acids Res.*, **17**, 147.
21. Sauer, B. and Henderson, N. (1988). *Proc. Natl Acad. Sci. USA* **85**, 5166.
22. Sauer, B. and Henderson, N. (1990). *New Biol.*, **2**, 441.
23. O'Gorman, S., Fox, D.T. and Wahl, G.M. (1991). *Science* **251**, 1351.
24. Snaith, M.R., Murray, J.A.H. and Boulter, C.A. (1995). *Gene* **166**, 173.
25. Dymecki, S.M. (1996). *Gene* **171**, 197.
26. Fiering, S., Epner, E., Robinson, K., Zhuang, Y., Telling, A., Hu, M., Martin, D.I.K., Enver, T., Ley, T.J. and Groudine, M. (1995). *Genes Dev.* **9**, 2203.
27. Meyers, E.N., Lewandoski, M. and Martin, G.R. (1998). *Nature Genet.* **18**, 136.
28. Olson, E.N., Arnold, H.-H., Rigby, P.W.J. and Wold, B.J. (1996). *Cell* **85**, 1.
29. Gu, H., Zou, Y.-R. and Rajewsky, K. (1993). *Cell* **73**, 1155.
30. Gu, H., Marth, J.D., Orban, P.C., Mossmann, H. and Rajewsky, K. (1994). *Science* **265**, 103.
31. Abuin, A. and Bradley, A. (1996). *Mol. Cell. Biol.* **16**, 1851.
32. Taniguchi, M., Sanbo, M., Watanabe, S., Naruse, I., Mishina, M. and Yagi, T. (1998). *Nucleic Acids Res.* **26**, 679.
33. Gagneten, S., Le, Y., Miller, J. and Sauer, B. (1997). *Nucleic Acids Res.* **25**, 3326.
34. Jung, S., Rajewsky, K. and Radbruch, A. (1993). *Science* **259**, 984.
35. Schwenk, F., Baron, U. and Rajewsky, K. (1995). *Nucleic Acids Res.* **23**, 5080.
36. Lakso, M., Pichel, J.G., Gorman, J.R., Sauer, B., Okamoto, Y., Lee, E., Alt, F.W. and Westphal, H. (1996). *Proc. Natl Acad. Sci. USA* **93**, 5860.
37. O'Gorman, S., Dagenais, N.A., Qian, M. and Marchuk, Y. (1997). *Proc. Natl Acad. Sci. USA* **94**, 14602.
38. Lewandoski, M., Wassarman, K.M. and Martin, G.R. (1997). *Curr. Biol.* **7**, 148.
39. Araki, K., Araki, M., Miyazaki, J.-I. and Vassalli, P. (1995). *Proc. Natl Acad. Sci. USA* **92**, 160.
40. de Wit, T., Drabek, D. and Grosveld, F. (1998). *Nucleic Acids Res.* **26**, 676.
41. Tsien, J.Z., Huerta, P.T. and Tonegawa, S. (1996). *Cell* **87**, 1327.
42. Tsien, J.Z., Chen, D.F., Gerber, D., Tom, C., Mercer, E.H., Anderson, D.J., Mayford, M., Kandel, E.R. and Tonegawa, S. (1996). *Cell* **87**, 1317.
43. Zhang, Y., Riesterer, C., Ayrall, A.-M., Sablitzky, F., Littlewood, T.D. and Reth, M. (1996). *Nucleic Acids Res.* **24**, 543.
44. Feil, R., Brocard, J., Mascrez, B., LeMeur, M., Metzger, D. and Chambon, P. (1996). *Proc. Natl Acad. Sci. USA* **93**, 10887.
45. Kellendonk, C., Tronche, F., Monaghan, A.-P., Angrand, P.-O., Stewart, F. and Schütz, G. (1996). *Nucleic Acids Res.* **24**, 1404.
46. Gossen, M. and Bujard, H. (1992). *Proc. Natl Acad. Sci. USA* **89**, 5547.
47. Furth, P.A., St-Onge, L., Böger, H., Gruss, P., Gossen, M., Kistner, A., Bujard, H. and Hennighausen, L. (1994). *Proc. Natl Acad. Sci. USA* **91**, 9302.
48. Kistner, A., Gossen, M., Zimmermann, F., Jerecic, J., Ullmer, C., Lübbert, H. and Bujard, H. (1996). *Proc. Natl Acad. Sci. USA* **93**, 10933.
49. St-Onge, L. Furth, P.A. and Gruss, P. (1996). *Nucleic Acids Res.* **24**, 3875.
50. Gossen, M., Freundlieb, S., Bender, G., Müller, G., Hillen, W. and Bujard, H. (1995). *Science* **268**, 1766.
51. Schwenk, F., Kühn, R., Angrand, P.-O., Rajewsky, K. and Stewart, A.F. (1998). *Nucleic Acids Res.* **26**, 1427.

52. Anton, M. and Graham, F.L. (1995). *J. Virol.* **69**, 4600.
53. Rohlmann, A., Gotthardt, M., Willnow, T.E., Hammer, R.E. and Herz, J. (1996). *Nature Biotechnol.* **14**, 1562.
54. Akagi, K., Sandig, V., Vooijs, M., Van der Valk, M., Giovannini, M., Strauss, M. and Berns, A. (1997). *Nucleic Acids Res.* **25**, 1766.
55. Fukushige, S. and Sauer, B. (1992). *Proc. Natl Acad. Sci. USA* **89**, 7905.
56. Smith, A.J.H., De Sousa, M.A., Kwabi-Addo, B., Heppell-Parton, A., Impey, H. and Rabbits, P. (1995). *Nature Genet.* **9**, 376.
57. Araki, K., Araki, M. and Yamamura, K.-I. (1997). *Nucleic Acids Res.* **25**, 868.
58. Bethke, B. and Sauer, B. (1997). *Nucleic Acids Res.* **25**, 2828.
59. Askew, G.R., Doetschman, T. and Lingrel, J.B. (1993). *Mol. Cell. Biol.* **13**, 4115.
60. Wu, H., Liu, X. and Jaenisch, R. (1994). *Proc. Natl Acad. Sci. USA* **91**, 2819.
61. Stacey, A., Schnieke, A., McWhir, J., Cooper, J., Colman, A. and Melton, D.W. (1994). *Mol. Cell. Biol.* **14**, 1009.
62. Hasty, P., Ramirez-Solis, R., Krumlauf, R. and Bradley, A. (1991). *Nature* **350**, 243.
63. Valancius, V. and Smithies, O. (1991). *Mol. Cell. Biol.* **11**, 1402.
64. Ramirez-Solis, R., Liu, P. and Bradley, A. (1995). *Nature* **378**, 720.
65. Brook, F.A. and Gardner, R.L (1997). *Proc. Natl Acad. Sci. USA* **94**, 5709.
66. Dean, W., Bowden, L., Aitchison, A., Klose, J., Moore, T., Meneses, J.J., Reik, W. and Feil, R. (1998). *Development* **125**, 2273.
67. Liu, X., Wu, H., Loring, J., Hormuzdi, S., Disteche, C.M., Bornstein, P. and Jaenisch, R. (1997). *Dev. Dyn.* **209**, 85.
68. Nagy, A., Rossant, J., Nagy, R., Abramow-Newerly, W. and Roder, J.C. (1993). *Proc. Natl Acad. Sci. USA* **90**, 8424.
69. Lemckert, F.A., Sedgwick, J.D. and Körner, H. (1997). *Nucleic Acids Res.* **25**, 917.
70. Kovacs, M.S., Lowe, L. and Kuehn, M.R. (1993). *Lab. Anim. Sci.* **43**, 91.
71. Lallemand, Y. and Brulet, P. (1990). *Development* **110**, 1241.
72. Saburi, S., Azuma, S., Sato, E., Toyoda, Y. and Tachi, C. (1997). *Differentiation* **62**, 1.
73. Wood, S.A., Allen, N.D., Rossant, J., Auerbach, A. and Nagy, A. (1993). *Nature* **365**, 87.
74. Banbury Conference on Genetic Background in Mice (1997). *Neuron* **19**, 755.
75. Markel, P., Shu, P., Ebeling, C., Carlson, G.A., Nagle, D.L. Smutko, J.S. and Moore, K.J. (1997). *Nature Genet.* **17**, 280.
76. Nagy, A. and Rossant, J. (1993). In *Gene Targeting: A Practical Approach* (ed. A.L. Joyner), p. 145. IRL Press, Oxford.
77. Chen, J., Lansford, R., Stewart, V., Young, F. and Alt, F.W. (1993). *Proc. Natl Acad. Sci. USA* **90**, 4528
78. Dymecki, S.M. (1996). *Proc. Natl Acad. Sci. USA* **93**, 6191.

A1

List of suppliers

A&J Beveridge Ltd, 5 Bonnington Road Lane, Leith, Edinburgh EH6 5BP.
Altro Floors, Works Road, Letchworth, Herts SG6 1NW.
Amersham
Amersham International plc., Lincoln Place, Green End, Aylesbury, Buckinghamshire HP20 2TP, UK.
Amersham Corporation, 2636 South Clearbrook Drive, Arlington Heights, IL 60005, USA.
Animal Identification and Marking Systems (AIMS), 13 Winchester Avenue, Budd Lake, New Jersey, 07828, USA.
Arco Ltd, PO Box 21, Waverley Street, Hull HU1 2SJ
Arrowmight Biosciences, Campwood Road, Rotherwas Industrial Estate, Hereford HR2 6JD.
B&K Universal Ltd, The field station, Grimston, Aldbrough, Hull HU11 4BQ
Bayer: Pharmaceutical Div, 400 Morgan Ln., West Haven, CT 06516, USA
Beaudouin Microforge, UK supplier: Micro Instruments Ltd., 18 Hangborough Park, Long Hanborough, Witney, Oxfordshire, OX8 8LH
Bio/medical Specialties, Box 1687, Santa Monica, CA 90406, USA.
Bio-Rad Laboratories
Bio-Rad Laboratories Ltd., Bio-Rad House, Maylands Avenue, Hemel Hempstead HP2 7TD, UK.
Bio-Rad Laboratories, Division Headquarters, 3300 Regatta Boulevard, Richmond, CA 94804, USA.
Brindus (industrial services) Ltd, Unit 67, Butterfly Avenue, Dartford Trade Park, Dartford, Kent DA1 1JG
Cambio Ltd., 34 Newnham Road, Cambridge, U.K. CB3 9EY.
Charles River UK Ltd, Manston Road, Margate, Kent CT9 4LT
Clark Electromedical Instruments, PO Box 8, Pangbourne, Reading RG8 7HU.
Difco Laboratories
Difco Laboratories Ltd., P.O. Box 14B, Central Avenue, West Molesey, Surrey KT8 2SE, UK.
Difco Laboratories, P.O. Box 331058, Detroit, MI 48232–7058, USA.
Dynal Inc., 5 Delaware Dr., Lake Success , NY 11042, USA
Epicentre Technologies:1402 Emil St., Madison, WI 53713, USA
Eppendorf (Germany): Barkhausenweg 1, Hamburg 22339, Germany

Ethicon GmbH & Co.: Essener Str. 99, 22419 Hamburg, Germany
Falcon (Falcon is a registered trademark of Becton Dickinson and Co.).
Fine Science Tools: 373-G Vintage Park Dr., Foster City, CA 94404, USA
Fluka
Fluka-Chemie AG, CH-9470, Buchs, Switzerland.
Fluka Chemicals Ltd., The Old Brickyard, New Road, Gillingham, Dorset SP8 4JL, UK.
FMC Bioproducts
FMC Bioproducts, 191 Thomaston Street, Rockland, ME 04841, USA
FMC Bioproducts Europe, Risingevej 1, DK-2665 Vallensbaek Strand, Denmark
FTS Systems, Inc., Stone Ridge, New York, 12484, USA
Genome Systems: 8620 Pennell Dr., St. Louis, MO 63114, USA, http://www.genomesystems.com/
Genosys Biotechnologies
Genosys Biotechnologies (Europe) Ltd., London Road, Pampisford, Cambridge, CB2 4EF, UK
Genosys Biotechnologies, Inc., 1442 Lake Front Circle, Suite 185, The Woodlands, Texas 77380, USA, http://www.genosys.com/
Gibco BRL
Gibco BRL (Life Technologies Ltd.), Trident House, Renfrew Road, Paisley PA3 4EF, UK.
Gibco BRL (Life Technologies Inc.), 3175 Staler Road, Grand Island, NY 14072–0068, USA.
Hamilton Co. 4970 Energy Way, Reno NV, 89502, USA
Hamilton needles distributed in the UK by V. A. Howe and Co. Ltd., Beaumont Close, Banbury, OX16 7RG, UK
Harlan UK Ltd, Shaw's Farm, Blackthorn, Bicester, Oxon OX6 0TP.
Harlan Teklad, Shaw's Farm, Blackthorn, Bicester, Oxon OX6 0TP.
Hawksley Ltd., Marlborough Road, Lancing, Sussex., UK
Heraeus Instruments: 111-A Corporate Blvd., South Plainfield, NJ 07080, USA
Hospital Management and Supplies Ltd, Carnegie Road, Hillington, Glasgow, G52 4NY.
H.W. Anderson Products Ltd, Davy Road, Gorse Lane Industrial Estate, Clacton-on-Sea, Essex CO15 4XA.
I.M.V. International Corp., 6870 Shingle Creek Parkway, Suite 100, Minneapolis, MN, 55430, USA
I.M.V., 10, Rue Clémenceau, BP 76, 61300 L'Aigle, France
I.M.V. Products distributed in the U.K. by Planer, (see above)
Interfocus Ltd, 14-15 Spring Rise, Falconer Road, Haverhill, Suffolk, CB9 7XU, UK.
International Market Supply (IMS), Dane Mill, Broadhurst Lane, Congleton, Cheshire CW12 1LA
Intervet: 405 State Street, P.O. Box 318, Millsboro, Delaware 19966, USA

Kopf Instruments: 7324 Elmo St., Tujanga, CA 91042, USA
Leica Inc./Leitz: 111 Deer Lake Road, Deerfield, Illinois 60015, USA
Leica Microsystems, UK Limited, Davy Avenue, Milton Keynes, MK5 8LB
Lillico, Wonham Hill Ltd, Betchworth. Surrey RH13 7AD
Locus Technologies, 910 Castine Road, Orland, ME 04472, USA, http://www.locustechnology.com
Marsh Biomedical Products, Inc., 565 Blossom Road, Rochester, NY 14610, USA
Merck
Merck Industries Inc., 5 Skyline Drive, Nawthorne, NY 10532, USA.
Merck, Frankfurter Strasse, 250, Postfach 4119, D-64293, Germany.
Micro Instruments Ltd., 18 Hanborough Park, Long Hanborough, Witney, Oxon OX8 8LH, U.K.
Microsoft Corporation, One Microsoft Way, Redmond, WA 98052-6399, USA
Miles Scientific, UK, Evans House, Hamolton Close, Hand Mills, Basingstoke, Hants, RG21 2YE
Millipore
Millipore (UK) Ltd., The Boulevard, Blackmoor Lane, Watford, Herts WD1 8YW, UK.
Millipore Corp./Biosearch, P.O. Box 255, 80 Ashby Road, Bedford, MA 01730, USA.
MJ Research Inc.,149 Grove St., Watertown , MA 02172, USA
Moredun Isolators, Pentlands Science Park, Bush Loan, Penicuik, Midlothian EH26 0PZ.
Murex Biotech, LTD., Central Road, Temple Hill, Dartford, Kent DA15LR, UK.
Murex Diagnostics LTD., 3075 Northwood Circle, Norcross, GA 30071, USA.
Nalge Nunc International, P.O. Box 20365, Rochester, New York 14602, USA
Narishige International: 1710 Hempstead Tpk., East Meadow, NY 11554, USA
Nikon/Image Systems Inc., Nikon Imaging Div., 8835 Columbia 100 Pkwy., Ste. A, Columbia , MD 21045, USA
Perkin-Elmer
Perkin-Elmer Ltd., Maxwell Road, Beaconsfield, Bucks. HP9 1QA, UK.
Perkin-Elmer Ltd., Post Office Lane, Beaconsfield, Bucks, HP9 1QA, UK.
Perkin-Elmer-Cetus (The Perkin-Elmer Corporation), 761 Main Avenue, Norwalk, CT 0689, USA.
Pharmacia Biosystems
Pharmacia Biosystems Ltd. (Biotechnology Division), Davy Avenue, Knowlhill, Milton Keynes MK5 8PH, UK.
Pharmacia LKB Biotechnology AB, Björngatan 30, S-75182 Uppsala, Sweden.
Pharmacia Biotech Europe Procordia EuroCentre, Rue de la Fuse-e 62, B-1130 Brussels, Belgium.

Planer Products Ltd, Windmill Road, Sunbury,Middlesex,TW16 7HD, UK

Planer distributed in the United States by T. S. Scientific Inc., PO Box 198, Perkasie, PA 18944, USA

Polysciences. 400 Valley Road, Warrington, PA 18976, USA

Precision Dynamics Corp., 13880 Del Sur St., San Fernando, CA 91340-3490, USA

Promega

Promega Ltd., Delta House, Enterprise Road, Chilworth Research Centre, Southampton, UK.

Promega Corporation, 2800 Woods Hollow Road, Madison, WI 53711–5399, USA.

Qiagen

Qiagen Inc., c/o Hybaid, 111–113 Waldegrave Road, Teddington, Middlesex, TW11 8LL, UK.

Qiagen Inc., 9259 Eton Avenue, Chatsworth, CA 91311, USA.

Raymond Lamb, 6 Sunbeam Road, London NW10 6JL, tel: 0181 965 18 34

Research Genetics, Inc., 2130 Memorial Pkwy., SW Huntsville, AL 35801, USA

Roche Diagnostics

Roche Diagnostics UK (Diagnostics and Biochemicals) Ltd, Bell Lane, Lewes, East Sussex BN17 1LG, UK.

Roche Diagnostics Corporation, Biochemical Products, 9115 Hague Road, P.O. Box 504 Indianapolis, IN 46250–0414, USA.

Roche Diagnostics, GmbH, Sandhofer Str. 116, Postfach 310120 D-6800 Ma 31, Germany.

Roth: Schoemperlenstr. 1-5, 76185 Karlsruhe, Germany

RS Biotech, Tower Works, Well Street, Finedon, Northants NN9 5JP.

Sigma Chemical Company

Sigma Chemical Company (UK), Fancy Road, Poole, Dorset BH17 7NH, UK.

Sigma Chemical Company, 3050 Spruce Street, P.O. Box 14508, St. Louis, MO 63178–9916.

Sorvall DuPont Company, Biotechnology Division, P.O. Box 80022, Wilmington, DE 19880–0022, USA.

Special Diets Services (SDS), PO BOX 705, Witham, Essex CM8 3AB

Spindler & Hoyer GmbH (Germany)/ Technical Manufacturing Corporation (USA): 15 Centennial Drive, Peabody, MA 01960, USA

Sterilising Equipment Company (SEC), Orchard Way, Calladine Park, Sutton In Ashfield, Notts NG17 1JU.

Stratagene

Stratagene Ltd., Unit 140, Cambridge Innovation Centre, Milton Road, Cambridge CB4 4FG, UK.

Strategene Inc., 11011 North Torrey Pines Road, La Jolla, CA 92037, USA.

Sutter Instrument Co.: 40 Leveroni Ct., Novato, CA 94949, USA

Techniplast UK Ltd, 2240 Parkway, Kettering Venture Park, Kettering, Northants NN15 6XR.

United States Biochemical, P.O. Box 22400, Cleveland, OH 44122, USA.

Vector Laboratories Inc, 30 Ingold Road, Burlingame, California, CA 94010, USA

Vector Laboratories Ltd., 16 Wulfric Square, Bretton, Peterborough, U.K. PE3 8RF.

Vysis Inc., 3100, Woodcreek Drive, Downers Grove, Illinois, IL 60515, USA.

Wellcome Reagents, Langley Court, Beckenham, Kent BR3 3BS, UK.

W. H. Brady Co., Identification Products Division, 6555 W. Good Hope Road, P.O. Box 571, Milwaukee, WI 53201-0571, USA

W. H. Brady Co products distributed in the UK by Cablectrix Ltd, 9 & 10 James Watt Close, Drayton Fields Industrial Estate, Daventry, Northampton, NN11 5QU, UK

WM Lillico & Son, Wonham Mill, Betchworth, Surrey RH3 7 AD

W.P. (World-Precision) Instruments:175 Saratosa Center Boulevard, Sarasota, FL 34240-9258, USA

Williton Box Company Ltd, Williton Industrial Estate, Williton, Somerset TA4 4RF

Yamanouchi Pharma: Breitspiel 19, 69126 Heidelberg, Germany

Zeiss Inc. Microscope Division, One Zeiss Drive, Thornwood, NY 10594, USA

Index

acetic-saline-Giemsa (ASG), G-bands 149
agouti, target size 208
alizarin, whole-mount skeletal analysis 84–5
alkaline phosphatase
  and β-galactosidase, double staining 78
  development of signal 71
  human placental (PLALP) 74–5
  staining procedure for human placental ALP 75
animal resources lists 180–1
  European Mouse Mutant Archive (EMMA) 181
  Induced Mutant Resource (IMR) 179–80
  International Mouse Strain Resource (IMSR) 180–1
  Jackson Laboratory:JAX mice IMR, MMR, DNA resource 180–1
  MGI-list, mouse community electronic bulletin board 181
3D Atlas of Mouse Development 172, 178
axenic and gnotobiotic accommodation 2

β-galactosidase
  activity, *lacZ* staining 73–4
  and ALP, double staining 78
B-lymphocyte specific promoter 260
BAC vector pBeloBAC11 136
backcross
  European Collaborative Interspecific Backcross (EUCIB) 122
  vs intercross 103–4
    recombination and segregation **90**
  progeny 92–3
    genotyping asubset for selected markers 95–6
  size of cross 99–100
backcross mapping experiment model 89–100
  see also phenotypic trait loci mapping
bacterial artificial chromosomes (BACs), isolation of DNA 219, 226–8
balancer chromosomes, genetic screens to isolate recessive mutations 198–9
barrier (or specified pathogen-free) housing 1–2
Big Blue transgenic assay 186
biotin, labelling, FISH 160
biotin-UTP 65
blastocyst, injection of ES cells 272–5
bone, staining 84–5
breeding records 16
breeding systems 16–20
  generating knock-out mice 268–82
  pseudopregnancy and vasectomy 18–19
  superovulation 19–20
  timed mating 17–18
  using deletions
    2-generation 199–200
    3-generation 197–200
  Whitten effect 17

Caesarian section, rederivation 22–4
cancer susceptibility, and lifespan 208–9
care issues
  Humane Society of US 212
  see also mouse care
cartilage, staining 84–5
cell volume changes
  controlled freezing **31**
  vitrification **33**
centimorgan (cM), defined 88
cervical dislocation 24
chimeras
  analysis 281–2
  host embryos 271–2
  tetraploid embryos 269
chlorambucil, radiation mutagenesis 210
chromosomal rearrangements 144–53
  G-bands 149–51
  identifying and karyotyping mouse chromosomes 151–2
  metaphase chromosomes 144–8
    mitotic chromosomes from mouse peripheral blood 145–7
  stimulation of peripheral mouse lymphocytes by phytohaemagglutinin and lipopolysaccharide 148–9
chromosome classification 151
chromosome engineering in embryonic stem cells 265–8
  deletions, inversions and duplicatiions 265–7
chromosome identification, FISH 154–5
chromosome paints, and direct labelling 155, 159–60
chromosome walking/gap filling, genome mapping 136–8
clone identification, using hybridization assay 134–5
CNS, detection of reporters 78–9
colchicine use 146
computed information *see* electronic tools
conconavalin A (con-A) 144
construct design, elements 249
contigs
  clone, validation 138–40
  primary generation, genome mapping 132–6

controlled freezing 30–2
Copeland–Jenkins interspecific backcross, specific data set 176
crosses, mapping, names, strains, websites 123
cryoprotectants
  glycerol 40–2
  propylene glycol 36–8
cryostat and paraffin sections, multiple detection 78–83
Cyanin3 159
cytogenetics 143–53

DAPI, chromosome staining 155
databases of genomic information 171–5
  electronic bulletin board, MGI-list 181
  major websites **172–3**
  transgenics, knock-outs and other induced mutations 179–80
deletions 265–7
diet 9–10
diethyl pyocarbonate (DEPC) treatment 63–4
digoxygenin, labelling, FISH 159–60
digoxygenin-UTP 65
dilute gene
  coat colour mutation 116–17
  target size 208
DNA
  BACs 219, 226–8
  excision of heterologous DNA from targeted locus **255**
  isolation, transgenesis 221–8
  pooling, genotyping 114–18
  probe, obtaining 158
  repair, future prospects in mutagenesis 211
  sampling 93
  YACs 219, 223–5
DNA repeats, LINES and SINES 154
dominant-lethal test (DLT) 188
double replacement, gene targeting 248, 264–5
duplications 265–7

ear punch system for identification 11–12, 243
ear tags 11
electronic tags 14
electronic tools for mouse genome assessment 171–83
  animal resources lists 180–1
  databases of genomic information 171–5
    major websites **172–3**
    MGI-list, mouse community electronic bulletin board 181
    mouse genome database (MGD) 171–5
    MRC Mammalian Genetics Unit 175
    primary and mirror sites for MGD **174**
    transgenics, knock-outs and other induced mutations 179
  gene expression data 178
    Gene Expression Database (GXD) 178
    mouse 3D atlas 178
  Induced Mutant Resource (IMR) 179–80
  Mouse Knock-out and Mutation Database (MKMD) 179
  physical mapping data 177–8
    genome-wide physical map 177
    X-chromosome physical map 177–8
  radiation hybrid data 176–7
    Jackson Laboratory mouse radiation hybrid database 177
  specific data sets 175–8
    EUCIB 175–6
    genetic data 175
    Jackson Laboratory backcross 176
    Seldin interspecific backcross 176
    Whitehead Institute/MIT map 175, 177
  transgenic/targeted mutation database (TBASE) 179
embryo cryopreservation 27–59
  collection of embryos for freezing 35–6
  controlled freezing 54
  cryoprotectants
    glycerol 40–2
    propylene glycol 36–8
  equipment 34
  expected overall survival 55–7
  genetic resource banks 57–8
  germplasm banking 28–9
  long-term maintenance of mouse models 28
  long-term storage 55
  MGU Harwell method 36–40
  pre-implantation, rederivation 20–2
  principles of cryopreservation 29–30
    controlled freezing 30–2
    vitrification 32–3
  proliferation of mouse models 27
  recovery of embryos 34–5
    by rapid warming 39–40
    live mice from PL-2 39–40
    slow frozen or vitrified in glycerol 44–6
  skills required 33–4
  source of embryos and preferred developmental stages 35–6
  vitrification, solution VS3a 42–4
embryo transfer 20–2, 237–43
embryonic stem (ES) cells *see* mutagenesis, in embryonic stem cells
enhancer trap 73
ethylene oxide, fumigation of housing 8
European Bioinformatics Institute radiation hybrid database 177
European Collaborative Interspecific Backcross (EUCIB) 122
  labelling PCR products by fluorescent dCTP incorporation 126–7

European Collaborative Interspecific Backcross Group (EUCIB) 175–6
European Mouse Mutant Archive (EMMA) 173, 181
euthanasia 24

fine mapping QTLs 107–9
FISH to mouse chromosomes 154–69
  analysis and microscopy 167–8
  chromosome identification 154–5
    mouse chromosome paints 155
    P1 probes 155
    repeat DNA 154–5
  human CD2 gene construct **167**
  hybridization 162–7
  lambda clone on chr-7 **166**
  mouse spleen culture 156–8
  probes and labelling 158–60
    chromosome paints and direct labelling 159–60
    denaturation 162–3
    DNA 158
    hybridization mix 162
    labelling by nick translation 158–9
    labelling DNA with biotin 160
    labelling DNA with digoxygenin 159–60
    preparation and precipitation 161–2
  slide pre-treatment and hybridization 163–4
  transgenic mice, results **168–9**
  washes and detection 164–5
FITC, chromosome paints 159
FITC–avidin 163, 164
fluorescein isothiocyanate-UTP 65
formaldehyde, fumigation of housing 6–7
founder analysis, transgenic mice 240–5
fumigation 5–8

G-banding 149–51
  acetic-saline-Giemsa (ASG), aa 149
  nomenclature 149–50
gamete cryopreservation 46–55
  expected overall survival 55–7
    effect of genotype **57**
  oocytes
    in vitro fertilization 51–3
    mouse metaphase II oocytes 47–8
    recovery of oocytes 48–9
  recovery 48–53
  sealing of straws 53–4
  sperm 49–51, 28–9
    in vitro fertilization 54–5
    recovery 52–3
  vitrification 32–3
gamete sampling, spermatogonial stem-cell mutagenesis 203–6

gene banks
  embryo cryopreservation 57–8
  germplasm 28–9
Gene Expression Database (GXD) 172, 178
gene expression *see* spatial analysis of gene expression
gene insertion 248
gene replacement 248
  vs insertion 249, **250**
gene targeting strategies **248**
gene trap 73
genetic mapping 124–31
  applications 122–4
  concentration and purification of fluorescently labelled PCR products before electrophoresis 127–8
  criteria for construction of physical genome maps 131
  detecting PCR products in SSCP native acrylamide gels by silver staining 125–6
  direct selection of microsatellite-containing sequences 129–31
  generating new genetic markers in specific genomic regions 129–31
  labelling PCR products by fluorescent dCTP incorporation 126–7
  linking different maps together 128–9
  marker types 124
  silver staining of SSCP gels 124–6
  see also genome mapping
genetic markers
  generation in specific genomic regions 129–31
  microsatellite markers 112–13, **117**
  phenotypic trait loci mapping 112–13
  simple sequence length polymorphism (SSLP) markers 124
  single nucleotide polymorphism (SNP) markers 124
  spacing, lod scores 105
genetic resource banks, embryo cryopreservation 57–8
genome, standard karyotype 89
genome assessment
  electronic tools 171–83
  specific data sets, Copeland–Jenkins interspecific backcross 176
genome database (MGD) 171–5
genome mapping 121–41
  applications of genetic and physical mapping 122–4
    mapping crosses **123**
  chromosome walking/gap filling 136–8
    isolating insert end sequences 136–8
    thermal asymmetric interlaced (TAIL) PCR 137–8
  marker generation in specific genomic regions 129–31

genome mapping (*cont.*)
primary contig generation 132–36
clone identification using hybridization assay 134–35
hybridization of overgos to high-density filter grids 135–6
large insert genomic library resources **134**
see also genetic mapping; genetic mapping; physical mapping
genotyping
DNA pooling 114–18
selective 105
germplasm banks 28–9
gestation, prolonged 16
glycerol as cryoprotectant 40–2
green fluorescent protein (GFP) 75–6
in transgenesis, observation and imaging of transgenic embryos 76

Haldane function, non-applicability 99
Haldane–Muller principle 188
haptens 158–9
health screening, pathogen-free animals 3
heritable trnslation test (HTT) 188
high efficiency particulate (HEPA) filtered air supply 4
'hit and run' gene targeting 248, 265, **266**
housing
bedding 9
caging 8–9
conventional vs barrier accommodation 1–2
diet 9–10
environmental control 10
fumigation 5–8
ethylene oxide 8
formaldehyde 6–7
germ-free (axenic) and gnotobiotic accommodation 2
health screening 3
pathogen-free maintenance of animals
breeding performance and production 2
collaborative experiments 2–3
interpretation of data 2
quarantine 3–4
flexible film isolators 4
individually ventilated cages 4–5
human disease, model development 189–90
human placental alkaline phosphatase 75
Humane Society of US 212
hybridization, overgos, to high-density filter grids 135–6
hybridization mix, preparation 162

identification
ear punch 11
ear tags 11
electronic tags 14
tail and paw tattoo 11–15
toe clip 11–12
in situ hybridization 63–71
immunodetection of signals 69–71
and *lacZ* staining 77–8
post-hybridization washes 68–9
removal of non-specific hybrids 68–9
whole-mount non-radioactive 63–71
in vitro fertilization (IVF) 51–3, 54–5
in–out ('hit and run') gene targeting 248, 265, **266**
incomplete penetrance, phenocopies and reduced viability 100–1
Induced Mutant Resource (IMR) 172, 179–80
injection equipment **272**
insert end sequences 136–7
intercross
vs backcross 103–4
and backcross, recombination and segregation **90**
linkage detection of recessive mutant gene **116**
mapping 101–2
International Committee for Standardized Nomenclature for Mice 172
International Mouse Strain Resource (IMSR) 173, 180–1
inversion polymorphisms 92
inversions 265–7
isolators, flexible film isolators 4

Jackson Laboratory
backcross 176
JAX mice IMR, MMR, DNA resource 180–1
mouse radiation hybrid database 177

karyotyping 151, **152**
normal karyotype **152**
recognizing aberrant chromosomes 152–3
knock-out mice 279–82
conditional somatic 257–62
generating, analysing and maintaining 268–82
summary **268**
host embryos 271–9
and other induced mutations 179
spatially restricted **258**, 259
temporally regulated, using inducible recombinases 259–62
knocking-in 262–3
Mouse Knock-out and Mutation Database (MKMD) 172, 179

lacI reporter gene, Big Blue transgenic assay 186
lacZ
embryos, embedding, sectioning, counterstaining 81–3
gene activation assay 260
reporter gene, MutaMouse transgenic assay 186
staining
counterstaining 81–3
detection of β-galactosidase activity 73–4
and *in situ* hybridization 77–8
large insert genomic library resources **134**
LET radiation 210
linkage
critical number of recombinants and swept radius **94**
data reporting 98–9
detection of recessive mutant gene by intercross **116**
phenocopies and reduced viability 101
scanning data 96
testing plan 93–5
estimating recombination frequency **96**
lipopolysaccharide, stimulation of peripheral mouse lymphocytes 148–9
loci, target size, *dilute* vs *agouti* 208
lod scores 97
decline 104–5
marker spacing 105
loxP gene, knocking-in 262–3
loxP sites, NMDA receptor, aa 259
LPS *see* lipopolysaccharide
lymphocytes, stimulation by phytohaemagglutinin and lipopolysaccharide 148–9

Mammalian Genetics Unit, MRC 172, 175
mapping crosses, names, strains, websites 123
markers *see* genetic markers
meiotic mapping 87–8
mesoderm, detection of reporters 78–9
metaphase chromosomes
G-banding 149–51
mitogens 144
preparation 144–8
MGI-list, mouse community electronic bulletin board 173, 181
microinjection set-up, for transgenesis 228–32
micromanipulators for transgenesis **229**, 230
microsatellite markers 112–13
analysis of DNA pools **117**
FLPs 113
microsatellite-containing sequences, direct selection 129–31
mitotic chromosomes *see* metaphase
models, proliferation and maintenance 27–8
modifier genes 91
morgan, defined 88
morula
and blastocyst transfer to host mothers 278–9
injection of ES cells 275–7
morula aggregation 277–8
mouse care and husbandry 1–25
care issues, Humane Society of US 212
environmental enhancement and stimulation 10–11
housing 1–8
legislation 24
rederivation 20–4
transportation 24
water 10
mouse genome assessment, electronic tools 171–83
Mouse Genome Database (MGD) 171–5
primary and mirror sites 174
Mouse Genome Informatics 99
Mouse Knock-out and Mutation Database (MKMD) 172, 179
MRC
3D Atlas of Mouse Development 172, 178
Mammalian Genetics Unit, genomic information 172, 175
Mus m. castaneus (CAST strain) 91
CAST/Ei x MEV 116
Mus m. molossinus (MOLD strain) 91
classification of progeny 93
Mus spretus 91
EUCIB data 175–6
mutagenesis, in embryonic stem cells 247–84
chromosome engineering
deletions inversions and duplicatiions 265–7
translocations 268
construct design 249–54
alternative approaches 254
mutation 251
regions of homology 250–1
replacement vs insertion 249–50
screening strategies 253–4
targeted events enrichment 252–3
targeting cassette components 251–2
gene targeting strategies **248**
generating
analysing and maintaining knock-out mice 268–82
chimera analysis 281–2
equipment for injection of ES-cells into embryos 269–71
ES-cell culture in chimeras with tetraploid embryos 269
ES-cell culture and pluripotency 268–9
genetic background 279–81

mutagenesis, in embryonic stem cells (*cont.*)
generating (*cont.*)
host embryos 271–9
blastocyst injection 272–5
embryo preparation 273
embryo transfer to uterus 278–9
ES-cell injection into blastocysts 274–5
ES-cell preparation 273–4
injection of ES cells into morulae 276
morula aggregation 277–8
morula injection 275–6
morulae and blastocyst transfer 278–9
maintenance of mutations 280–1
microinstrument preparation 270–1
site-specific recombinases in targeting 254–63
excision from ES-cell clones with recombinase expression plasmids 256
excision of heterologous DNA from targeted locus 255–7
excision in mice using transgenic mice expressing recombinase 257
knocking in: recombinase mediated integration into target sites 262–3
oocyte microinjection of recombinase 257
recombinases to generate conditional somatic knock-outs 257–62
spatially restricted knock-outs **258**, 259
temporally regulated knock-outs using inducible recombinases 259–62
subtle mutations without site-specific recombinases 263–5
double replacement 264–5
hit and run 265
mutagenesis of germline 185–215
development of resource 188–9
allelic series 190
biochemical or developmental pathways 190
functional complexity of genomic regions 189
human disease models 189–90
future prospects 211–12
DNA repair 211
mutagenesis future 211–12
sequence-based screening for lesions 211
mutation rate assessment systems
dominant phenotype assays and others 188
reporter genes 186–7
specific locus test 187–8, 208
N-ethyl-*N*-nitrosourea (ENU) 191–201
optimizing dose for inbred strains 206–7
practical breeding considerations 201–9
gamete sampling for spermatogonial stem-cell mutagenesis 203–6
male rotations 201–3
strain background effects 206–9
radiation mutagenesis 209–10
chlorambucil 210
other types of radiation 210
X-rays, treatment and mutations recovered 209–10
rate of recovery to fertility 208
spontaneous mutations 185–6
strain background effects
cancer susceptibility and lifespan 208–9
ENU optimizing dose for inbred strains 206–7
F1 hybrids 207
mutation rates for different loci 208
MutaMouse transgenic assay 186
mutations, maintenance as co-isogenic and congenic lines 280–1

N-ethyl-*N*-nitrosourea (ENU)
chemical properties, stability and half life 191
DNA lesions 191–2
doses and treatment protocols 192–6
F1 hybrids, toleration of treatment 207
genetic screens to isolate mutations 196–201
2- and 3-generation breeding scheme using deletions 197–200
balancer chromosomes 198–9
dominant mutations 196–7
modifiers and sensitized pathways 200–1
recessive mutations 197–9
inactivating and disposal of ENU 195–6
induction of mutations 192–6
mutation rates **193**
practical breeding considerations 201–9
protein product effects 192
sequenced mutations, summary **187**
spectrophotometry 194
structure **191**
treatment of male mice with ENU 195
NMDA receptor, *loxP* sites 259

oestrous cycle 17
oligonucleotides
fill-in probes 134–5
overgo probes **134**
oocytes
cryopreservation 47–8
recovery, rapid warming 49
microinjection of recombinase 257
'overgos' 134
oviduct transfer 236–40

P1 probes 155
pathogen-free animals 2–3

paw tattoo 11–15
PCR
  acrylamide concentrations **125**
  concentration and purification of fluorescently labelled PCR products before electrophoresis 127–8
  detecting products in SSCP native acrylamide gels by silver staining 125–6
  labelling products by fluorescent dCTP incorporation 126–7
  thermal asymmetric interlaced (TAIL) 137–8
  see also genetic mapping
PCR screens 253
PHA *see* phytohaemagglutinin
phenocopies 101
'phenotype gap' 27
phenotypic trait loci mapping 87–120
  backcross mapping experiment model 89–100
    analysing data to determine gene order and distance 97–8
    choosing a tester strain 90–2
    classifying progeny for target locus 93
    extracting a DNA sample 93
    finding markers flanking target 97
    genotyping backcross progeny for selected markers 95–6
    identifying the phenotypic variant 89–90
    linkage testing plan 93–5
    markers for fine mapping 100
    mating mutant/tester stock to produce F1 progeny 92
    producing backcross progeny 92–3
    recording data 96
    reporting linkage data 98–9
    scanning data for linkage 96–7
    size of cross 99–100
  background information 88–9
    definitions 88
    genome 89
    mapping basics 89
  DNA pooling 114–18
  incomplete penetrance, phenocopies and reduced viability 100–1
  mapping by intercross 101–2
  markers 112–13
  QTL mapping using derivative strains 102–12
    congenic strains 111
    consomic strains 112
    fine mapping 107–9
      advanced intercross populations 108–9
      congenic strain construction 108
      progeny testing 108
      selective phenotyping 107–8
    recombinant congenic strains 111
    recombinant inbred strains 109–12
    strain crosses 103–7
      backcross vs intercross 103–4
      confidence intervals 107
      marker spacing 104–5
      number of progeny 104
      selection 103
      selective genotyping 105
      significance thresholds 106–7
  rationale 87–8
    genetic definition of trait locus 87
    genetic manipulation 87–8
    identification of candidate genes 87
  see also QTL mapping
phenylalanine biochemical pathway 190
physical mapping 131–41
  applications 122–4
  criteria for construction 131
  data 177–8
  primary contig generation 132–6
  validation of clone contigs 138–40
  website URLs **140**
phytohaemagglutinin
  and lipopolysaccharide, stimulation of peripheral mouse lymphocytes 148–9
  strains and responses **148**
plasmid DNA, isolation 221–3
Poisson formula 205
post-implantation embryos, harvest and dissection 62–3
primers
  arbitrary degenerate **139**
  vector-specific for pBeloBAC11 **139**
probes
  denaturation 162–3
  hybridization mix 162
  and labelling 158–60
  P1 155
  precipitation 161–2
  preparation 160–2
promoter trap 73
promoters, mutagenesis in embryonic stem cells 252
propylene glycol as cryoprotecant 36–8
pseudopregnancy 18–19
  oviduct transfer 236–7

QTL mapping
  confidence intervals 107
  derivative inbred strains 109–12
  fine mapping 107–9
  significance thresholds 106–7
  strain crosses 103–7
    marker spacing 104–5
  using derivative strains 102–12
quarantine, housing 3–4

Index

radiation hybrid data 176–7
radiation mutagenesis 209–10
  X-rays, treatment and mutations recovered 209–10
recessive mutations, genetic screens to isolate using *N*-ethyl-*N*-nitrosourea (ENU), aa 197–9
recombinant inbred strains 109–10
recombinase expression plasmids, excision from ES-cell clones 256
recombinase systems, gene targeting 248
recombinases
  generation of conditional somatic knock-outs 257–62
  inducible
    temporally regulated knock-outs 259–62
    tetracyclin-specific **261**
  oocyte microinjection 257
  site-specific, in targeting 254–63
  supplying **256**
recombination frequency 88
  conversion to cM 99
  estimating **96**
  incomplete penetrance 100–1
  UCL[U]95u] 98–9
recombination and segregation, backcross and intercross **90**
record keeping 16
  breeding records 16
  experimental returns 16
  general housekeeping 16
rederivation 27–59
  Caesarian section 22–4
  defined 20–4
  embryos and gametes 27–59
  pre-implantation embryo transfer 20–2
reporter transgenes
  applications 72–6
  detection in CNS and mesoderm 78–9
  lacZ 73–4
resorption, moles 188
restraint of animal 15
RFLPs
  markers 124
  microsatellite markers, mapping 113
riboprobes
  generation 65–6
  immunodetection of signals 69–71
  in vitro transcription 65–6
Robertsonian translocations 152–3

Seldin interspecific backcross 176
sequence-based screening for lesions, future prospects 211
SHIRPA protocol 196
simple sequence length polymorphism (SSLP) markers 124
single nucleotide polymorphism (SNP) markers 124
single strand conformation polymorphism (SSCP), silver staining of SSCP gels 124–6
site-specific recombinases, in targeting 254–63
skeletal analysis, whole-mount 83–5
Southern blot analysis 245
Southern screens 253
spatial analysis of gene expression 61–86
  detection of alkaline phosphatase activity 74–5
  detection of β-galactosidase activity 73–6
    lacZ staining 73–4
  dissection of post-implantation embryos 62–3
  green fluorescent protein (GFP) in transgenesis 75–6
  immunodetection and development of signal 69–71
  in situ hybridization
    and *lacZ* staining 77–8
    and post-hybridization washes 68–9
    and removal of non-specific hybrids 68–9
  lacZ embryos, embedding, sectioning and counterstaining 81–3
  multiple and combined detection systems 77–8
    double staining for β-galactosidase and alkaline phosphatase 78
    whole mount staining of early (8.0-12.5 dpc) embryos 77–8
  multiple staining on cryosections 81
  pre-treatment of slides for frozen and wax sections 80
  reporter transgenes 72–6
  staining of cartilage and bones 84–5
  whole-mount non-radioactive *in situ* hybridization 63–71
    fixation and permeabilization 67–8
    generation of riboprobe 65–6
    immunodetection of signals 69–71
    in vitro transcription of riboprobe 65–6
    precautions for RNA work 63–4
    theory of *in situ* hybridization 64
    troubleshooting 71
  whole-mount skeletal analysis 83–5
specific locus test (SLT) 187–8, 208
specified pathogen-free (SPF) housing 1–2
sperm cryopreservation 49–51
  and *in vitro* fertilization 54–5
  and rederivation 49–51
spermatogonia
  N-ethyl-*N*-nitrosourea (ENU) treatment 192
  stem-cell mutagenesis, practical breeding considerations 203–6

spleen culture, obtaining chromosomes for FISH 156–7
sucrose dilution procedure, embryo cryopreservation 40–2
superovulation 19–20
  and isolation of fertilized oocytes 232–4
swept radius, linkage, defined 94

tail biopsies, isolation of genomic DNA 243–4
tail and paw tattoo 11–15
TBASE 172, 179
tester strain, choice 90–2
tetracyclin-specific inducible recombinases **261**
thermal asymmetric interlaced (TAIL) PCR
  chromosome walking/gap filling 137–8
  thermocycling conditions **139**
toe clip 11–12
transgenesis 217–45
  DNA isolation 221–8
    BAC DNA 226–7
    plasmid DNA 221–3
    testing DNA concentration and quality 227–8
    YAC DNA 223–5
  flow chart **232**
  knock-outs and other induced mutations 179
  microinjection experiment 232–40
    oviduct transfer 236–40
      protocol 238–40
      pseudopregnant females 236–7
      transfer pipette preparation 237
    superovulation and isolation of fertilized oocytes 232–4
    timing and methods 234–6
  microinjection set-up 228–32
    holding pipette 231
    injection table, location and design 229–30
    microinjection needles 231–2
    micromanipulators **229**, 230
    microscope 230
  principles and general considerations 217–21
    choice of mouse strains 221
    construct design 219–20
    difficulties and limitations 218–19
    efficiency factors 220–1
    perspectives 220
  transgenic/targeted mutation database (TBASE) 172, 179
transgenic founder analysis 240–5
  isolation of genomic DNA from tail biopsies 243–4
  Southern blot analysis 245
  tail tipping and ear punching 243
translocations, chromosome engineering **267**, 268

vasectomy 18–19
vitrification 32–3
  3-step equilibration 32–3
  solution VS3a 42–4

water provision 10
weaning 17
website URLs, physical mapping, resources **140**
websites, databases of genomic information **172–3**
Whitehead Institute/MIT map 175, 177
Whitten effect 17
whole embryo fixation, and pre-treatment 67–8
whole embryo staining
  double reporter staining 77–8
  multiple detection on cryostat and paraffin sections 78–83
Whole Mouse Catalog 173
whole-mount skeletal analysis 83–5

X-chromosome physical mapping, data 177–8
X-rays, treatment and mutations recovered 209–10

yeast artificial chromosomes (YACs), isolation of DNA 219, 223–5